A Radical Approach to Real Analysis

Second Edition

Originally published by
The Mathematical Association of America, 2007.

Softcover ISBN: 978-1-4704-6904-7
LCCN: 2006933946

10 9 8 7 6 5 4 3 2   27 26 25 24 23 22

AMS/MAA | TEXTBOOKS

VOL **10**

A Radical Approach to Real Analysis

Second Edition

David M. Bressoud

Providence, Rhode Island

to the memory of my mother
Harriet Carnrite Bressoud

Preface

The task of the educator is to make the child's spirit pass again where its forefathers have gone, moving rapidly through certain stages but suppressing none of them. In this regard, the history of science must be our guide.

—Henri Poincaré

This course of analysis is radical; it returns to the roots of the subject. It is not a history of analysis. It is rather an attempt to follow the injunction of Henri Poincaré to let history inform pedagogy. It is designed to be a first encounter with real analysis, laying out its context and motivation in terms of the transition from power series to those that are less predictable, especially Fourier series, and marking some of the traps into which even great mathematicians have fallen.

This is also an abrupt departure from the standard format and syllabus of analysis. The traditional course begins with a discussion of properties of the real numbers, moves on to continuity, then differentiability, integrability, sequences, and finally infinite series, culminating in a rigorous proof of the properties of Taylor series and perhaps even Fourier series. This is the right way to view analysis, but it is not the right way to teach it. It supplies little motivation for the early definitions and theorems. Careful definitions mean nothing until the drawbacks of the geometric and intuitive understandings of continuity, limits, and series are fully exposed. For this reason, the first part of this book follows the historical progression and moves backwards. It starts with infinite series, illustrating the great successes that led the early pioneers onward, as well as the obstacles that stymied even such luminaries as Euler and Lagrange.

There is an intentional emphasis on the mistakes that have been made. These highlight difficult conceptual points. That Cauchy had so much trouble proving the mean value theorem or coming to terms with the notion of uniform convergence should alert us to the fact that these ideas are not easily assimilated. The student needs time with them. The highly refined proofs that we know today leave the mistaken impression that the road of discovery in mathematics is straight and sure. It is not. Experimentation and misunderstanding have been essential components in the growth of mathematics.

Exploration is an essential component of this course. To facilitate graphical and numerical investigations, *Mathematica* and *Maple* commands and programs as well as investigative projects are available on a dedicated website at **www.macalester.edu/aratra**.

The topics considered in this book revolve around the questions raised by Fourier's trigonometric series and the restructuring of calculus that occurred in the process of answering them. Chapter 1 is an introduction to Fourier series: why they are important and why they met with so much resistance. This chapter presupposes familiarity with partial differential equations, but it is purely motivational and can be given as much or as little emphasis as one wishes. Chapter 2 looks at the background to the crisis of 1807. We investigate the difficulties and dangers of working with infinite summations, but also the insights and advances that they make possible. More of these insights and advances are given in Appendix A. Calculus would not have revolutionized mathematics as it did if it had not coupled with infinite series. Beginning with Newton's *Principia*, the physical applications of calculus rely heavily on infinite sums. The chapter concludes with a closer look at the understandings of late eighteenth century mathematicians: how they saw what they were doing and how they justified it. Many of these understandings stood directly in the way of the acceptance of trigonometric series.

In Chapter 3, we begin to find answers to the questions raised by Fourier's series. We follow the efforts of Augustin Louis Cauchy in the 1820s to create a new foundation to the calculus. A careful definition of differentiability comes first, but its application to many of the important questions of the time requires the mean value theorem. Cauchy struggled—unsuccessfully—to prove this theorem. Out of his struggle, an appreciation for the nature of continuity emerges.

We return in Chapter 4 to infinite series and investigate the question of convergence. Carl Friedrich Gauss plays an important role through his complete characterization of convergence for the most important class of power series: the hypergeometric series. This chapter concludes with a verification that the Fourier cosine series studied in the first chapter does, in fact, converge at every value of x.

The strange behavior of infinite sums of functions is finally tackled in Chapter 5. We look at Dirichlet's insights into the problems associated with grouping and rearranging infinite series. We watch Cauchy as he wrestles with the problem of the discontinuity of an infinite sum of continuous functions, and we discover the key that he was missing. We begin to answer the question of when it is legitimate to differentiate or integrate an infinite series by differentiating or integrating each summand.

Our story culminates in Chapter 6 where we present Dirichlet's proof of the validity of Fourier series representations for all "well behaved" functions. Here for the first time we encounter serious questions about the nature and meaning of the integral. A gap remains in Dirichlet's proof which can only be bridged after we have taken a closer look at integration, first using Cauchy's definition, and then arriving at Riemann's definition. We conclude with Weierstrass's observation that Fourier series are indeed strange creatures. The function represented by the series

$$\cos(\pi x) + \frac{1}{2}\cos(13\pi x) + \frac{1}{4}\cos(169\pi x) + \frac{1}{8}\cos(2197\pi x) + \cdots$$

converges and is continuous at every value of x, but it is never differentiable.

The material presented within this book is not of uniform difficulty. There are computational inquiries that should engage all students and refined arguments that will challenge the best. My intention is that every student in the classroom and each individual reader striking out alone should be able to read through this book and come away with an understanding of analysis. At the same time, they should be able to return to explore certain topics in greater depth.

Historical Observations

In the course of writing this book, unexpected images have emerged. I was surprised to see Peter Gustav Lejeune Dirichlet and Niels Henrik Abel reveal themselves as the central figures of the transformation of analysis that fits into the years from 1807 through 1872. While Cauchy is associated with the great theorems and ideas that launched this transformation, one cannot read his work without agreeing with Abel's judgement that "what he is doing is excellent, but very confusing." Cauchy's seminal ideas required two and a half decades of gestation before anyone could begin to see what was truly important and why it was important, where Cauchy was right, and where he had fallen short of achieving his goals.

That gestation began in the fall of 1826 when two young men in their early 20s, Gustav Dirichlet and Niels Henrik Abel, met to discuss and work out the implications of what they had heard and read from Cauchy himself. Dirichlet and Abel were not alone in this undertaking, but they were of the right age to latch onto it. It would become a recurring theme throughout their careers. By the 1850s, the stage was set for a new generation of bright young mathematicians to sort out the confusion and solidify this new vision for mathematics. Riemann and Weierstrass were to lead this generation. Dirichlet joined Gauss as teacher and mentor to Riemann. Abel died young, but his writings became Weierstrass's inspiration.

It was another twenty years before the vision that Riemann and Weierstrass had grasped became the currency of mathematics. In the early 1870s, the general mathematical community finally understood and accepted this new analysis. A revolution had taken place. It was not an overthrow of the old mathematics. No mathematical truths were discredited. But the questions that mathematicians would ask and the answers they would accept had changed in a fundamental way. An era of unprecedented power and possibility had opened.

Changes to the Second Edition

This second edition incorporates many changes, all with the aim of aiding students who are learning real analysis. The greatest conceptual change is in Chapter 2 where I clarify that the Archimedean understanding of infinite series *is* the approach that Cauchy and the mathematical community has adopted. While this chapter still has a free-wheeling style in its use of infinite series—the intent being to convey the power and importance of infinite series—it also begins to introduce rigorous justification of convergence. A new section devoted entirely to geometric series has been added. Chapter 4, which introduces tests of convergence, has been reorganized.

I have also trimmed some of the digressions that I found led students to lose sight of my intent. In particular, the section on the Newton–Raphson method and the proof of Gauss's test for convergence of hypergeometric series have been taken out of the text. Because I feel that this material is still important, though not central, these sections and much more are available on the web site dedicated to this book.

> **Web Resource:** When you see this box with the designation "Web Resource", more information is available in a pdf file, *Mathematica* notebook, or *Maple* worksheet that can be downloaded at **www.macalester.edu/aratra**. The box is also used to point to additional information available in **Appendix A**.

I have added many new exercises, including many taken from *Problems in Mathematical Analysis* by Kaczor and Nowak. Problems taken from this book are identified in Appendix C. I wish to acknowledge my debt to Kaczor and Nowak for pulling together a beautiful collection of challenging problems in analysis. Neither they nor I claim that they are the original source for all of these problems.

All code for *Mathematica* and *Maple* has been removed from the text to the website. Exercises for which these codes are available are marked with the symbol $\boxed{\text{M&M}}$. The appendix with selected solutions has been replaced by a more extensive appendix of hints.

I considered adding a new chapter on the structure of the real numbers. Ultimately, I decided against it. That part of the story properly belongs to the second half of the nineteenth century when the progress described in this book led to a thorough reappraisal of integration. To everyone's surprise this was not possible without a full understanding of the real numbers which were to reveal themselves as far more complex than had been thought. That is an entirely other story that will be told in another book, *A Radical Approach to Lebesgue's Theory of Integration*.

Acknowledgements

Many people have helped with this book. I especially want to thank the NSA and the MAA for financial support; Don Albers, Henry Edwards, and Walter Rudin for their early and enthusiastic encouragement; Ray Ayoub, Allan Krall, and Mark Sheingorn for helpful suggestions; and Ivor Grattan-Guinness who was extremely generous with his time and effort, suggesting historical additions, corrections, and references. The epilogue is among the additions that were made in response to his comments. I am particularly indebted to Meyer Jerison who went through the manuscript of the first edition very carefully and pointed out many of the mathematical errors, omissions, and questionable approaches in the early versions. Some was taken away and much was added as a result of his suggestions. I take full responsibility for any errors or omissions that remain. Susan Dziadosz assisted with the exercises. Her efforts helped weed out those that were impossible or incorrectly stated. Beverly Ruedi helped me through many aspects of production and has shepherded this book toward a speedy publication. Most especially, I want to thank the students who took this course at Penn State in the spring of 1993, putting up with a very preliminary edition and helping to identify its weaknesses. Among those who suggested improvements were Ryan Anthony, Joe Buck, Robert Burns, Stephanie Deom, Lisa Dugent, David Dunson, Susan

Dziadosz, Susan Feeley, Rocco Foderaro, Chris Franz, Karen Lomicky, Becky Long, Ed Mazich, Jon Pritchard, Mike Quarry, Curt Reese, Brad Rothenberger, Chris Solo, Randy Stanley, Eric Steel, Fadi Tahan, Brian Ward, Roger Wherley, and Jennifer White.

Since publication of the first edition, suggestions and corrections have come from many people including Dan Alexander, Bill Avant, Robert Burn, Dennis Caro, Colin Denis, Paul Farnham II, Julian Fleron, Kristine Fowler, Øistein Gjøvik, Steve Greenfield, Michael Kinyon, Mary Marion, Betty Mayfield, Mi-Kyong, Helen Moore, Nick O'Neill, David Pengelley, Mac Priestley, Tommy Ratliff, James Reber, Fred Rickey, Wayne Roberts, Cory Sand, Karen Saxe, Sarah Spence, Volker Strehl, Simon Terrington, and Stan Wagon. I apologize to anyone whose name I may have forgotten.

David M. Bressoud
bressoud@macalester.edu
October 20, 2006

Contents

1

Crisis in Mathematics: Fourier's Series

The crisis struck four days before Christmas 1807. The edifice of calculus was shaken to its foundations. In retrospect, the difficulties had been building for decades. Yet while most scientists realized that something had happened, it would take fifty years before the full impact of the event was understood. The nineteenth century would see ever expanding investigations into the assumptions of calculus, an inspection and refitting of the structure from the footings to the pinnacle, so thorough a reconstruction that calculus was given a new name: *Analysis*. Few of those who witnessed the incident of 1807 would have recognized mathematics as it stood one hundred years later. The twentieth century was to open with a redefinition of the integral by Henri Lebesgue and an examination of the logical underpinnings of arithmetic by Bertrand Russell and Alfred North Whitehead, both direct consequences of the events set in motion in that critical year. The crisis was precipitated by the deposition at the Institut de France in Paris of a manuscript, *Theory of the Propagation of Heat in Solid Bodies*, by the 39-year old prefect of the department of Isère, Joseph Fourier.

1.1 Background to the Problem

Fourier began his investigations with the problem of describing the flow of heat in a very long and thin rectangular plate or lamina. He considered the situation where there is no heat loss from either face of the plate and the two long sides are held at a constant temperature which he set equal to 0. Heat is applied in some known manner to one of the short sides, and the remaining short side is treated as infinitely far away (Figure 1.1). This sheet can be represented in the x, w plane by a region bounded below by the x-axis, on the left by $x = -1$, and on the right by $x = 1$. It has a constant temperature of 0 along the left and right edges so that if $z(x, w)$ represents the temperature at the point (x, w), then

$$z(-1, w) = z(1, w) = 0, \quad w > 0. \tag{1.1}$$

1

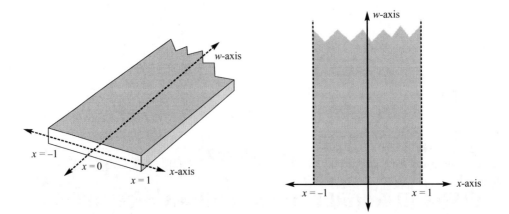

FIGURE 1.1. Two views of Fourier's thin plate.

The known temperature distribution along the bottom edge is described as a function of x:

$$z(x, 0) = f(x). \tag{1.2}$$

Fourier restricted himself to the case where f is an *even* function of x, $f(-x) = f(x)$. The first and most important example he considered was that of a constant temperature normalized to

$$z(x, 0) = f(x) = 1. \tag{1.3}$$

The task was to find a stable solution under these constraints. Trying to apply a constant temperature across the base of this sheet raises one problem: what is the value at $x = 1$, $w = 0$? The temperature along the edge $x = 1$ is 0. On the other hand, the temperature across the bottom where $w = 0$ is 1. Whatever value we try to assign here, there will have to be a discontinuity.

But Joseph Fourier did find a solution, and he did it by looking at situations where the temperature does drop off to zero as x approaches 1 along the bottom edge. What he found is that if the original temperature distribution along the bottom edge $-1 \leq x \leq 1$ and $w = 0$ can be written in the form

$$f(x) = a_1 \cos\left(\frac{\pi x}{2}\right) + a_2 \cos\left(\frac{3\pi x}{2}\right) + a_3 \cos\left(\frac{5\pi x}{2}\right) + \cdots + a_n \cos\left(\frac{(2n-1)\pi x}{2}\right), \tag{1.4}$$

where $a_1, a_2, \ldots, a_n$ are arbitrary constants, then the temperature of the sheet will drop off exponentially as we move away from the x-axis,

$$z(x, w) = a_1 e^{-\pi w/2} \cos\left(\frac{\pi x}{2}\right) + a_2 e^{-3\pi w/2} \cos\left(\frac{3\pi x}{2}\right) + \cdots$$

$$+ a_n e^{-(2n-1)\pi w/2} \cos\left(\frac{(2n-1)\pi x}{2}\right). \tag{1.5}$$

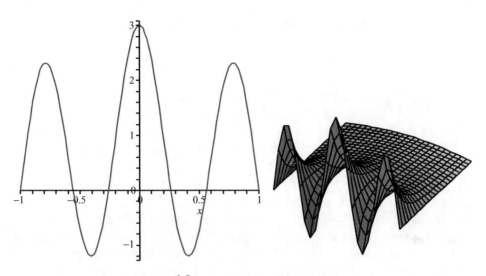

FIGURE 1.2. The functions $f(x)$ and $z(x, w)$.

> **Web Resource:** To see how Fourier found this solution, go to **The Derivation of Fourier's Solution**.

For example (see Figure 1.2), if the temperature along the bottom edge is given by the function $f(x) = \cos(\pi x/2) + 2\cos(5\pi x/2)$, then the temperature at the point (x, w), $-1 \le x \le 1$, $w \ge 0$, is given by

$$z(x, w) = e^{-\pi w/2} \cos\left(\frac{\pi x}{2}\right) + 2e^{-5\pi w/2} \cos\left(\frac{5\pi x}{2}\right). \tag{1.6}$$

The problem with the solution in equation (1.5) is that it assumes that the distribution of heat along the bottom edge is given by a formula of the form found in equation (1.4). Any function that can be written in this way must be continuous and equal to 0 at $x = \pm1$. The constant function $f(x) = 1$ cannot be written in this form. One possible interpretation is that there simply is no solution when $f(x) = 1$. That possibility did not sit well with Fourier. After all, it *is* possible to apply a constant temperature to one end of a metal bar.

Fourier observed that as we take larger values of n, the number of summands in equation (1.4), we can get functions that more closely approximate $f(x) = 1$. If we could take infinitely many terms, then we should be able to get a function of this form that is exactly $f(x) = 1$. Fourier was convinced that this would work and boldly proclaimed his solution. For $-1 < x < 1$, he asserted that

$$1 = \frac{4}{\pi}\left[\cos\left(\frac{\pi x}{2}\right) - \frac{1}{3}\cos\left(\frac{3\pi x}{2}\right) + \frac{1}{5}\cos\left(\frac{5\pi x}{2}\right) - \cdots\right]$$

$$= \frac{4}{\pi}\sum_{n=1}^{\infty}\frac{(-1)^{n-1}}{2n-1}\cos\left(\frac{(2n-1)\pi x}{2}\right). \tag{1.7}$$

If true, then this implies that the temperature in the plate is given by

$$z(x, w) = \frac{4}{\pi} \left[e^{-\pi w/2} \cos \left(\frac{\pi x}{2} \right) - \frac{1}{3} e^{-3\pi w/2} \cos \left(\frac{3\pi x}{2} \right) + \cdots \right]$$

$$= \frac{4}{\pi} \sum_{n=1}^{\infty} \frac{(-1)^{n-1}}{2n - 1} e^{-(2n-1)\pi w/2} \cos \left(\frac{(2n - 1)\pi x}{2} \right). \tag{1.8}$$

Web Resource: To explore graphs of approximations to Fourier series go to **Approximating Fourier's Solution**.

Here was the heart of the crisis. Infinite sums of trigonometric functions had appeared before. Daniel Bernoulli (1700–1782) proposed such sums in 1753 as solutions to the problem of modeling the vibrating string. They had been dismissed by the greatest mathematician of the time, Leonhard Euler (1707–1783). Perhaps Euler scented the danger they presented to his understanding of calculus. The committee that reviewed Fourier's manuscript: Pierre Simon Laplace (1749–1827), Joseph Louis Lagrange (1736–1813), Sylvestre François Lacroix (1765–1843), and Gaspard Monge (1746–1818), echoed Euler's dismissal in an unenthusiastic summary written by Siméon Denis Poisson (1781–1840). Lagrange was later to make his objections explicit. In section 2.6 we shall investigate the specific objections to trigonometric series that were raised by Lagrange and others. Well into the 1820s, Fourier series would remain suspect because they contradicted the established wisdom about the nature of functions.

Fourier did more than suggest that the solution to the heat equation lay in his trigonometric series. He gave a simple and practical means of finding those coefficients, the a_i, for any function. In so doing, he produced a vast array of verifiable solutions to specific problems. Bernoulli's proposition could be debated endlessly with little effect for it was only theoretical. Fourier was modeling actual physical phenomena. His solution could not be rejected without forcing the question of why it seemed to work.

Web Resource: To see Fourier's method for finding the values of a_i and to see how to determine values for the function $f(x) = 1$, go to **The General Solution**.

There are problems with Fourier series, but they are subtler than anyone realized in that winter of 1807–08. It was not until the 1850s that Bernhard Riemann (1826–1866) and Karl Weierstrass (1815–1897) would sort out the confusion that had greeted Fourier and clearly delineate the real questions.

1.2   Difficulties with the Solution

Fourier realized that equation (1.7) is only valid for $-1 < x < 1$. If we replace x by $x + 2$ in the nth summand, then it changes sign:

$$\cos \left(\frac{(2n - 1)\pi (x + 2)}{2} \right) = \cos \left(\frac{(2n - 1)\pi x}{2} + (2n - 1)\pi \right)$$

$$= -\cos \left(\frac{(2n - 1)\pi x}{2} \right).$$

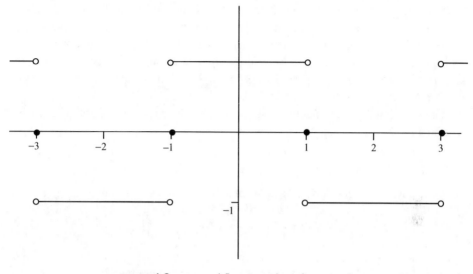

FIGURE 1.3. $f(x) = \frac{4}{\pi} \left[\cos \frac{\pi x}{2} - \frac{1}{3} \cos \frac{3\pi x}{2} + \cdots \right]$.

It follows that for x between 1 and 3, equation (1.7) becomes

$$f(x) = -1 = \frac{4}{\pi} \left[\cos \frac{\pi x}{2} - \frac{1}{3} \cos \frac{3\pi x}{2} + \frac{1}{5} \cos \frac{5\pi x}{2} - \frac{1}{7} \cos \frac{7\pi x}{2} + \cdots \right]. \quad (1.9)$$

In general, $f(x + 2) = -f(x)$. The function represented by this cosine series has a graph that alternates between -1 and $+1$ as shown in Figure 1.3.

This is very strange behavior. Equation (1.7) seems to be saying that our cosine series is the constant function 1. Equation (1.9) says that our series is not constant. Moreover, to the mathematicians of 1807, Figure 1.3 did not look like the graph of a function. Functions were polynomials; roots, powers, and logarithms; trigonometric functions and their inverses; and whatever could be built up by addition, subtraction, multiplication, division, or composition of these functions. Functions had graphs with unbroken curves. Functions had derivatives and Taylor series. Fourier's cosine series flew in the face of everything that was *known* about the behavior of functions. Something must be very wrong.

Fourier was aware that his justification for equation (1.7) was not rigorous. It began with the assumption that such a cosine series should exist, and in a crucial step he assumes that the integral of such a series can be obtained by integrating each summand. In fact, strange things happen when you try to integrate or differentiate this series by integrating or differentiating each term.

Term-by-term Integration and Differentiation

Term-by-term integration, the ability to find the integral of a sum of functions by integrating each summand, works for finite sums,

$$\int_a^b \left(f_1(x) + f_2(x) + \cdots + f_n(x) \right) dx$$

$$= \int_a^b f_1(x)dx + \int_a^b f_2(x)dx + \cdots + \int_a^b f_n(x)dx.$$

It is not surprising that Fourier would assume that it also works for any infinite sum of functions. After all, this lay behind one of the standard methods for finding integrals.

Pressed for a definition of integration, mathematicians of Fourier's time would have replied that it is the inverse process of differentiation: to find the integral of $f(x)$, you find a function whose derivative is $f(x)$. This definition has its limitations: what is the integral of e^{-x^2}?

There is no simple function with this derivative, but the integral can be found explicitly by using power series. Using the fact that

$$e^{-x^2} = 1 - x^2 + \frac{x^4}{2} - \frac{x^6}{3!} + \frac{x^8}{4!} - \cdots,$$

and the fact that a power series can be integrated by integrating each summand, we see that

$$\int e^{-x^2} dx = C + x - \frac{x^3}{3} + \frac{x^5}{2 \cdot 5} - \frac{x^7}{3! \cdot 7} + \frac{x^9}{4! \cdot 9} - \cdots. \tag{1.10}$$

Mathematicians knew that as long as you stayed inside the interval of convergence there was never any problem integrating a power series term-by-term. The worst that could go wrong when differentiating term-by-term was that you might lose convergence at the endpoints. Few mathematicians even considered that switching to an infinite sum of trigonometric functions would create problems. But you did not have to press Fourier's solution very far before you started to uncover real difficulties.

> **Web Resource:** To see how complex analysis can shed light on why Fourier series are problematic, go to **Fourier Series as Complex Power Series**.

Looking at the graph of

$$f(x) = \frac{4}{\pi} \left[\cos\left(\frac{\pi x}{2}\right) - \frac{1}{3} \cos\left(\frac{3\pi x}{2}\right) + \frac{1}{5} \cos\left(\frac{5\pi x}{2}\right) - \cdots \right]$$

shown in Figure (1.3), it is clear that the derivative, $f'(x)$, is zero for all values of x other than odd integers. The derivative is not defined when x is an odd integer. But if we try to differentiate this function by differentiating each summand, we get the series

$$-2 \left[\sin\left(\frac{\pi x}{2}\right) - \sin\left(\frac{3\pi x}{2}\right) + \sin\left(\frac{5\pi x}{2}\right) - \sin\left(\frac{7\pi x}{2}\right) + \cdots \right] \tag{1.11}$$

which only converges when x is an even integer.

Many mathematicians of the time objected to even considering infinite sums of cosines. These infinite summations cast doubt on what scientists thought they knew about the nature of functions, about continuity, about differentiability and integrability. If Fourier's disturbing series were to be accepted, then all of calculus needed to be rethought.

Lagrange thought he found the flaw in Fourier's work in the question of convergence: whether the summation approaches a single value as more terms are taken. He asserted that the cosine series,

$$\cos \frac{\pi x}{2} - \frac{1}{3} \cos \frac{3\pi x}{2} + \frac{1}{5} \cos \frac{5\pi x}{2} - \frac{1}{7} \cos \frac{7\pi x}{2} + \cdots,$$

does not have a well-defined value for all x. His reason for believing this was that the series consisting of the absolute values of the coefficients,

$$1 + \frac{1}{3} + \frac{1}{5} + \frac{1}{7} + \frac{1}{9} + \cdots,$$

grows without limit (see exercise 1.2.3). In fact, Fourier's cosine expansion of $f(x) = 1$ *does* converge for any x, as Fourier demonstrated a few years later. The complete justification of the use of these infinite trigonometric series would have to wait twenty-two years for the work of Peter Gustav Lejeune Dirichlet (1805–1859), a young German who, in 1807 when Fourier deposited his manuscript, was two years old.

Exercises

The symbol (**M&M**) indicates that *Maple* and *Mathematica* codes for this problem are available in the **Web Resources** at **www.macalester.edu/aratra**.

1.2.1. (**M&M**) Graph each of the following partial sums of Fourier's expansion over the interval $-1 \le x \le 3$.

a. $\frac{4}{\pi} \cos(\pi x/2)$

b. $\frac{4}{\pi} \left[\cos(\pi x/2) - \frac{1}{3} \cos(3\pi x/2) \right]$

c. $\frac{4}{\pi} \left[\cos(\pi x/2) - \frac{1}{3} \cos(3\pi x/2) + \frac{1}{5} \cos(5\pi x/2) \right]$

d. $\frac{4}{\pi} \left[\cos(\pi x/2) - \frac{1}{3} \cos(3\pi x/2) + \frac{1}{5} \cos(5\pi x/2) - \frac{1}{7} \cos(7\pi x/2) \right]$

1.2.2. (**M&M**) Let $F_n(x)$ denote the sum of the first n terms of Fourier's series evaluated at x:

$$F_n(x) = \frac{4}{\pi} \left[\cos \frac{\pi x}{2} - \frac{1}{3} \cos \frac{3\pi x}{2} + \cdots + \frac{(-1)^{n-1}}{2n-1} \cos \frac{(2n-1)\pi x}{2} \right].$$

a. Evaluate $F_{100}(x)$ at $x = 0, 0.5, 0.9, 0.99, 1.1$, and 2. Is this close to the expected value?

b. Evaluate $F_n(0.99)$ at $n = 100, 200, 300, \ldots, 2000$ and plot these successive approximations.

c. Evaluate $F_n(0.999)$ at $n = 100, 200, 300, \ldots, 2000$ and plot these successive approximations.

d. What is the value of this infinite series at $x = 1$?

1.2.3. (**M&M**) Evaluate the partial sums of the series

$$1 + \frac{1}{3} + \frac{1}{5} + \frac{1}{7} + \frac{1}{9} + \cdots$$

for the first 10, 20, 40, 80, 160, 320, and 640 terms. Does this series appear to approach a value? If so, what value is it approaching?

1.2.4. (**M&M**) Graph the surfaces described by the partial sums consisting of the first term, the first two terms, the first three terms, and the first four terms of Fourier's solution over

$0 \le w \le 0.6, \ -1 \le x \le 1:$

$$z(x, w) = \frac{4}{\pi} \left[e^{-\pi w/2} \cos \frac{\pi x}{2} - \frac{1}{3} e^{-3\pi w/2} \cos \frac{3\pi x}{2} \right.$$

$$\left. + \frac{1}{5} e^{-5\pi w/2} \cos \frac{5\pi x}{2} - \frac{1}{7} e^{-7\pi w/2} \cos \frac{7\pi x}{2} + \cdots \right].$$

1.2.5. (**M&M**) Consider the series

$$1 + \frac{1}{5} - \frac{1}{7} - \frac{1}{11} + \frac{1}{13} + \frac{1}{17} - \frac{1}{19} - \frac{1}{23} + \cdots .$$

Prove that the partial sums are always greater than or equal to 1 once we have at least five terms. What number does this series appear to approach?

1.2.6. Fourier series illustrate the dangers of trying to find limits by simply substituting the value that x approaches. Consider Fourier's series:

$$f(x) = \frac{4}{\pi} \left[\cos \frac{\pi x}{2} - \frac{1}{3} \cos \frac{3\pi x}{2} + \frac{1}{5} \cos \frac{5\pi x}{2} - \frac{1}{7} \cos \frac{7\pi x}{2} + \cdots \right]. \quad (1.12)$$

a. What value does this approach as x approaches 1 from the left?

b. What value does this approach as x approaches 1 from the right?

c. What is the value of $f(1)$?

These three answers are all different.

1.2.7. (**M&M**) Consider the function that we get if we differentiate each summand of the function $f(x)$ defined in equation (1.12),

$$g(x) = -2 \left(\sin \frac{\pi x}{2} - \sin \frac{3\pi x}{2} + \sin \frac{5\pi x}{2} - \sin \frac{7\pi x}{2} + \cdots \right).$$

a. For $-1 < x < 3$, graph the partial sums of this series consisting of the first 10, 20, 30, 40, and 50 terms. Does it appear that these graphs are approaching the constant function 0?

b. Evaluate the partial sums up to at least 20 terms when $x = 0, 0.2, 0.3,$ and 0.5. Does it appear that this series is approaching 0 at each of these values of x?

c. What *is* happening at $x = 0, 0.2, 0.3, 0.5$? What can you prove?

2

Infinite Summations

The term *infinite summation* is an oxymoron. *Infinite* means without limit, nonterminating, never ending. *Summation* is the act of coming to the highest point (*summus*, summit), reaching the totality, achieving the conclusion. How can we conclude a process that never ends? The phrase itself should be a red flag alerting us to the fact that something very subtle and nonintuitive is going on. It is safer to speak of an **infinite series** for a summation that has no end, but we shall use the symbols of addition, the $+$ and the $\sum$. We need to remember that they no longer mean quite the same thing.

In this chapter we will see why infinite series are important. We will also see some of the ways in which they can behave totally unlike finite summations. The discovery of Fourier series accelerated this recognition of the strange behavior of infinite series. We will learn more about why they were so disturbing to mathematicians of the early 19th century. We begin by learning how Archimedes of Syracuse dealt with infinite processes. While he may seem to have been excessively cautious, ultimately it was his approach that mathematicians would adopt.

2.1 The Archimedean Understanding

The Greeks of the classical era avoided such dangerous constructions as infinite series. An illustration of this can be found in the quadrature of the parabola by Archimedes of Syracuse (287–212 B.C.). To make the problem concrete, we state it as one of finding the area of the region bounded below by the x-axis and above by the curve $y = 1 - x^2$ (Figure 2.1), but Archimedes actually showed how to find the area of any segment bounded by an arc of a parabola and a straight line.

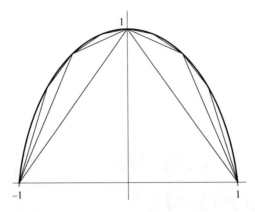

FIGURE 2.1. Archimedes' triangulation of a parabolic region.

> **Web Resource:** To see what Archimedes actually did to find the area of any segment bounded by a parabola and a straight line, go to **The quadrature of a parabolic segment**.

The triangle with vertices at $(\pm 1, 0)$ and $(0, 1)$ has area 1. The two triangles that lie above this and have vertices at $(\pm 1/2, 3/4)$ have a combined area of $1/4$. If we put four triangles above these two, adding vertices at $(\pm 1/4, 15/16)$ and $(\pm 3/4, 7/16)$, then these four triangles will add a combined area of $1/16$. In general, what Archimedes showed is that no matter how many triangles we have placed inside this region, we can put in two new triangles for each one we just inserted and increase the total area by one-quarter of the amount by which we last increased it.

As we take more triangles, we get successive approximations to the total area:

$$1, \quad 1 + \frac{1}{4}, \quad 1 + \frac{1}{4} + \frac{1}{16}, \quad 1 + \frac{1}{4} + \frac{1}{16} + \frac{1}{64}, \quad \ldots, \quad 1 + \frac{1}{4} + \frac{1}{16} + \cdots + \frac{1}{4^n}.$$

Archimedes then makes the observation that each of these sums brings us closer to $4/3$:

$$1 = \frac{4}{3} - \frac{1}{3},$$

$$1 + \frac{1}{4} = \frac{4}{3} - \frac{1}{4 \cdot 3},$$

$$1 + \frac{1}{4} + \frac{1}{16} = \frac{4}{3} - \frac{1}{16 \cdot 3},$$

$$1 + \frac{1}{4} + \frac{1}{16} + \frac{1}{64} = \frac{4}{3} - \frac{1}{64 \cdot 3},$$

$$\vdots$$

$$1 + \frac{1}{4} + \cdots + \frac{1}{4^k} = \frac{4}{3} - \frac{1}{4^k \cdot 3}. \tag{2.1}$$

A modern reader is inclined to make the jump to an infinite summation at this point and say that the actual area is

$$1 + \frac{1}{4} + \frac{1}{16} + \cdots = \frac{4}{3}.$$

This is precisely what Archimedes did *not* do. He proceeded very circumspectly, letting K denote the area to be calculated and demonstrating that K could not be larger than 4/3 nor .less than 4/3.

Archimedes' Argument

Let K denote the area bounded by the parabolic arc and the line segment. Archimedes showed that each time we add new triangles, the area of the region inside the parabolic arc that is not covered by our triangles is reduced by more than half (see exercises 2.1.2–2.1.3). It follows that we can make this error as small as we want by taking enough inscribed triangles. If K were larger than 4/3, then we could inscribe triangles until their total area was more than 4/3. This would contradict equation (2.1) which says that the sum of the areas of the inscribed triangles is always strictly less than 4/3. If K were smaller than 4/3, then we could find a k for which $4/3 - 1/(4^k \cdot 3)$ is larger than K. But then equation (2.1) tells us that the sum of the areas of the corresponding inscribed triangles is strictly larger than K. This contradicts the fact that the sum of the areas of *inscribed* triangles cannot exceed the total area.

This method of calculating areas by summing inscribed triangles is often referred to as the "method of exhaustion." E. J. Dijksterhuis has pointed out that this is "the worst name that could have been devised." As Archimedes or Eudoxus of Cnidus (*ca.* 408–355 B.C.) (the first to employ this method) would have insisted, you never exhaust the area. You only get arbitrarily close to it.

Archimedes argument is important because it points to our modern definition of the infinite series $1 + 1/4 + 1/16 + \cdots + 1/4^n + \cdots$. Just as Archimedes handled his infinite process by producing a value and demonstrating that the answer could be neither greater nor less than this produced value, so Cauchy and others of the early nineteenth century would handle infinite series by producing the desired value and demonstrating that the series could not have a value either greater or less than this.

To a modern mathematician, an infinite series *is* the succession of approximations by finite sums. Our finite sums may not close in quite as nicely as Archimedes' series $1 + 1/4 + 1/16 + \cdots + 1/4^n + \cdots$, but the idea stays the same. We seek a target value T so that for each $M > T$, the finite sums eventually will all be below M, and for all real numbers $L < T$, the finite sums eventually will all be above L. In other words, given any open interval (L, M) that contains T, all of the partial sums are inside this interval once they have enough terms. If we can find such a target value T, then it is the value of the infinite series. We shall call this the *Archimedean understanding of an infinite series*.

For example, the Archimedean understanding of $1 + 1/4 + 1/16 + \cdots + 1/4^n + \cdots$ is that it is the sequence $1, 1 + 1/4, 1 + 1/4 + 1/16, \ldots$. All of the partial sums are less than 4/3, and so they are less than any $M > 4/3$. For any $L < 4/3$, from some point on all of the partial sums will be strictly larger than L. Therefore, the value of the series is 4/3.

> **Definition: Archimedean understanding of an infinite series**
>
> The **Archimedean understanding** of an infinite series is that it is shorthand for the sequence of finite summations. The **value** of an infinite series, if it exists, is that number T such that given any $L < T$ and any $M > T$, all of the finite sums from some point on will be strictly contained in the interval between L and M. More precisely, given $L < T < M$, there is an integer n, whose value depends on the choice of L and M, such that every partial sum with at least n terms lies inside the interval (L, M).

In the seventeenth and eighteenth centuries, there was a free-wheeling style in which it appeared that scientists treated infinite series as finite summations with a very large number of terms. In fact, scientists of this time were very aware of the distinction between series with a large number of summands and infinite series. They knew you could get into serious trouble if you did not make this distinction. But they also knew that treating infinite series as if they really were summations led to useful insights such as the fact that the integral of a power series could be found by integrating each term, just as in a finite summation. They developed a sense for what was and was not legitimate. But by the early 1800s, the sense for what should and should not work was proving insufficient, as exemplified by the strange behavior of Fourier's trigonometric series. Cauchy and others returned to Archimedes' example of how to handle infinite processes.

It may seem the Archimedean understanding creates a lot of unnecessary work simply to avoid infinite summations, but there is good reason to avoid infinite summations for they are manifestly *not* summations in the usual sense.

> **Web Resource:** To learn about the Archimedean principle and why it is essential to the Archimedean understanding of an infinite series, go to **The Archimedean principle**.

The Oddity of Infinite Sums

Ordinary sums are very well behaved. They are associative, which means that it does not matter how we group them:

$$(2 + 3) + 5 = 2 + (3 + 5),$$

and they are commutative, which means that it does not matter how we order them:

$$2 + 3 + 5 = 3 + 5 + 2.$$

These simple facts do not always hold for infinite sums. If we could group an infinite sum any way we wanted, then we would have that

$$
\begin{aligned}
& 1 - 1 + 1 - 1 + 1 - 1 + 1 - 1 + \cdots \\
& = (1 - 1) + (1 - 1) + (1 - 1) + (1 - 1) + \cdots \\
& = 0,
\end{aligned}
$$

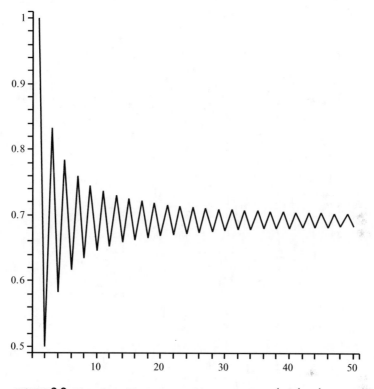

FIGURE 2.2. Plot of partial sums up to fifty terms of $1 - \frac{1}{2} + \frac{1}{3} - \frac{1}{4} + \cdots$.

whereas by regrouping we obtain

$$1 - 1 + 1 - 1 + 1 - 1 + 1 - 1 + \cdots$$
$$= 1 + (-1 + 1) + (-1 + 1) + (-1 + 1) + (-1 + 1) + \cdots$$
$$= 1.$$

It takes a little more effort to see that rearrangements are not always allowed, but the effort is rewarded in the observation that some very strange things are happening here. Consider the alternating harmonic series

$$1 - \frac{1}{2} + \frac{1}{3} - \frac{1}{4} + \frac{1}{5} - \frac{1}{6} + \cdots .$$

A plot of the partial sums of this series up to the sum of the first fifty terms is given in Figure 2.2. The partial sums are narrowing in on a value near 0.7 (in fact, this series converges to $\ln 2$).

If we rearrange the summands in this series, taking two positive terms, then one negative term, then the next two positive terms, then the next negative term:

$$1 + \frac{1}{3} - \frac{1}{2} + \frac{1}{5} + \frac{1}{7} - \frac{1}{4} + \frac{1}{9} + \frac{1}{11} - \frac{1}{6} + \cdots ,$$

we obtain a series whose partial sums are plotted in Figure 2.3. The partial sums are now approaching a value near 1.04. Rearranging the summands has changed the value.

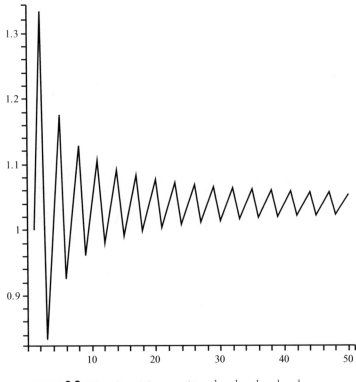

FIGURE 2.3. Plot of partial sums of $1 + \frac{1}{3} - \frac{1}{2} + \frac{1}{5} + \frac{1}{7} - \frac{1}{4} + \cdots$.

Web Resource: To explore the alternating harmonic series and its rearrangements, go to **Explorations of the alternating harmonic series**.

Exercises

The symbol $\boxed{\textbf{M\&M}}$ indicates that *Maple* and *Mathematica* codes for this problem are available in the **Web Resources** at **www.macalester.edu/aratra**.

2.1.1.

a. Show that the triangle with vertices at (a_1, b_1), (a_2, b_2), and (a_3, b_3) has area equal to

$$\frac{1}{2}\left|(a_2 - a_1)(b_3 - b_1) - (b_2 - b_1)(a_3 - a_1)\right|. \tag{2.2}$$

One approach is to use the fact that the area of the parallelogram defined by $(a_2 - a_1)\vec{\imath} + (b_2 - b_1)\vec{\jmath}$ and $(a_3 - a_1)\vec{\imath} + (b_3 - b_1)\vec{\jmath}$ is

$$\left|\left((a_2 - a_1)\vec{\imath} + (b_2 - b_1)\vec{\jmath} + 0\,\vec{k}\right) \times \left((a_3 - a_1)\vec{\imath} + (b_3 - b_1)\vec{\jmath} + 0\,\vec{k}\right)\right|.$$

b. Use the area formula in line (2.2) to prove that the area of the triangle with vertices at (a, a^2), $(a + \delta, (a + \delta)^2)$, $(a + 2\delta, (a + 2\delta)^2)$ is $|\delta|^3$.

c. Use these results to prove that the area of the polygon with vertices at $(-1, 0)$, $(-1 + 2^{-n}, 1 - (-1 + 2^{-n})^2)$, $(-1 + 2 \cdot 2^{-n}, 1 - (-1 + 2 \cdot 2^{-n})^2)$, $(-1 + 3 \cdot 2^{-n}, 1 - (-1 + 3 \cdot 2^{-n})^2)$, ..., $(1, 0)$ is $1 + 4^{-1} + 4^{-2} + \cdots + 4^{-n}$.

2.1.2. Archimedes' formula for the area of a parabolic region is obtained by constructing triangles where the base is the line segment that bounds the region and the apex is located at the point where the tangent line to the parabola is parallel to the base. Show that the tangent to $y = 1 - x^2$ at $((k + 1/2)2^{-n}, 1 - (k + 1/2)^2 2^{-2n})$ has the same slope as the line segment connecting the two endpoints: $(k\,2-n, 1 - k^2 2^{-2n})$ and $((k + 1)2^{-n}), 1 - (k + 1)^2 2^{-2n})$.

2.1.3. Show that if we take a parabolic region and inscribe a triangle whose base is the line segment that bounds the region and whose apex is located at the point where the tangent line to the parabola is parallel to the base, then the area of the triangle is more than half the area of the parabolic region.

2.1.4. Archimedes' method to find the area under the graph of $y = 1 - x^2$ is equivalent to using trapezoidal approximations to the integral of this function from $x = -1$ to $x = 1$, first with steps of size 1, then size $1/2$, $1/4$, $1/8$, ...

a. Verify that the trapezoidal approximation to $\int_{-1}^{1} 1 - x^2\,dx$ with steps of size $1/2$ is equal to $4/3 - 1/3 \cdot 4 = 5/4$.

b. Verify that the trapezoidal approximation to $\int_{-1}^{1} 1 - x^2\,dx$ with steps of size $1/4$ is equal to $4/3 - 1/3 \cdot 4^2 = 21/16$.

c. Verify that the trapezoidal approximation to $\int_{-1}^{1} 1 - x^2\,dx$ with steps of size $1/8$ is equal to $4/3 - 1/3 \cdot 4^3 = 85/64$.

2.1.5. Explain each step in the following evaluation of the trapezoidal approximation to $\int_{-1}^{1}(1 - x^2)\,dx$ with steps of size 2^{-k}:

$$\int_{-1}^{1}(1 - x^2)\,dx \approx \frac{1}{2^k} \sum_{i=1-2^k}^{2^k-1}\left(1 - \left(\frac{i}{2^k}\right)^2\right) \tag{2.3}$$

$$= \frac{1}{2^k}\left[2^{k+1} - 1 - 2\sum_{i=1}^{2^k-1}\frac{i^2}{2^{2k}}\right] \tag{2.4}$$

$$= 2 - \frac{1}{2^k n} - \frac{1}{2^{3k-1}}\left(\frac{2^{3k}}{3} - \frac{2^{2k}}{2} + \frac{2^k}{6}\right) \tag{2.5}$$

$$= \frac{4}{3} - \frac{1}{3 \cdot 4^k}. \tag{2.6}$$

It follows that the sum of the areas of the last 2^k triangles is

$$\left(\frac{4}{3} - \frac{1}{3 \cdot 4^k}\right) - \left(\frac{4}{3} - \frac{1}{3 \cdot 4^{k-1}}\right) = \frac{4 - 1}{3 \cdot 4^k} = \frac{1}{4^k}.$$

2.1.6. Consider the series

$$1 + \frac{1}{2} + \frac{1}{4} + \frac{1}{8} + \cdots + \frac{1}{2^k} + \cdots .$$

Find the target value, T, of the partial sums. How do you know that for any M greater than your target value, all of the partial sums are strictly less than M? How many terms do you have to take in order to guarantee that all of the partial sums from that point on will be larger than $L = T - 1/10$?

2.1.7. Consider the series

$$3 - \frac{3}{2} + \frac{3}{4} - \frac{3}{8} + \cdots + \frac{(-1)^k 3}{2^k} + \cdots , \quad k \geq 0.$$

Find the target value, T, of the partial sums. How many terms do you have to take in order to guarantee that all of the partial sums from that point on will be smaller than $M = T + 1/10$? How many terms do you have to take in order to guarantee that all of the partial sums from that point on will be larger than $L = T - 1/10$? How many terms do you have to take in order to guarantee that all of the partial sums from that point on will be within $1/100$ of T?

2.1.8. Consider the series

$$1 - \frac{1}{2} + \frac{1}{3} - \frac{1}{4} + \cdots + \frac{(-1)^{k-1}}{k} + \cdots .$$

Explain why there should be a target value. You may not be able to prove that the target value is $T = \ln 2$, but you should still be able to explain why there should be one. How many terms will be enough to guarantee that all of the partial sums from that point on will be within $1/10$ of T? Explain the reasoning that leads to your answer.

2.1.9. What is the Archimedean understanding of the infinite series $1 - 1 + 1 - 1 + \cdots$? Explain why this series *cannot* have a value under this understanding.

2.1.10. $\boxed{\textbf{M\&M}}$

a. Calculate the first $2n$ terms of the alternating harmonic series with the summands in the usual order. Check that it gets close to the target value of $T = \ln 2$ as n gets large. How large does n have to be before the partial sums are all with 10^{-6} of $\ln 2$?

b. Find the value that the series approaches when you take two positive summands for every negative summand,

$$1 + \frac{1}{3} - \frac{1}{2} + \frac{1}{5} + \frac{1}{7} - \frac{1}{4} + \frac{1}{9} + \frac{1}{11} - \frac{1}{6} + \cdots .$$

c. Find the value that the series approaches when you take one positive summand for every two negative summands,

$$1 - \frac{1}{2} - \frac{1}{4} + \frac{1}{3} - \frac{1}{6} - \frac{1}{8} + \frac{1}{5} - \frac{1}{10} - \frac{1}{12} + \cdots .$$

d. Take your decimal answers from parts (b) and (c). For each decimal, d, calculate e^{2d}. Guess the true values of these rearranged series. Explore the decimal values you get with other integers for r and s. Guess the general value formula. Keep notes of your exploration and explain the process that led to your guess.

2.1.11. (**M&M**) Explore what happens if you rearrange the series

$$1 - \frac{1}{2^2} + \frac{1}{3^2} - \frac{1}{4^2} + \cdots .$$

Compare the values that you get with the original series, taking two positive summands for every negative summand, and taking two negative summands for every positive summand. Explore what happens with other values for r and s.

2.2 Geometric Series

By the fourteenth century, the Scholastics in Oxford and Paris, people such as Richard Swineshead (*fl. c.* 1340–1355) and Nicole Oresme (1323–1382), were using and assigning values to infinite series that arose in problems of motion. They began with series for which each pair of consecutive summands has the same ratio, such as the summation used by Archimedes,

$$1 + \frac{1}{4} + \frac{1}{16} + \cdots + \frac{1}{4^n} + \cdots .$$

Any series such as this for which there is a constant ratio between successive summands is called a **geometric series**.

For many values of x, the infinite geometric series can be summed using the identity

$$1 + x + x^2 + x^3 + x^4 + \cdots = \frac{1}{1 - x}. \tag{2.7}$$

Examples of this are

$$1 + \frac{1}{3} + \frac{1}{9} + \frac{1}{27} + \frac{1}{81} + \cdots = \frac{3}{2} = \frac{1}{1 - 1/3}$$

and

$$1 - \frac{1}{2} + \frac{1}{4} - \frac{1}{8} + \frac{1}{16} - \cdots = \frac{2}{3} = \frac{1}{1 - (-1/2)}.$$

One has to be very careful with equation (2.7). If we set $x = 2$, we get a very strange equality:

$$1 + 2 + 4 + 8 + 16 + \cdots = \frac{1}{1 - 2} = -1. \tag{2.8}$$

We need to decide what we mean by an infinite summation. We could define $1 + x + x^2 + x^3 + \cdots$ to mean $1/(1 - x)$, in which case equation (2.8) is correct. We would be in good company. Leonhard Euler accepted this definition. It yields many other interesting results, for example:

$$1 - 2 + 4 - 8 + 16 - \cdots = \frac{1}{1 - (-2)} = \frac{1}{3}.$$

In the exhaustive and fascinating account, *Convolutions in French Mathematics, 1800–1840*, Ivor Grattan-Guinness writes, "Some modern appraisals of the cavalier style of 18th-century mathematicians in handling infinite series convey the impression that these poor men set their brains aside when confronted by them." They did not. Certainly Euler had not set his brain aside. He rather viewed infinite series in a larger context, a context that he makes clear in his article "On divergent series" published in 1760. Euler illustrates his understanding with the series $1 - 1 + 1 - 1 + \cdots$ which he asserts to be equal to 1/2, obtained by setting $x = -1$ in equation (2.7).

> Notable enough, however, are the controversies over the series $1 - 1 + 1 - 1 + 1-$ *etc.* whose sum was given by Leibniz as 1/2, although others disagree.... Understanding of this question is to be sought in the word "sum"; this idea, if thus conceived—namely, the sum of a series is said to be that quantity to which it is brought closer as more terms of the series are taken—has relevance only for the convergent series, and we should in general give up this idea of sum for divergent series. On the other hand, as series in analysis arise from the expansion of fractions or irrational quantities or even of transcendentals, it will in turn be permissible in calculation to substitute in place of such series that quantity out of whose development it is produced.

Here is the point we have been making: for *any* infinite summation we need to stretch our definition of sum. Euler merely asks that in the case of a series that does not converge, we allow a value determined by the genesis of the series.

As we shall see in section 2.6, Euler's approach raises more problems than it settles. Eventually, mathematicians would be forced to allow divergent series to have values. Such values are too useful to abandon completely. But using these values must be done with great delicacy. The scope of this book will only allow brief glimpses of how this can be done safely. The Archimedean understanding is the easiest and most reliable way of assigning values to infinite series.

Web Resource: To learn more about divergent series, go to **Assigning values to divergent series**.

When an infinite series has a target value in the sense of Archimedes' understanding, we say that our series converges. For our purposes, it will be safest not to assign a value to an infinite series unless it converges.

Definition: convergence of an infinite series

An infinite series **converges** if there is a target value T so that for any $L < T$ and any $M > T$, all of the partial sums from some point on are strictly between L and M.

Cauchy's Approach

Returning to equation (2.7), it is tempting to try to prove this result using precisely the associative law that we saw does not work:

$$1 = 1 - x + x - x^2 + x^2 - x^3 + x^3 - \cdots$$
$$= (1 - x) + x(1 - x) + x^2(1 - x) + x^3(1 - x) + \cdots$$
$$= (1 + x + x^2 + x^3 + \cdots)(1 - x),$$
$$\frac{1}{1 - x} = 1 + x + x^2 + x^3 + \cdots . \tag{2.9}$$

In 1821, Augustin Louis Cauchy published his *Cours d'analyse de l'École Royale Polytechnique* (*Course in Analysis of the Royal Institute of Technology*). One of his intentions in writing this book was to put the study of infinite series on a solid foundation. In his introduction, he writes,

> As for the methods, I have sought to give them all of the rigor that one insists upon in geometry, in such manner as to never have recourse to explanations drawn from algebraic technique. Explanations of this type, however commonly admitted, especially in questions of convergent and divergent series and real quantities that arise from imaginary expressions, cannot be considered, in my opinion, except as heuristics that will sometimes suggest the truth, but which accord little with the accuracy that is so praised in the mathematical sciences.

When Cauchy speaks of "algebraic technique," he is specifically referring to the kind of technique employed in equation (2.9). While this argument is suggestive, we cannot rely upon it.

Cauchy shows how to handle a result such as equation (2.7). We need to restrict our argument to the safe territory of finite summations:

$$1 = 1 - x + x - x^2 + x^2 - \cdots - x^n + x^n$$
$$= (1 - x) + x(1 - x) + x^2(1 - x) + \cdots + x^{n-1}(1 - x) + x^n,$$
$$\frac{1}{1 - x} = 1 + x + x^2 + \cdots + x^{n-1} + \frac{x^n}{1 - x},$$
$$1 + x + x^2 + \cdots + x^{n-1} = \frac{1}{1 - x} - \frac{x^n}{1 - x}. \tag{2.10}$$

Cauchy follows the lead of Archimedes. What we call the infinite series is really just the sequence of values obtained from these finite sums. Approaching the problem in this way, we can see exactly how much the finite geometric series differs from the target value, $T = 1/(1 - x)$. The difference is

$$\frac{x^n}{1 - x}.$$

If we take a value larger than T, is this finite sum eventually below it? If we take a value smaller than T, is this finite sum eventually above it? The value of this series is $1/(1 - x)$ if and only if we can make the difference as close to 0 as we wish by putting a lower bound on n. This happens precisely when $|x| < 1$.

Cauchy's careful analysis shows us that equation (2.7) needs to carry a restriction:

$$1 + x + x^2 + x^3 + \cdots = \frac{1}{1-x}, \qquad \text{provided that } |x| < 1. \qquad (2.11)$$

We have stumbled across a curious and important phenomenon. Ordinary equalities do not carry restrictions like this. A statement such as

$$1 + x = \frac{1 - x^2}{1 - x}$$

is valid for any x, as long as the denominator on the right is not 0. Equation (2.11) is something very different. It is a statement about successive approximations. The equality does not mean what it usually does. The symbol $+$ no longer means quite the same. The Archimedean understanding, cumbersome as it may seem, has become essential.

Exercises

The symbol $\boxed{\textbf{M\&M}}$ indicates that *Maple* and *Mathematica* codes for this problem are available in the **Web Resources** at **www.macalester.edu/aratra**.

2.2.1. Find the target value of the series

$$1 + \frac{1}{3} + \frac{1}{9} + \cdots + \frac{1}{3^k} + \cdots .$$

Find a value of n so that any partial sum with at least n terms is within 0.001 of the target value. Justify your answer.

2.2.2. Find the target value of the series

$$1 - \frac{3}{4} + \frac{9}{16} - \frac{27}{64} + \cdots + (-1)^k \frac{3^k}{4^k} + \cdots .$$

Find a value of n so that any partial sum with at least n terms is within 0.001 of the target value. Justify your answer.

2.2.3. Find the target value of the series

$$\frac{1}{5} - \frac{1}{6} + \frac{5}{36} - \frac{25}{216} + \cdots + (-1)^k \frac{5^{k-1}}{6^k} + \cdots .$$

Find a value of n so that any partial sum with at least n terms is within 0.001 of the target value. Justify your answer.

2.2.4. Find the target value of the series

$$1 + \frac{1}{2} - \frac{1}{4} + \frac{1}{8} + \frac{1}{16} - \frac{1}{32} + \cdots + \frac{1}{2^{3k}} + \frac{1}{2^{3k+1}} - \frac{1}{2^{3k+2}} + \cdots .$$

Find a value of n so that any partial sum with at least n terms is within 0.001 of the target value. Justify your answer.

2.2.5. It is tempting to differentiate each side of equation (2.11) with respect to x and to assert that

$$1 + 2x + 3x^2 + 4x^3 + \cdots = \frac{1}{(1-x)^2}. \tag{2.12}$$

Following Cauchy's advice, we know we need to be careful. Differentiate each side of equation (2.10). What is the difference between $1 + 2x + 3x^2 + \cdots + nx^{n-1}$ and $(1 - x)^{-2}$? For which values of x will this difference approach 0 as n increases?

2.2.6. Find the target value of the series

$$1 - \frac{2}{3} + \frac{3}{9} - \frac{4}{27} + \cdots + (-1)^{k-1}\frac{k}{3^{k-1}} + \cdots.$$

Find a value of n so that any partial sum with at least n terms is within 0.001 of the target value. Justify your answer.

2.2.7. Find the target value of the series

$$2 - \frac{1}{3} + \frac{4}{9} - \frac{3}{27} + \frac{6}{81} - \frac{5}{243} + \cdots + \frac{2k+2}{3^{2k}} - \frac{2k+1}{3^{2k+1}} + \cdots.$$

Find a value of n so that any partial sum with at least n terms is within 0.001 of the target value. Justify your answer.

2.2.8. (**M&M**) Explore the rearrangements of $1 - 1/2 + 1/4 - 1/8 + 1/16 - 1/32 + \cdots$. Explain why all rearrangements of this series must have the same target value.

2.2.9. (**M&M**) It is tempting to integrate each side of equation (2.11) with respect to x and to assert that

$$x + \frac{x^2}{2} + \frac{x^3}{3} + \frac{x^4}{4} + \cdots = -\ln(1 - x). \tag{2.13}$$

Following Cauchy's advice, we know we need to be careful, but now we run into trouble. What happens when we try to integrate $x^n/(1 - x)$? Fortunately, we do not have to find the exact value of the difference between $x + x^2/2 + x^3/3 + \cdots + x^n/n$ and $-\ln(1 - x)$. All we have to show is that we can make this difference as small as we wish by taking enough terms. We can do this by bounding the integral of $x^n/(1 - x)$.

We use the fact that if

$$|f(x)| < g(x) \quad \text{for all } x, \quad \text{then} \quad \left| \int_a^b f(x)\, dx \right| < \int_a^b g(x)\, dx.$$

If $0 < x < 1$, then we can find a number a so that $0 < x < a < 1$ and

$$\left| \frac{x^n}{1 - x} \right| < x^n(1 - a)^{-1}.$$

Integrate this bounding function with respect to x, and show that if $0 < x < 1$, then the partial sums of the series in equation (2.13) approach the target value of $-\ln(1-x)$ as n increases. Explain what happens for $-1 < x < 0$. Justify your answer.

2.2.10. Find the target value of the series

$$\frac{1}{2} - \frac{1}{2 \cdot 2^2} + \frac{1}{3 \cdot 2^3} - \frac{1}{4 \cdot 2^4} + \cdots + (-1)^{k-1} \frac{1}{k \cdot 2^k} + \cdots .$$

Find a value of n so that any partial sum with at least n terms is within 0.001 of the target value. Justify your answer.

2.2.11. Find the target value of the series

$$1 - 1 + \frac{3}{2 \cdot 2^2} - \frac{4}{3 \cdot 2^3} + \frac{5}{4 \cdot 2^4} - \cdots + (-1)^k \frac{k+1}{k \cdot 2^k} + \cdots .$$

Find a value of n so that any partial sum with at least n terms is within 0.001 of the target value. Justify your answer.

2.3 Calculating π

Beginning in the Middle Ages, at first hesitantly and then with increasing confidence, mathematicians plunged into the infinite. They resurfaced with treasures that Archimedes could never have imagined. The true power of calculus lies in its coupling with infinite processes. Mathematics as we know it and as it has come to shape modern science could never have come into being without some disregard for the dangers of the infinite.

As we saw in the last section, the dangers are real. The genius of the early explorers of calculus lay in their ability to sense when they could treat an infinite summation according to the rules of the finite and when they could not. Such intuition is a poor foundation for mathematics. By the time Fourier proposed his trigonometric series, it was recognized that a better understanding of what was happening—what was legitimate and what would lead to error—was needed. The solution that was ultimately accepted looks very much like what Archimedes was doing, but it would be a mistake to jump directly from Archimedes to our modern understanding of infinite series, for it would miss the point of that revolution in mathematics that occured in the late seventeenth century and that was so powerful precisely because it dared to treat the infinite as if it obeyed the same laws as the finite.

The time will come when we will insist on careful definitions, when we will concentrate on potential problems and learn how to avoid them. But the problems will not be meaningful unless we first appreciate the usefulness of playing with infinite series as if they really are summations. We begin by seeing what we can accomplish if we simply assume that infinite series behave like finite sums.

Much of the initial impetus for using the infinite came from the search for better approximations to π, the ratio of the circumference of a circle to its diameter. In this section we will describe several different infinite series as well as an infinite product that can be used to approximate π.

The Arctangent Series

One of the oldest and most elegant series for computing π is usually attributed to Gottfried Leibniz (1646–1716) but was also known to Isaac Newton (1642–1727) and to James Gregory (1638–1675). Almost two centuries earlier, it was known to Nilakantha (*ca.* 1450–1550) of Kerala in southwest India where the power series for the sine and cosine probably had been discovered even earlier by Madhava (*ca.* 1340–1425). It is

$$\frac{\pi}{4} = 1 - \frac{1}{3} + \frac{1}{5} - \frac{1}{7} + \frac{1}{9} - \frac{1}{11} + \cdots . \tag{2.14}$$

This is the special case $x = 0$ of Fourier's equation (1.7). It was discovered by integrating a geometric series.

We use the fact that the derivative of the arctangent function is $1/(1 + x^2) = 1 - x^2 + x^4 - x^6 + \cdots$. If we integrate this series, we should get the arctangent:

$$x - \frac{x^3}{3} + \frac{x^5}{5} - \frac{x^7}{7} + \cdots = \arctan x. \tag{2.15}$$

Equation (2.14) is the special case $x = 1$.

The series in equation (2.14) converges very slowly, but we have at our disposal the series for the arctangent of any value between 0 and 1. The convergence becomes much faster if we take a value of x close to 0. Around 1706, John Machin (1680–1751) calculated the first 100 digits of π using the identity

$$\pi = 16 \arctan\left(\frac{1}{5}\right) - 4 \arctan\left(\frac{1}{239}\right)$$

$$= 16\left(\frac{1}{5} - \frac{1}{3 \cdot 5^3} + \frac{1}{5 \cdot 5^5} - \frac{1}{7 \cdot 5^7} + \cdots\right)$$

$$- 4\left(\frac{1}{239} - \frac{1}{3 \cdot 239^3} + \frac{1}{5 \cdot 239^5} - \frac{1}{7 \cdot 239^7} + \cdots\right). \tag{2.16}$$

> **Web Resource:** To investigate series that converge to π, go to **More pi**.

Wallis's Product

John Wallis (1616–1703) considered the integral

$$\int_0^1 (1 - t^{1/p})^q \, dt.$$

When $p = q = 1/2$, this is the area in the first quadrant bounded by the graph of $y = \sqrt{1 - x^2}$, the upper half circle. It equals $\pi/4$. Wallis knew the binomial theorem for integer exponents,

$$(1 + x)^n = 1 + \binom{n}{1}x + \binom{n}{2}x^2 + \cdots + \binom{n}{k}x^k + \cdots + x^n, \tag{2.17}$$

and he knew how to integrate a rational power of x. Relying on what happens at integer values of q, he was able to extrapolate to other values. From the patterns he observed, he discovered remarkable bounds for $\sqrt{\pi/2}$:

$$\frac{2 \cdot 4 \cdot 6 \cdots (2n - 2)\sqrt{2n}}{3 \cdot 5 \cdot 7 \cdots (2n - 1)} > \sqrt{\frac{\pi}{2}} > \frac{2 \cdot 4 \cdot 6 \cdots (2n - 2)(2n)}{3 \cdot 5 \cdot 7 \cdots (2n - 1)\sqrt{2n + 1}}, \qquad (2.18)$$

valid for any $n \geq 2$. This implies that

$$\frac{\pi}{2} = \frac{2}{1} \cdot \frac{2}{3} \cdot \frac{4}{3} \cdot \frac{4}{5} \cdot \frac{6}{5} \cdot \frac{6}{7} \cdot \frac{8}{7} \cdots . \qquad (2.19)$$

> To learn how John Wallis discovered equation (2.19), go to **Appendix A.1, Wallis on π**.

Newton's Binomial Series

In 1665, Isaac Newton read Wallis's *Arithmetica infinitorum* in which he explains how to derive his product identity. This led Newton to an even more important discovery. He described the process in a letter to Leibniz written on October 24, 1676.

The starting point was both to generalize and to simplify Wallis's integral. Newton looked at

$$\int_0^x (1 - t^2)^{m/2} \, dt.$$

When m is an even integer, we can use the binomial expansion in equation (2.17) to produce a polynomial in x:

$$\int_0^x (1 - t^2)^0 \, dt = x,$$

$$\int_0^x (1 - t^2)^1 \, dt = x - \frac{1}{3}x^3,$$

$$\int_0^x (1 - t^2)^2 \, dt = x - \frac{2}{3}x^3 + \frac{1}{5}x^5,$$

$$\int_0^x (1 - t^2)^3 \, dt = x - \frac{3}{3}x^3 + \frac{3}{5}x^5 - \frac{1}{7}x^7,$$

$$\vdots$$

What happens when m is an odd integer? Is it possible to interpolate between these polynomials? If it is, then we could let $m = 1$ and $x = 1$ and obtain an expression for $\pi/4$.

Newton realized that the problem comes down to expanding $(1 - t^2)^{m/2}$ as a polynomial in t^2 and then integrating each term. Could this be done when m is an odd integer? Playing with the patterns that he discovered, he stumbled upon the fact that not only could he find an expansion for the binomial when the exponent is $m/2$, m odd, he could get the expansion with any exponent. Unless the exponent is a positive integer (or zero), the expansion is an infinite series.

> **Newton's Binomial Series**
>
> For *any* real number a and any x such that $|x| < 1$, we have that
>
> $$(1+x)^a = 1 + ax + \frac{a(a-1)}{2!}x^2 + \frac{a(a-1)(a-2)}{3!}x^3 + \cdots. \qquad (2.20)$$

> **Web Resource:** To learn how Newton discovered his binomial series, go to **Newton's formula**. To explore the convergence of the series for π that arises from the binomial series, go to **More pi**.

Equipped with equation (2.20) and assuming that there are no problems with term-by-term integration, we can find another series that approaches $\pi/4$:

$$\frac{\pi}{4} = \int_0^1 (1 - t^2)^{1/2}\, dt$$

$$= \int_0^1 \left(1 - \frac{1}{2}t^2 + \frac{(1/2)(-1/2)}{2!}t^4 - \frac{(1/2)(-1/2)(-3/2)}{3!}t^4 + \cdots\right) dt$$

$$= 1 - \frac{1}{2\cdot 3} - \frac{1}{4\cdot 2!\cdot 5} - \frac{3}{8\cdot 3!\cdot 7} - \frac{3\cdot 5}{16\cdot 4!\cdot 9} - \cdots. \qquad (2.21)$$

This series is an improvement over equation (2.14), but Newton showed how to use his binomial series to do much better. He considered the area of the shaded region in Figure 2.4. On the one hand, this area is represented by the series:

$$\text{Area} = \int_0^{1/4} \sqrt{x - x^2}\, dx$$

$$= \int_0^{1/4} x^{1/2}(1 - x)^{1/2}\, dx$$

$$= \int_0^{1/4} \left(x^{1/2} - \frac{1}{2}x^{3/2} + \frac{(1/2)(-1/2)}{2!}x^{5/2} - \frac{(1/2)(-1/2)(-3/2)}{3!}x^{7/2} + \cdots\right) dx$$

$$= \frac{2}{3}\left(\frac{1}{4}\right)^{3/2} - \frac{2}{5\cdot 2}\left(\frac{1}{4}\right)^{5/2} - \frac{2}{7\cdot 2^2\cdot 2!}\left(\frac{1}{4}\right)^{7/2} - \frac{2\cdot 3}{9\cdot 2^3\cdot 3!}\left(\frac{1}{4}\right)^{9/2}$$

$$- \frac{2\cdot 3\cdot 5}{11\cdot 2^4\cdot 4!}\left(\frac{1}{4}\right)^{11/2} - \cdots$$

$$= \frac{1}{3\cdot 2^2} - \frac{1}{5\cdot 2^5} - \frac{1}{7\cdot 2^8\cdot 2!} - \frac{3}{9\cdot 2^{11}\cdot 3!} - \frac{3\cdot 5}{11\cdot 2^{14}\cdot 4!} - \cdots. \qquad (2.22)$$

On the other hand, this area is one-sixth of a circle of radius 1/2 minus a right triangle whose base is 1/4 and whose hypotenuse is 1/2:

$$\text{Area} = \frac{\pi}{24} - \frac{\sqrt{3}}{32}. \qquad (2.23)$$

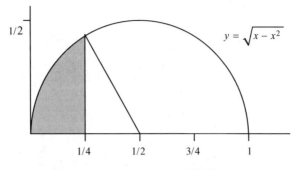

FIGURE 2.4. The area under $y = \sqrt{x - x^2}$ from 0 to 1/4.

The square root of 3 can be expressed using the binomial series:

$$\sqrt{3} = 2\sqrt{3/4}$$

$$= 2\left(1 - \frac{1}{4}\right)^{1/2}$$

$$= 2\left(1 - \frac{1}{2 \cdot 4} - \frac{1}{2^2 \cdot 4^2 \cdot 2!} - \frac{3}{2^3 \cdot 4^3 \cdot 3!} - \frac{3 \cdot 5}{2^4 \cdot 4^4 \cdot 4!} - \cdots\right). \quad (2.24)$$

Putting these together, we see that

$$\pi = 24\left(\frac{1}{3 \cdot 2^2} - \frac{1}{5 \cdot 2^5} - \frac{1}{7 \cdot 2^8 \cdot 2!} - \frac{3}{9 \cdot 2^{11} \cdot 3!} - \frac{3 \cdot 5}{11 \cdot 2^{14} \cdot 4!} - \cdots\right)$$

$$+ \frac{3}{2}\left(1 - \frac{1}{2^3} - \frac{1}{2^6 \cdot 2!} - \frac{3}{2^9 \cdot 3!} - \frac{3 \cdot 5}{2^{12} \cdot 4!} - \cdots\right). \quad (2.25)$$

All of this work is fraught with potential problems. We have simply assumed that we may integrate the infinite summations by integrating each term. In fact, here it works. That will not always be the case.

Newton's discovery was more than a means of calculating π. The binomial series is one that recurs repeatedly and has become a critical tool of analysis. It is a simple series that raises difficult questions. In Chapter 4, we will return to this series and determine the values of a for which it converges at one or both of the endpoints, $x = \pm 1$.

Ramanujan's Series

The calculation of π was and continues to be an important source of interesting infinite series. Modern calculations to over two billion digits are based on far more complicated series such as the one published by S. Ramanujan (1887–1920) in 1915:

$$\frac{1}{\pi} = \frac{\sqrt{8}}{9801} \sum_{n=0}^{\infty} \frac{(4n)!}{(n!)^4} \frac{(1103 + 26390n)}{396^{4n}}.$$

> **Web Resource:** To learn more about approximations to π and to find links and references, go to **More pi**.

Exercises

The symbol (**M&M**) indicates that *Maple* and *Mathematica* codes for this problem are available in the **Web Resources** at **www.macalester.edu/aratra**.

2.3.1. Find a value of n so that any partial sum with at least n terms is within $1/100$ of the target value $\pi/4$ of the series $1 - 1/3 + 1/5 - 1/7 + \cdots$. Justify your answer.

2.3.2. Find a value of n so that any partial sum with at least n terms is within 0.001 of the target value for the series expansion of $\arctan(1/2)$. Justify your answer.

2.3.3. Use the method outlined in exercise 2.2.9 to show that for $|x| < 1$ the partial sums of $x - x^3/3 + x^5/5 - x^7/7 + \cdots$ can be forced arbitrarily close to the target value of $\arctan x$ by taking enough terms.

2.3.4. Use equation (2.14) to prove that

$$\frac{\pi}{8} = \frac{1}{1\cdot 3} + \frac{1}{5\cdot 7} + \frac{1}{9\cdot 11} + \cdots.$$

2.3.5. Prove Machin's identity:

$$\frac{\pi}{4} = 4\arctan\left(\frac{1}{5}\right) - \arctan\left(\frac{1}{239}\right).$$

2.3.6. How many terms of each series in equation (2.16) did Machin have to take in order to calculate the first 100 digits of π? Specify how many terms are needed so that the series for $16\arctan(1/5)$ and for $4\arctan(1/239)$ are each within 2.5×10^{-100} of their target values.

2.3.7. (**M&M**) Use your answer to exercise 2.3.6 to find the first 100 digits of π.

2.3.8. Explain why the geometric series is really just a special case of Newton's binomial series.

2.3.9. What happens to Newton's binomial series when a is a positive integer? Explain why it turns into a polynomial.

2.3.10. When $a = 1/2$, Newton's binomial series becomes the series expansion for $\sqrt{1+x}$. Find a value of n so that any partial sum with at least n terms is within 0.001 of the target value $\sqrt{3/2}$. Justify your answer.

2.3.11. It may appear that Newton's binomial series can only be used to find approximations to square roots of numbers between 0 and 2, but once you can do this, you can find a series for the square root of any positive number. If $x \geq 2$, then find the integer n so that $n^2 \leq x < (n+1)^2$. It follows that $\sqrt{x} = n\sqrt{x/n^2}$ and $1 \leq x/n^2 < 2$. Use this idea to find a series expansion for $\sqrt{13}$. Find a value of n so that any partial sum with at least n terms is within 0.001 of the target value $\sqrt{13}$. Justify your answer.

2.3.12. (**M&M**) Evaluate the partial sum of at least the first hundred terms of the binomial series expansion of $(1 + x)^2$ at $x = 0.5$ and for each of the following values of a: -2, -0.4, 1/3, 3, and 5.2. In each case, does the numerical evidence suggest that you are converging to the true value of $(1 + .5)^a$? Describe and comment on what you see happening.

2.3.13. (**M&M**) Evaluate the partial sum of at least the first hundred terms of the binomial series expansion of $\sqrt{1 + x}$ at $x = -2$, -1, 0.9, 0.99, 1, 1.01, 1.1, and 2. In each case, does the numerical evidence suggest that you are converging to the true value of $\sqrt{1 + x}$? Describe and comment on what you see happening.

2.3.14. (**M&M**) Graph $y = \sqrt{1 + x}$ for $-1 \leq x \leq 2$ and compare this with the graphs over the same interval of the polynomial approximations of degrees 2, 5, 8 and 11 obtained from the binomial series

$$1 + \frac{1/2}{1} x + \frac{(1/2)(-1/2)}{2!} x^2 + \frac{(1/2)(-1/2)(-3/2)}{3!} x^3 + \cdots .$$

Describe what is happening in these graphs. For which values of x is each polynomial a good approximation to $\sqrt{1 + x}$?

2.3.15. Using the methods of this section, find an infinite series that is equal to

$$\int_0^1 \left(1 - t^3\right)^{1/3} dt.$$

2.4   Logarithms and the Harmonic Series

In exercise 2.2.9 on page 21, we saw how to justify integrating each side of

$$\frac{1}{1 - x} = 1 + x + x^2 + x^3 + x^4 + \cdots$$

to get the series expansion, valid for $|x| < 1$,

$$-\ln(1 - x) = x + \frac{x^2}{2} + \frac{x^3}{3} + \frac{x^4}{4} + \frac{x^5}{5} + \cdots . \tag{2.26}$$

Replacing x by $-x$ and multiplying through by -1, we get the series expansion for the natural logarithm in its usual form,

$$\ln(1 + x) = x - \frac{x^2}{2} + \frac{x^3}{3} - \frac{x^4}{4} + \frac{x^5}{5} - \cdots . \tag{2.27}$$

Around 1667, this identity was independently discovered by Isaac Newton and by Nicolaus Mercator (?–1687). Mercator was the first to publish it. Though we have only proved its validity for $-1 < x < 1$, it also holds for $x = 1$ where it yields the target value for the alternating harmonic series,

$$\ln 2 = 1 - \frac{1}{2} + \frac{1}{3} - \frac{1}{4} + \frac{1}{5} - \frac{1}{6} + \cdots . \tag{2.28}$$

What about the harmonic series,

$$1 + \frac{1}{2} + \frac{1}{3} + \frac{1}{4} + \frac{1}{5} + \cdots?$$

It is not hard to see that under the Archimedean understanding, this does not have a value. Consider the partial sums of the first 2^n terms:

$$n = 1: \quad 1 + \frac{1}{2} = \frac{3}{2},$$

$$n = 2: \quad 1 + \frac{1}{2} + \left(\frac{1}{3} + \frac{1}{4}\right) > 1 + \frac{1}{2} + \frac{2}{4} = 2,$$

$$n = 3: \quad 1 + \frac{1}{2} + \left(\frac{1}{3} + \frac{1}{4}\right) + \left(\frac{1}{5} + \cdots + \frac{1}{8}\right) > 1 + \frac{1}{2} + \frac{2}{4} + \frac{4}{8} = \frac{5}{2},$$

$$n = 4: \quad 1 + \frac{1}{2} + \left(\frac{1}{3} + \frac{1}{4}\right) + \left(\frac{1}{5} + \cdots + \frac{1}{8}\right) + \left(\frac{1}{9} + \cdots + \frac{1}{16}\right)$$

$$> 1 + \frac{1}{2} + \frac{2}{4} + \frac{4}{8} + \frac{8}{16} = 3.$$

In general, we see that

$$1 + \frac{1}{2} + \frac{1}{3} + \cdots + \frac{1}{2^n} > 1 + \frac{1}{2} + \frac{2}{4} + \frac{4}{8} + \cdots + \frac{2^{n-1}}{2^n} = 1 + n \cdot \frac{1}{2} = \frac{n+2}{2}.$$

No matter what number we pick, we can find an n so that all of the partial sums with at least 2^n terms will exceed that number. There is no target value.

What about ∞? The problem is that ∞ is not a number, so it cannot be a target value. Nevertheless, there is something special about the way in which this series diverges. No matter how large a number we pick, all of the partial sums beyond some point will be larger than that number.

Definition: divergence to infinity

When we write that an infinite series equals ∞, we mean that no matter what number we pick, we can find an n so that all of the partial sums with at least n terms will exceed that number.

We write

$$1 + \frac{1}{2} + \frac{1}{3} + \frac{1}{4} + \frac{1}{5} + \cdots = \infty,$$

but the fact that we have set this series equal to infinity does not mean that it has a value. This series does not have a value under the Archimedean understanding. What we have written is shorthand for the fact that this series diverges in a special way.

Euler's Constant

How large is the nth partial sum of the harmonic series? This is not an idle question. It arises in many interesting and important problems. We are going to find a simple approximation, in terms of n, to the value of the partial sum of the first $n - 1$ terms of the harmonic series, $1 + 1/2 + 1/3 + \cdots + 1/(n - 1)$.

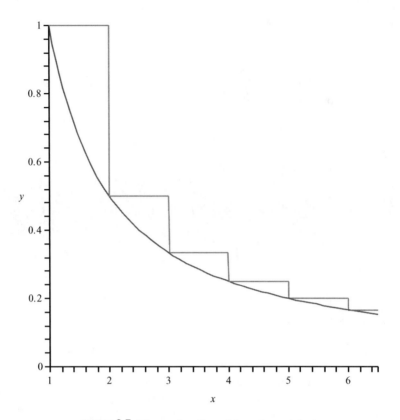

FIGURE 2.5. The graphs of $y = 1/x$ and $y = 1/\lfloor x \rfloor$.

Web Resource: For problems that require finding the values of the partial sums of the harmonic series, go to **Explorations of the Harmonic Series**.

The key to getting started is to think of this value as an area. It is the area under $y = 1/\lfloor x \rfloor$ from $x = 1$ to n—this is why we only took the first $n - 1$ terms of the harmonic series. (The symbol $\lfloor x \rfloor$ denotes the greatest integer less than or equal to x, read as the **floor** of x.) Our area is slightly larger than that under the graph of $y = 1/x$ from $x = 1$ to $x = n$ (see Figure 2.5). The area under the graph of $y = 1/x$ is

$$\int_1^n \frac{1}{x}\,dx = \ln n.$$

How much larger is the area we want to find?

The missing areas can be approximated by triangles. The first has area $(1 - 1/2)/2$, the second $(1/2 - 1/3)/2$, and so on. The sum of the areas of the triangles is

$$\frac{1}{2}\left(1 - \frac{1}{2}\right) + \frac{1}{2}\left(\frac{1}{2} - \frac{1}{3}\right) + \cdots + \frac{1}{2}\left(\frac{1}{n-1} - \frac{1}{n}\right) = \frac{1}{2} - \frac{1}{2n}.$$

This sum approaches $1/2$ as n gets larger. This is not big enough. We have missed part of the area between the curves. But it gives us some idea of the probable size of this missing area.

The value that this missing area approaches as n increases is denoted by the greek letter γ, read *gamma*, and is called Euler's constant, since it was Leonard Euler (1707–1783) who discovered this constant and established the exact connection between the harmonic series and the natural logarithm in 1734.

Definition: Euler's constant, γ

Euler's constant is defined as the limit between the partial sum of the harmonic series and the natural logarithm,

$$\gamma = \lim_{n \to \infty} \left(1 + \frac{1}{2} + \frac{1}{3} + \cdots + \frac{1}{n-1} - \ln n \right). \tag{2.29}$$

Estimating Euler's Gamma

We define

$$x_n = 1 + \frac{1}{2} + \frac{1}{3} + \cdots + \frac{1}{n-1} - \ln n. \tag{2.30}$$

This sequence records the accumulated areas between $1/\lfloor x \rfloor$ and $1/x$ for $1 \leq x \leq n$, so it is increasing. By definition, γ is the value of its limit, but how do we know that this sequence has a limit? If it does, how large is γ? We will answer these questions by finding another sequence that decreases toward γ, enabling us to squeeze the true value between these two sequences.

We define

$$y_n = 1 + \frac{1}{2} + \frac{1}{3} + \cdots + \frac{1}{n-1} + \frac{1}{n} - \ln n = x_n + \frac{1}{n}. \tag{2.31}$$

We can use the series expansion for $\ln(1 + 1/n)$ to show that this sequence is decreasing (see exercise 2.4.3 for a geometric proof that the sequence $(y_1, y_2, \ldots)$ decreases):

$$y_n - y_{n+1} = \ln(n+1) - \ln(n) - \frac{1}{n+1}$$

$$= \ln\left(1 + \frac{1}{n}\right) - \frac{1}{n+1}$$

$$= \frac{1}{n} - \frac{1}{n+1} - \frac{1}{2n^2} + \frac{1}{3n^3} - \frac{1}{4n^4} + \frac{1}{5n^5} - \cdots$$

$$= \frac{2n(n+1) - 2n^2 - (n+1)}{2n^2(n+1)} + \frac{1}{3n^3} - \frac{1}{4n^4} + \frac{1}{5n^5} - \cdots$$

$$= \frac{n-1}{2n^2(n+1)} + \left(\frac{1}{3n^3} - \frac{1}{4n^4}\right) + \left(\frac{1}{5n^5} - \frac{1}{6n^6}\right) + \cdots$$

$$> 0.$$

This implies that

$$y_1 > y_2 > y_3 > y_4 > \cdots.$$

Now $y_n = x_n + 1/n$, and so y_n is always larger than x_n. As n gets larger, x_n and y_n get closer. For example:

$$x_{10} = 0.526\ldots < \gamma < y_{10} = 0.626\ldots,$$
$$x_{100} = 0.5722\ldots < \gamma < y_{100} = 0.5822\ldots,$$
$$x_{1000} = 0.57671\ldots < \gamma < y_{1000} = 0.57771\ldots.$$

Since we know that x_n and y_n differ by $1/n$, we can bring them as close together as we want, and find γ to whatever accuracy we desire.

But as we narrow in, is there really something there?

The Nested Interval Principle

What may seem to be a ridiculous question is actually very profound. It gets to the heart of what we mean by the real number line. It would not be until the second half of the 19th century that anyone seriously asked this question: If we have two sequences approaching each other, one always increasing, the other always decreasing, and so that the distance between the sequences can be made as small as we wish by going out far enough on both sequences, do these sequences have a limit?

We cannot prove that such a limit must exist. The existence of the limit must be wrapped up in the definition of what we mean by the real number line, stated as an **axiom** or fundamental assumption.

Definition: nested interval principle

Given an increasing sequence, $x_1 \leq x_2 \leq x_3 \leq \cdots$, and a decreasing sequence, $y_1 \geq y_2 \geq y_3 \geq \cdots$, such that y_n is always larger than x_n but the difference between y_n and x_n can be made arbitrarily small by taking n sufficiently large, there is *exactly* one real number that is greater than or equal to every x_n and less than or equal to every y_n.

The important part of this principle is that there is *at least* one number that is greater than or equal to every x_n and less than or equal to every y_n. We cannot have more than one such number. If

$$x_n \leq a < b \leq y_n$$

for all n, then $y_n - x_n$ would have to be at least as large as $b - a$. But our assumption is that we can make $y_n - x_n$ as small as we want.

The conclusion that there is at least one such number is something we cannot prove, not without making some other assumption that is equivalent to the nested interval principle. We have reached one of the foundational assumptions on which a careful and rigorous treatment of calculus can be built. It took mathematicians a long time to realize this. In the early 1800s, the nested interval principle was used as if it was too obvious to bother justifying or even stating very carefully. What it guarantees is that the real number line has no holes in it. If two sequences are approaching each other from different directions, then where they "meet" there is always some number.

This principle will play an important role in future chapters when we enter the nineteenth century and begin to grapple with questions of continuity and convergence. It will be our primary tool for showing that a desired number actually exists even when we do not know what it is.

Approximating Partial Sums of the Harmonic Series

What is the first integer n for which

$$1 + \frac{1}{2} + \frac{1}{3} + \cdots + \frac{1}{n} > 10?$$

We know that the sum of the first n terms of the harmonic series is about $\ln n$, so $e^{10} \approx$ 22,000 should be roughly accurate. That is off by a factor close to 2 because we did not use γ. Our partial sum is closer to $\ln n + \gamma$, so we want $\ln n + \gamma \approx 10$ or $n \approx e^{10-\gamma} \approx 12{,}366.968$. This requires a fairly accurate approximation to γ. We can use

$$\gamma = 0.5772156649012.$$

Will 12,366 terms be enough, or do we need the 12,367th? Or is it the case that we are close, but not quite close enough to be able to determine the exact number of terms? Can we find out without actually adding 12,366 fractions?

We know that

$$1 + \frac{1}{2} + \cdots + \frac{1}{n-1} - \ln n < \gamma < 1 + \frac{1}{2} + \cdots + \frac{1}{n-1} + \frac{1}{n} - \ln n,$$

and so

$$\gamma + \ln(n-1) \; < \; 1 + \frac{1}{2} + \cdots + \frac{1}{n-1} \; < \; \gamma + \ln n \; < \; 1 + \frac{1}{2} + \cdots + \frac{1}{n}. \quad (2.32)$$

It follows that

$$9.9999217 < 1 + \frac{1}{2} + \cdots + \frac{1}{12366} < 10.00000258 < 1 + \frac{1}{2} + \cdots + \frac{1}{12367},$$

so the answer must be either 12,366 or 12,367.

In his original paper of 1734, Euler gave more information that will enable us to decide which is the correct answer. We observe that

$$\ln(n+1) - \ln n = \ln\left(1 + \frac{1}{n}\right) = \frac{1}{n} - \frac{1}{2n^2} + \frac{1}{3n^3} - \frac{1}{4n^4} + \cdots,$$

which we will write as

$$\frac{1}{n} = \ln(n+1) - \ln n + \frac{1}{2n^2} - \frac{1}{3n^3} + \frac{1}{4n^4} - \frac{1}{5n^5} + \cdots. \quad (2.33)$$

This implies the following identities:

$$1 = \ln 2 - \ln 1 + \frac{1}{2} - \frac{1}{3} + \frac{1}{4} - \frac{1}{5} + \cdots,$$

$$\frac{1}{2} = \ln 3 - \ln 2 + \frac{1}{2 \cdot 2^2} - \frac{1}{3 \cdot 2^3} + \frac{1}{4 \cdot 2^4} - \frac{1}{5 \cdot 2^5} + \cdots,$$

$$\frac{1}{3} = \ln 4 - \ln 3 + \frac{1}{2 \cdot 3^2} - \frac{1}{3 \cdot 3^3} + \frac{1}{4 \cdot 3^4} - \frac{1}{5 \cdot 3^5} + \cdots,$$

$$\frac{1}{4} = \ln 5 - \ln 4 + \frac{1}{2 \cdot 4^2} - \frac{1}{3 \cdot 4^3} + \frac{1}{4 \cdot 4^4} - \frac{1}{5 \cdot 4^5} + \cdots,$$

$$\vdots$$

$$\frac{1}{n-1} = \ln(n) - \ln(n-1) + \frac{1}{2 \cdot (n-1)^2} - \frac{1}{3 \cdot (n-1)^3}$$
$$+ \frac{1}{4 \cdot (n-1)^4} - \frac{1}{5 \cdot (n-1)^5} + \cdots.$$

Adding these equations (and recognizing that $\ln 1 = 0$), we see that

$$1 + \frac{1}{2} + \frac{1}{3} + \cdots + \frac{1}{n-1} = \ln n$$
$$+ \frac{1}{2}\left(1 + \frac{1}{2^2} + \frac{1}{3^2} + \cdots + \frac{1}{(n-1)^2}\right)$$
$$- \frac{1}{3}\left(1 + \frac{1}{2^3} + \frac{1}{3^3} + \cdots + \frac{1}{(n-1)^3}\right)$$
$$+ \frac{1}{4}\left(1 + \frac{1}{2^4} + \frac{1}{3^4} + \cdots + \frac{1}{(n-1)^4}\right)$$
$$- \frac{1}{5}\left(1 + \frac{1}{2^5} + \frac{1}{3^5} + \cdots + \frac{1}{(n-1)^5}\right) + \cdots. \quad (2.34)$$

This implies that

$$\gamma = \lim_{n \to \infty} \left(1 + \frac{1}{2} + \frac{1}{3} + \cdots + \frac{1}{n-1} - \ln n\right)$$
$$= \frac{1}{2}\sum_{m=1}^{\infty}\frac{1}{m^2} - \frac{1}{3}\sum_{m=1}^{\infty}\frac{1}{m^3} + \frac{1}{4}\sum_{m=1}^{\infty}\frac{1}{m^4} - \frac{1}{5}\sum_{m=1}^{\infty}\frac{1}{m^5} + \cdots. \quad (2.35)$$

We can now see exactly how far the partial sum of the harmonic series is from $\ln n + \gamma$:

$$1 + \frac{1}{2} + \frac{1}{3} + \cdots + \frac{1}{n-1} - \ln n - \gamma$$
$$= -\frac{1}{2}\sum_{m=n}^{\infty}\frac{1}{m^2} + \frac{1}{3}\sum_{m=n}^{\infty}\frac{1}{m^3} - \frac{1}{4}\sum_{m=n}^{\infty}\frac{1}{m^4} + \frac{1}{5}\sum_{m=n}^{\infty}\frac{1}{m^5} - \cdots. \quad (2.36)$$

It follows that

$$1 + \frac{1}{2} + \frac{1}{3} + \cdots + \frac{1}{n-1} < \ln n + \gamma - \frac{1}{2}\sum_{m=n}^{\infty}\frac{1}{m^2} + \frac{1}{3}\sum_{m=n}^{\infty}\frac{1}{m^3}.$$

We can use integrals to approximate these sums:

$$\sum_{m=n}^{\infty} \frac{1}{m^2} > \int_n^{\infty} \frac{dx}{x^2} = \frac{1}{n},$$

$$\sum_{m=n}^{\infty} \frac{1}{m^3} < \frac{1}{n^3} + \int_n^{\infty} \frac{dx}{x^3} = \frac{2+n}{2n^3}.$$

Finally, we see that

$$1 + \frac{1}{2} + \frac{1}{3} + \cdots + \frac{1}{12366} < \ln 12367 + \gamma - \frac{1}{2 \cdot 12367} + \frac{12369}{6 \cdot 12367^3} < 9.9999622.$$

The first time the partial sum of the harmonic series exceeds 10 is with the 12,367th summand.

Exercises

The symbol (**M&M**) indicates that *Maple* and *Mathematica* codes for this problem are available in the **Web Resources** at **www.macalester.edu/aratra**.

2.4.1. Give an example of a series that diverges to ∞ but whose partial sums do not form an increasing sequence.

2.4.2. Give an example of a series that does not diverge to ∞ but whose partial sums are increasing.

2.4.3. Show that the area above $1/\lfloor x + 1 \rfloor$ and below $1/x$ for $1 \leq x \leq n$ is equal to $\ln n - (1/2 + 1/3 + \cdots + 1/n)$. Use this fact to prove that $y_n = 1 + 1/2 + 1/3 + \cdots + 1/n - \ln n$ is a decreasing sequence.

2.4.4. (**M&M**) Evaluate the partial sum of the power series for $\ln(1 + x)$ with at least 100 terms at $x = -0.9, 0.9, 0.99, 0.999, 1, 1.001, 1.01$, and 1.1. Compare these approximations with the actual value of $\ln(1 + x)$. Describe and comment on what you see happening.

2.4.5. (**M&M**) In 1668, James Gregory came up with an improvement on the series for $\ln(1 + x)$. He started with the observation that

$$\ln\left(\frac{1+z}{1-z}\right) = 2\left(z + \frac{z^3}{3} + \frac{z^5}{5} + \cdots\right). \tag{2.37}$$

Using the series for $\ln(1 + x)$, prove this identity. Show that if $1 + x = (1 + z)/(1 - z)$ then $z = x/(x + 2)$ and therefore, for any $x > -1$,

$$\ln(1 + x) = 2\left(\frac{x}{x+2} + \frac{1}{3}\left(\frac{x}{x+2}\right)^3 + \cdots\right). \tag{2.38}$$

Explore the convergence of this series. How many terms are needed in order to calculate $\ln 5$ to within an accuracy of $1/100{,}000$ if we set $x = 4$ in equation (2.38)? How many terms are needed to get this same accuracy if we set $x = -4/5$ and calculate $-\ln(1/5)$? Justify your answers.

2.4.6. (**M&M**) Evaluate the partial sum of the series given in equation (2.38) using at least 100 terms at $x = -0.9, 0.9, 1, 1.1, 5, 20$, and 100. Describe and comment on these results and compare them with the results of exercise 2.4.4.

2.4.7. (**M&M**) Using equations (2.14), page 23, and (2.28), page 28, express the series

$$1 + \frac{1}{2} - \frac{1}{3} - \frac{1}{4} + \frac{1}{5} + \frac{1}{6} - \frac{1}{7} - \frac{1}{8} + \cdots$$

in terms of π and $\ln 2$. Check your result by calculating the sum of the first 1000 terms of this series.

2.4.8. Identify the point at which the following argument goes wrong. Which is the first equation that is not true? Explain why it is not true.

Let f be any given function. Then:

$$\int_1^2 f(x)\,dx = \int_0^2 f(x)\,dx - \int_0^1 f(x)\,dx. \tag{2.39}$$

Letting $x = 2y$ in the first integral on the right:

$$\int_0^2 f(x)\,dx = 2\int_0^1 f(2y)\,dy \tag{2.40}$$

$$= 2\int_0^1 f(2x)\,dx. \tag{2.41}$$

Take $f(x)$ such that $f(2x) = \frac{1}{2}f(x)$ for all values of x. Then

$$\int_1^2 f(x)\,dx = 2\int_0^1 \frac{1}{2}f(x)\,dx - \int_0^1 f(x)\,dx \tag{2.42}$$

$$= 0. \tag{2.43}$$

Now $f(2x) = \frac{1}{2}f(x)$ is satisfied by $f(x) = 1/x$. Thus, $\int_1^2 dx/x = 0$, so $\log 2 = 0$.

2.4.9. (**M&M**) The summations in equation (2.35) are well known as the zeta (greek letter ζ) functions:

$$\zeta(k) = \sum_{m=1}^{\infty} \frac{1}{m^k}. \tag{2.44}$$

When k is even, these are equal to a rational number times π^k. There is no simple formula when k is odd, but these values of $\zeta(k)$ are also known to very high accuracy. What happens to the values of $\zeta(k)$ as k increases? Assuming that we have arbitrarily good accuracy on the values of $\zeta(k)$, how many terms of the series in equation (2.35) are needed to calculate γ to within 10^{-6}?

> To see how to evaluate the zeta function at positive even integers, go to **Appendix A.3, Sums of Negative Powers**.

2.4.10. We know that the harmonic series does not converge. A result that is often seen as surprising is that if we eliminate those integers that contain the digit 9, the partial sums of the resulting series *do* stay bounded. Prove that the partial sums of the reciprocals of the integers that do not contain any 9's in their decimal representation are bounded.

2.4.11. Show that if we sum the reciprocals of the positive integers that contain neither an 8 nor a 9, the sum must be less than 35.

2.4.12. Find an upper bound for the sum of the reciprocals of the positive integers that do not contain the digit 1.

2.4.13. (**M&M**) The following procedure enables us to estimate the sum of

$$1 + \frac{1}{\sqrt{2}} + \frac{1}{\sqrt{3}} + \frac{1}{\sqrt{4}} + \cdots + \frac{1}{\sqrt{n}}.$$

a. Show that

$$\sqrt{k+1} - \sqrt{k} = \frac{1}{\sqrt{k+1} + \sqrt{k}},$$

and then explain how this implies that

$$2\sqrt{k+1} - 2\sqrt{k} < \frac{1}{\sqrt{k}} < 2\sqrt{k} - 2\sqrt{k-1}. \tag{2.45}$$

b. List the double inequalities of (2.45) for $k = 1, 2, 3, \ldots, n$, and then add up each column to prove that

$$2\sqrt{n+1} - 2\sqrt{1} < 1 + \frac{1}{\sqrt{2}} + \cdots + \frac{1}{\sqrt{n}} < 2\sqrt{n} - 2\sqrt{0}. \tag{2.46}$$

Show that this implies that

$$61.27 < 1 + \frac{1}{\sqrt{2}} + \frac{1}{\sqrt{3}} + \frac{1}{\sqrt{4}} + \cdots + \frac{1}{\sqrt{1000}} < 63.25.$$

c. Using a computer or calculator, what is the value to six significant digits of the series

$$1 + \frac{1}{\sqrt{2}} + \frac{1}{\sqrt{3}} + \frac{1}{\sqrt{4}} + \cdots + \frac{1}{\sqrt{1000}}?$$

d. Find bounds for

$$1 + \frac{1}{\sqrt{2}} + \frac{1}{\sqrt{3}} + \frac{1}{\sqrt{4}} + \cdots + \frac{1}{\sqrt{1,000,000,000}}.$$

e. Does the infinite series

$$1 + \frac{1}{\sqrt{2}} + \frac{1}{\sqrt{3}} + \frac{1}{\sqrt{4}} + \cdots$$

converge or does it diverge to ∞? Explain your answer.

2.4.14. Find a simple function of n in terms of $\ln n$, call it ω, so that

$$\lim_{n \to \infty} \left(1 + \frac{1}{3} + \frac{1}{5} + \cdots + \frac{1}{2n-1} - \omega(n) \right) = 0.$$

2.4.15. Consider the rearranged alternating harmonic series which takes the first r positive summands, then the first s negative summands, and then alternates r positive summands with s negative summands. If we take the partial sum with $n(r+s)$ terms, it is equal to

$$\left(1 + \frac{1}{3} + \frac{1}{5} + \cdots + \frac{1}{2nr-1} \right) - \left(\frac{1}{2} + \frac{1}{4} + \frac{1}{6} + \cdots + \frac{1}{2ns} \right).$$

Using the result from exercise 2.4.14, find a simple function that differs from this summation by an amount that approaches 0 as n gets larger. Show that this function does not depend on n. Explain why *every* partial sum of this rearranged alternating harmonic series will be as close as we wish to this target value if we have enough terms.

2.4.16. Use the fact that

$$1 + \frac{1}{2} + \frac{1}{3} + \cdots + \frac{1}{n-1} - \ln n - \gamma > \frac{-1}{2} \sum_{m=n}^{\infty} \frac{1}{m^2}$$

to find a lower bound for $1 + 1/2 + 1/3 + \cdots + 1/(n-1)$ in the form $\ln n + \gamma - R(n)$ where $R(n)$ is a **rational function** of n (a ratio of polynomials). Show that $\ln n + \gamma - R(n)$ is strictly larger than $\ln(n-1) + \gamma$.

2.4.17. (M&M) Use the result from exercise 2.4.16 to find the precise smallest integer n such that $1 + 1/2 + 1/3 + \cdots + 1/n$ is larger than 100. Show the work that leads to your answer.

2.4.18. You are asked to walk to the end of an infinitely stretchable rubber road that is one mile long. After each step, the road stretches uniformly so that it is one mile longer than it was before you took that step. Assuming that there are 2000 steps to a mile and that you are moving at the brisk pace of two steps per second, show that you will eventually reach the end of the road. Find the approximate time (in years) that it will take.

2.5 Taylor Series

Infinite series explode across the eighteenth century. They are discovered, investigated, and utilized. They are recognized as a central pillar of calculus, so much so that one of the most important books to be published in this century, Euler's *Introductio in analysin infinitorum* of 1748, is a primer on infinite series. There is no calculus in it in the sense that there are no derivatives, no integrals, only what Euler calls "algebra," but it is the algebra of the infinite: derivations of the power series for all of the common functions and some extraordinary manipulations of them. This is done not as a consequence of calculus but as a preparation for it. As he says in the *Preface*:

> Often I have considered the fact that most of the difficulties which block the progress of students trying to learn analysis stem from this: that although they understand little of ordinary algebra, still they attempt this more subtle art. From this it follows not only that they remain on the fringes, but in addition they entertain strange ideas about the concept of the infinite, which they must try to use ... I am certain that the material I have gathered in this book is quite sufficient to remedy that defect.

By the end of the seventeenth century, **power series**,

$$a_0 + a_1 x + a_2 x^2 + a_3 x^3 + a_4 x^4 + \cdots ,$$

had emerged as one of the primary tools of calculus. They were useful for finding approximations. They soon became indispensable for solving differential equations. As long as x is restricted to the interval where the power series are defined, they can be differentiated, integrated, added, multiplied, and composed as if they were ordinary polynomials.

One example of their utility can be found in Leonhard Euler's analysis of 1759 of the vibrations of a circular drumhead. Euler was led to the differential equation

$$\frac{d^2 u}{dr^2} + \frac{1}{r} \frac{du}{dr} + \left(\alpha^2 - \frac{\beta^2}{r^2} \right) u = 0, \tag{2.47}$$

where u (the vertical displacement) is a function of r (the distance from the center of the drum) and α and β are constants depending on the properties of the drumhead. There is no closed form for the solution of this differential equation, but if we assume that the solution can be expressed as a power series,

$$u = r^\lambda + a_1 r^{\lambda+1} + a_2 r^{\lambda+2} + a_3 r^{\lambda+3} + \cdots ,$$

then we can solve for λ and the a_i.

Web Resource: To learn more about the drumhead problem and to see how Euler solved it, go to **Euler's solution to the vibrating drumhead**.

Power series are *useful*. They are also ubiquitous. Every time a power series representation was sought, it was found. It might be valid for all x as with $\sin x$, or only for a restricted range of x as with $\ln(1 + x)$, but it was always there. In 1671, James Gregory wrote to John Collins and listed the first five or six terms of the power series for $\tan x$, $\arctan x$, $\sec x$, $\ln \sec x$, $\sec^{-1}(\sqrt{2}\, e^x)$, $\ln(\tan(x/2 + \pi/4))$, and $2\arctan(\tanh x/2)$. Clearly, he was drawing on some underlying machinery to generate these.

Everyone seemed to know about this power series machine. Gottfried Leibniz and Abraham de Moivre had each described it and explained the path of their discovery in separate letters to Jean Bernoulli, Leibniz in 1694, de Moivre in 1708. Newton had hinted at it in his geometric interpretation of the coefficients of a power series in Book II, Proposition X of the *Principia* of 1687. He elucidated it fully in an early draft of *De Quadratura* but removed it before publication. Johann Bernoulli published the general result

$$\int_0^x f(t)\, dt = xf(x) - \frac{x^2}{2!} f'(x) + \frac{x^3}{3!} f''(x) - \frac{x^4}{4!} f'''(x) + \cdots \qquad (2.48)$$

in the journal *Acta Eruditorum* in 1694. He would later point out that this is equivalent to the machine in question. Today, this machine is named for the first person to actually put it into print, Brook Taylor (1685–1731). It appeared in his *Methodus incrementorum* of 1715. His derivation is based on an interpolation formula discovered independently by James Gregory and Isaac Newton (Book III, Lemma V of the *Principia*).

Taylor's Formula

The machine described by Taylor expresses the coefficients of the power series in terms of the derivatives at a particular point.

Definition: Taylor Series

If all of the derivatives of the function f exist at the point a, then the Taylor series for f about a is the infinite series

$$f(x) = f(a) + f'(a)(x - a) + \frac{f''(a)}{2!} (x - a)^2 + \frac{f'''(a)}{3!} (x - a)^3$$

$$+ \frac{f^{(4)}(a)}{4!} (x - a)^4 + \cdots . \qquad (2.49)$$

This has as a special case ($a = 0$):

$$f(x) = f(0) + f'(0) x + \frac{f''(0)}{2!} x^2 + \frac{f'''(0)}{3!} x^3 + \frac{f^{(4)}(0)}{4!} x^4 + \cdots . \qquad (2.50)$$

All power series are special cases of equation (2.49). For example, if $f(x) = \ln(1 + x)$, we observe that

$$f(a) = \ln(1 + a),$$
$$f'(a) = (1 + a)^{-1},$$
$$f''(a) = -(1 + a)^{-2},$$
$$f'''(a) = 2(1 + a)^{-3},$$
$$\vdots$$
$$f^{(n)}(a) = (-1)^{n-1}(n - 1)!\,(1 + a)^{-n}.$$

Making this substitution into equation (2.50), we obtain

$$\ln(1 + x) = \ln 1 + 1 \cdot x + \frac{-1}{2!} x^2 + \frac{2}{3!} x^3 + \cdots + \frac{(-1)^{n-1}(n-1)!}{n!} x^n + \cdots$$

$$= x - \frac{x^2}{2} + \frac{x^3}{3} - \cdots + (-1)^{n-1} \frac{x^n}{n} + \cdots .$$

We are not yet ready to prove that equation (2.49) satisfies the Archimedean understanding of an infinite series. In fact, we know that we will have to restrict the values of x for which it is true, since even the series for $\ln(1 + x)$ is only valid for $-1 < x \leq 1$. But we can give a fast and dirty reason why it makes sense. If we assume that $f(x)$ can be represented as a power series,

$$f(x) = c_0 + c_1(x - a) + c_2(x - a)^2 + \cdots + c_k(x - a)^k + \cdots ,$$

and if we assume that we are allowed to differentiate this power series by differentiating each summand, then the kth derivative of f is equal to

$$f^{(k)}(x) = c_k k! + c_{k+1} \frac{(k+1)!}{1!}(x - a) + c_{k+2} \frac{(k+2)!}{2!}(x - a)^2 + \cdots .$$

Setting $x = a$ eliminates everything but the first term. If power series are as nice as we hope them to be, then we will have

$$f^{(k)}(a) = c_k k! \quad \Longrightarrow \quad c_k = \frac{f^{(k)}(a)}{k!}.$$

d'Alembert and the Question of Convergence

One of the first mathematicans to study the convergence of series was Jean Le Rond d'Alembert (1717–1783) in his paper of 1768, "Réflexions sur les suites et sur les racines imaginaires." d'Alembert was science editor for Diderot's *Encyclopédie* and contributed many of the articles across many different fields. He was born Jean Le Rond, a foundling whose name was taken from the church of Saint Jean Le Rond in Paris on whose steps he had been abandoned. "Le Rond" (the round or plump) refers to the shape of the church. Perhaps feeling that John the Round was a name lacking in dignity, he added d'Alembert, and occasionally signed himself Jean Le Rond d'Alembert et de la Chapelle.[1]

d'Alembert considered Newton's binomial series and asked when it is valid. In particular, he looked at the following series:

$$\sqrt{1 + \frac{200}{199}} \stackrel{?}{=} 1 + (1/2)\frac{200}{199} + \frac{(1/2)(1/2 - 1)}{2!}\left(\frac{200}{199}\right)^2$$

$$+ \frac{(1/2)(1/2 - 1)(1/2 - 2)}{3!}\left(\frac{200}{199}\right)^3 + \cdots . \tag{2.51}$$

As d'Alembert pointed out, the series begins well. The partial sums of the first 100 and the first 101 terms are, respectively, 1.416223987 and 1.415756552. It appears to be converging very quickly toward the correct value near 1.41598098.

---

[1] d'Alembert was not the only famous mathematician to create his own surname. James Joseph Sylvester was born James Joseph.

> **Web Resource:** To explore the convergence of d'Alembert's series, go to **Explorations of d'Alembert's series**.

Starting out well is not enough. d'Alembert analyzed this series by comparing it to the geometric series. What characterizes a geometric series is the fact that the ratio of any two consecutive summands is always the same. This suggests analyzing the binomial series by looking at the ratio of consecutive summands. We can then compare our series to a geometric series. The series in (2.51) has as the nth summand

$$a_n = \frac{(1/2)(1/2-1)\dots(1/2-n+2)}{(n-1)!}\left(\frac{200}{199}\right)^{n-1}.$$

The absolute value of the ratio of consecutive summands is

$$\left|\frac{a_{n+1}}{a_n}\right| = \left|\frac{\frac{(1/2)(1/2-1)\dots(1/2-n+2)(1/2-n+1)}{n!}\left(\frac{200}{199}\right)^n}{\frac{(1/2)(1/2-1)\dots(1/2-n+2)}{(n-1)!}\left(\frac{200}{199}\right)^{n-1}}\right|$$

$$= \left|\frac{(1/2-n+1)}{n}\frac{200}{199}\right|$$

$$= \left(1 - \frac{3}{2n}\right)\frac{200}{199}. \tag{2.52}$$

d'Alembert now observed that this ratio is larger than 1 whenever n is larger than 300:

$$\left(1 - \frac{3}{2n}\right)\frac{200}{199} > 1 \quad \text{if and only if} \quad n > 300.$$

At $n = 301$, the ratio is larger than 1.000016, and it approaches 200/199 as n gets larger. Once we pass $n = 300$, our summands will start to get larger. If the summands approach zero, we are not guaranteed convergence. On the other hand, if the summands do not approach zero, then the series cannot converge. Table 2.1. shows the partial sums for various values of n up to $n = 3000$. The partial sums are closest to the target value when $n = 300$, and then they move away with ever increasing error.

The general binomial series is

$$(1+x)^\alpha = 1 + \alpha x + \frac{\alpha(\alpha-1)}{2!}x^2 + \frac{\alpha(\alpha-1)(\alpha-2)}{3!}x^3 + \cdots. \tag{2.53}$$

A similar analysis can be applied. The absolute value of the ratio of the $n + 1$st to the nth summands is

$$\left|\frac{a_{n+1}}{a_n}\right| = \left|\frac{(\alpha-n+1)}{n}x\right| = \left|1 - \frac{1+\alpha}{n}\right||x|. \tag{2.54}$$

As n increases, this ratio approaches $|x|$. If $|x| > 1$, then the summands do not approach 0 and the series cannot converge to $(1+x)^\alpha$.

If $|x| < 1$, then the summands approach 0. Is this enough to guarantee that the binomial series converges to the desired value? d'Alembert did not answer this although he seemed to

Table 2.1. Binomial series approximations to $\sqrt{1 + 200/199}$.

n	sum of first n terms	sum of first $n + 1$ terms
100	1.416223987	1.415756552
200	1.416125419	1.415853117
300	1.416111363	1.415866832
400	1.416120069	1.415857961
500	1.416143716	1.415834169
600	1.416183194	1.415794514
700	1.416243295	1.415734170
800	1.416332488	1.415644629
900	1.416464086	1.415512518
1000	1.416658495	1.415317346
1500	1.420454325	1.411505919
2000	1.451536959	1.380289393
2500	1.727776909	1.102822490
3000	4.323452545	−1.504623925

imply it. Neither did he investigate what happens when $|x| = 1$ (a question with a delicate answer that depends on the value of α and the sign of x), or how far this approach can be extended to other series.

Lagrange's Remainder

Joseph Louis Lagrange (1736–1813) was born in Turin, Italy under the name Giuseppe Lodovico Lagrangia. When he published his first mathematics in 1754, he signed it Luigi De la Grange Tournier (the final name being a reference to his native city). Shortly thereafter, he adopted the French form of his name, Joseph Louis. Like many people of his time, he was not consistent in his signature. "J. L. de la Grange" was the most common. It was only after his death that "Joseph Louis Lagrange" became the common spelling. His mathematical reputation was established at an early age, and he was a frequent correspondent of Euler and d'Alembert. In 1766 he succeeded Euler at the Berlin Academy, and in 1787 he moved to Paris. He was clearly the dominant member of the committee that met, probably in early 1808, to reject Fourier's treatise on the propagation of heat.

In the revised edition of *Théorie des fonctions analytiques* published in the year of his death, Lagrange gives a means of estimating the size of the error that is introduced when any partial sum of a Taylor series is used to approximate the value of the original function. In other words, he finds a way of bounding the difference between the partial sums of the Taylor series and the function value that it approaches. This is exactly what we will need in order to prove that Taylor series satisfy the Archimedean understanding of infinite series. Proving Lagrange's Remainder Theorem will be one of the chief goals of the next chapter.

Theorem 2.1 (Lagrange's Remainder Theorem). *Given a function f for which all derivatives exist at $x = a$, let $D_n(a, x)$ denote the difference between the nth partial sum of the Taylor series for f expanded about $x = a$ and the target value $f(x)$,*

$$D_n(a, x) = f(x) - \left(f(a) + f'(a)(x - a) + \frac{f''(a)}{2!}(x - a)^2 \right.$$
$$\left. + \cdots + \frac{f^{(n-1)}(a)}{(n - 1)!}(x - a)^{n-1} \right). \tag{2.55}$$

There is at least one real number c strictly between a and x for which

$$D_n(a, x) = \frac{f^{(n)}(c)}{n!}(x - a)^n. \tag{2.56}$$

While we do not know the value of c, the fact that it lies between a and x is often enough to be able to bound the size of the error.

Web Resource: To explore the behavior of this difference function, go to **Explorations of Lagrange's Remainder** .

The Exponential, Sine, and Cosine Functions

The exponential, sine, and cosine functions all have particularly nice derivatives which, when evaluated at $x = 0$ always yield 1, -1, or 0, giving us simple Taylor series:

$$e^x = 1 + x + \frac{x^2}{2!} + \frac{x^3}{3!} + \frac{x^4}{4!} + \cdots, \tag{2.57}$$

$$\sin x = x - \frac{x^3}{3!} + \frac{x^5}{5!} - \frac{x^7}{7!} + \cdots, \tag{2.58}$$

$$\cos x = 1 - \frac{x^2}{2!} + \frac{x^4}{4!} - \frac{x^6}{6!} + \cdots. \tag{2.59}$$

For which values of x do the partial sums of these infinite series approach the target value, that is to say, when do they approach the value of the function at x?

Lagrange's remainder theorem tells us that the difference between the partial sum that ends with the term involving x^{n-1} and the target value is bounded by

$$\frac{\left| f^{(n)}(c) \right|}{n!} |x|^n,$$

for some c between 0 and x. For the exponential function, the nth derivative is e^x, and so the difference is $e^c x^n / n!$. The absolute value of this is bounded by

$$\frac{e^x}{n!} x^n \quad \text{for } x > 0, \qquad \frac{|x|^n}{n!} \quad \text{for } x < 0.$$

The situation is even simpler for the sine and cosine. Since the nth derivative is always a sine or cosine function, we know that $\left| f^{(n)}(c) \right| \le 1$ no matter what the value of c, and so the difference is bounded by $|x|^n / n!$.

What happens to $\frac{|x|^n}{n!}$ as n gets large? Stirling's formula gives us an easy answer.

Stirling's Formula

The factorial function $n!$ is well approximated by the function $(n/e)^n \sqrt{2\pi n}$. Specifically, we have that

$$\lim_{n \to \infty} \frac{n!}{(n/e)^n \sqrt{2\pi n}} = 1. \tag{2.60}$$

For a proof of Stirling's formula and for information on the accuracy of this approximation, go to **Appendix A.4, The Size of $n!$**.

With Stirling's formula on hand, we see that for *any* real number x, the difference between the exponential, sine, or cosine function and the partial sum of its Taylor series can be made arbitrarily small by taking n sufficiently large,

$$\lim_{n \to \infty} \frac{|x|^n}{n!} = \lim_{n \to \infty} \left(\frac{e|x|}{n} \right)^n \frac{1}{\sqrt{2\pi n}} = 0.$$

Lagrange and the Binomial Series

Lagrange's remainder enables us to answer three questions left open by d'Alembert's analysis of the binomial series:

a. What happens when $x = 1$?

b. If the series converges, how many terms must we take in order to obtain the desired degree of accuracy?

c. If the series diverges, how accurate can we be?

To simplify our calculations, we shall restrict our attention to Newton's original expansion:

$$\sqrt{1+x} = 1 + (1/2)x + \frac{(1/2)(-1/2)}{2!} x^2 + \cdots .$$

If we take the partial sum up to

$$\frac{(1/2)(-1/2) \cdots (-n + 1 + 3/2)}{(n-1)!} x^{n-1},$$

then the difference between this partial sum and $\sqrt{1+x}$ is

$$D_n(0, x) = \frac{(1/2)(-1/2) \cdots (-n + 3/2)(1 + c)^{-n+1/2}}{n!} x^n. \tag{2.61}$$

For $x > 0$, we find an upper bound on the absolute value of $D_n(0, x)$ by taking $c = 0$. The error that is introduced by using the polynomial approximation of degree $n - 1$ is

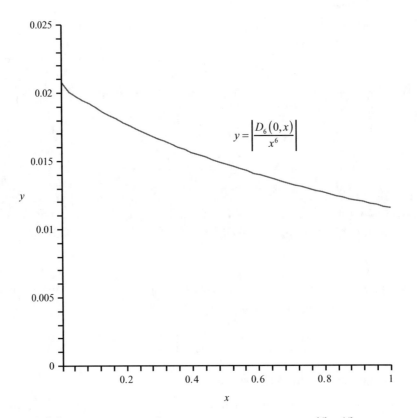

FIGURE 2.6. Plot of $|D_6(0, x)/x^6|$. Note that it stays well below $6^{-3/2}\pi^{-1/2} \approx 0.0384$.

bounded by

$$
\begin{aligned}
|D_n(0, x)| &\leq \left| \frac{(1/2)(-1/2)\cdots(-n + 3/2)}{n!} x^n \right| \\
&= \frac{1 \cdot 3 \cdot 5 \cdots (2n - 3)}{2 \cdot 4 \cdot 6 \cdots (2n)} |x|^n.
\end{aligned} \tag{2.62}
$$

Using Wallis's inequality (2.18) on page 24, we have that

$$
\frac{1 \cdot 3 \cdots (2n - 1)}{2 \cdot 4 \cdots (2n - 2)\sqrt{2n}} < \sqrt{\frac{2}{\pi}}, \tag{2.63}
$$

and so (Figure 2.6)

$$
\begin{aligned}
|D_n(0, x)| &\leq \frac{1 \cdot 3 \cdot 5 \cdots (2n - 3)}{2 \cdot 4 \cdot 6 \cdots (2n)} |x|^n \\
&< \frac{\sqrt{2/\pi}}{(2n - 1)\sqrt{2n}} |x|^n \\
&< |x|^n n^{-3/2} \pi^{-1/2}.
\end{aligned} \tag{2.64}
$$

When $x = 1$, the error term does approach zero as n gets larger. Given $|x| < 1$ and a limit on the size of the allowable error, inequality (2.64) can be used to see how large n must be. If $|x|$ is larger than 1, then the error will eventually grow without bound. This bound is minimized when

$$n = \frac{3}{2 \ln |x|}.$$

If $x = 200/199$, we want to choose $n = 300$. The resulting approximation will be within

$$\frac{(200/199)^{300}}{300^{3/2} \sqrt{\pi}} \approx 4.88 \times 10^{-4}.$$

The True Significance

The Lagrange remainder for the Taylor series is more than a tool for estimating errors. It makes precise the difference between the partial sum which is a polynomial and the target function that this polynomial approximates. This precision will come to play a critical role as we try to pin down the reasons why certain series behave well while others must be treated with great care.

Exercises

The symbol $\boxed{\text{M\&M}}$ indicates that *Maple* and *Mathematica* codes for this problem are available in the **Web Resources** at **www.macalester.edu/aratra**.

2.5.1. $\boxed{\text{M\&M}}$ Find the first five nonzero terms in the power series for each of Gregory's functions.

a. $\tan x$

b. $\arctan x$

c. $\sec x$

d. $\ln(\sec x)$

e. $\sec^{-1}(\sqrt{2}\, e^x)$, ($\sec^{-1}$ is the arc secant)

f. $\ln[\tan(x/2 + \pi/4)]$

g. $2 \arctan(\tanh x/2)$, [$\tanh x = (e^x - e^{-x})/(e^x + e^{-x})$ is the hyperbolic tangent]

2.5.2. $\boxed{\text{M\&M}}$ For each problem in exercise 2.5.1, graph the given function and compare it to the graph of the first five terms of its power series. For what values of x do you have a good approximation?

2.5.3. Prove Bernoulli's identity, equation (2.48), by using repeated integration by parts:

$$\int_0^x f(t)\, dt = x\, f(x) - \int_0^x t f'(t)\, dt$$

$$= x\, f(x) - \frac{x^2}{2} f'(x) + \int_0^x \frac{t^2}{2} f''(t)\, dt$$

$$= \dots .$$

2.5.4. Use Lagrange's remainder theorem to determine the number of terms of the partial sum for the power series expansion of the exponential, sine, or cosine function that are needed in order to guarantee that the partial sum is within $1/100$ of each of the following target values.

a. e^3

b. e^{10}

c. $\sin\left(\frac{\pi}{4}\right)$

d. $\sin\left(\frac{2\pi}{3}\right)$

e. $\cos\left(\frac{\pi}{2}\right)$

2.5.5. Show that Taylor's series implies Bernoulli's identity by first using Taylor's series to prove that

$$f(x) - f(0) = f'(0)x + \frac{f''(0)}{2!}x^2 + \frac{f'''(0)}{3!}x^3 + \cdots,$$

$$f'(x) - f'(0) = f''(0)x + \frac{f'''(0)}{2!}x^2 + \frac{f^{(4)}(0)}{3!}x^3 + \cdots,$$

$$f''(x) - f''(0) = f'''(0)x + \frac{f^{(4)}(0)}{2!}x^2 + \frac{f^{(5)}(0)}{3!}x^3 + \cdots,$$

$$f'''(x) - f'''(0) = f^{(4)}(0)x + \frac{f^{(5)}(0)}{2!}x^2 + \frac{f^{(6)}(0)}{3!}x^3 + \cdots,$$

$$\vdots$$

Now eliminate $f'(0)$, $f''(0)$, $f'''(0)$, ... to obtain

$$f(x) = f(0) + f'(x)x - \frac{f''(x)}{2!}x^2 + \frac{f'''(x)}{3!}x^3 - \frac{f^{(4)}(x)}{4!}x^4 + \cdots.$$

2.5.6. Use Taylor series to find the power series for $(1+x)^\alpha$. What happens when α is a positive integer? What happens when $\alpha = 0$?

2.5.7. (**M&M**) Consider the binomial series for the reciprocal of the square root,

$$(1+x)^{-1/2} = 1 + (-1/2)x + \frac{(-1/2)(-3/2)}{2!}x^2 + \frac{(-1/2)(-3/2)(-5/2)}{3!}x^3 + \cdots.$$

$$(2.65)$$

Calculate the partial sums as n goes from 100 to 1000 in steps of 100 when $x = 200/199$. Describe what you see.

2.5.8. (**M&M**) Calculate the partial sum of the series in equation (2.65) as n goes from 100 to 1000 in steps of 100 when $x = 1$ and when $x = -1$. Describe what you see. Make a guess of whether or not this series converges for these values of x. Explain your reasoning.

2.5.9. Find a good bound for the absolute value of the Lagrange remainder for the series in equation (2.65). What happens to this bound when $|x| = 1$ and n gets large? What can you say about the convergence of this series at $x = \pm 1$?

2.5.10. For the series in equation (2.65) with $x = 1/2$, how many terms in the partial sum are needed in order to guarantee that the partial sum is within 0.01 of the target value? Use the Lagrange remainder to answer this question and show the work that leads to your answer.

2.5.11. Prove that

$$\left(1 - \frac{3}{2n}\right)\frac{200}{199} > 1 \qquad \text{if and only if} \qquad n > 300.$$

In general, prove that when $n \geq 1 + \alpha > 0$, then

$$\left|\frac{(\alpha - n + 1)x}{n}\right| > 1$$

if and only if $|x| > 1$ and

$$n > \frac{1 + \alpha}{1 - |x|^{-1}}.$$

What happens if $1 + \alpha \leq 0$?

2.5.12. (M&M) Calculate the partial sum for $\sqrt{1 + (200/199)}$:

$$1 + (1/2)\left(\frac{200}{199}\right) + \cdots + \frac{(1/2)(-1/2)\dots(-297.5)}{299!}\left(\frac{200}{199}\right)^{299}$$

and compare your result to the calculator value of $\sqrt{399/199}$. Is the error within the predicted bounds? How close are you to the outer bound?

2.5.13. Find the Lagrange remainder for an approximation to an arbitrary binomial series:

$$(1 + x)^{\alpha} \approx 1 + \alpha x + \frac{\alpha(\alpha - 1)}{2!}x^2 + \cdots + \frac{\alpha(\alpha - 1)\dots(\alpha - n + 2)}{(n - 1)!}x^{n-1}.$$

2.5.14. Simplify the Lagrange remainder of the previous exercise when $\alpha = -1$. What happens to this remainder when $x = 1$ and n increases? Does it approach 0?

2.5.15. What is wrong with the following argument? Using the Lagrange remainder, we know that

$$(1 + x)^{-1} = 1 - x + x^2 - x^3 + \cdots + (-1)^{n-1}x^{n-1} + D_n(0, x),$$

where $D_n(0, x) = (-1)^n x^n/(1 + c)^{n+1}$ for some c between 0 and x. If $x = 1$, then c is positive and the absolute value of the Lagrange remainder is $1/(1 + c)^{n+1}$ which approaches 0 as n increases.

2.5.16. $\boxed{\textbf{M\&M}}$ Experiment with different values of α between -1 and $1/2$ in the Lagrange remainder for the binomial series (exercise 2.5.13). Does the remainder approach zero when $x = 1$ and n increases? Describe and discuss the results of your experiments.

2.5.17. Find the Lagrange form of the remainder for the partial sum approximation to $\ln(1 + x)$:

$$\ln(1 + x) \approx x - \frac{x^2}{2} + \cdots + (-1)^{n-2}\frac{x^{n-1}}{n-1}. \tag{2.66}$$

Use this error bound to prove that when $x = 1$ the partial sums approach the target value $\ln 2$ as n increases. How large must n be in order to guarantee that the partial sum is within 0.001 of the target value? Show the work that leads to your answer.

2.5.18. Continuing exercise 2.5.17, find the value of c, $0 \le c \le 200/199$, that maximizes the absolute value of the Lagrange form of the remainder for the partial sum approximation to $\ln(1 + x)$ given in equation (2.66) and evaluated at $x = 200/199$. Find the value of n that minimizes this bound.

2.5.19. Consider Lagrange's form of the remainder for $D_6(0, x)$ when the function is $f(x) = \sqrt[3]{1 + x} = (1 + x)^{1/3}$, $x > 0$. Find the value of c, $0 \le c \le x$, that maximizes the absolute value of $D_6(0, x)$. Graph the resulting function of x, and find the largest value of x for which this bound is less than or equal to 0.5.

2.5.20. $\boxed{\textbf{M\&M}}$ For the function $f(x) = \ln(1 + x)$, graph x^{-7} times each of the two functions, the greater of which must bound the remainder when $n = 6$:

$$y = \frac{f^{(7)}(0)}{7!} \quad \text{and} \quad y = \frac{f^{(7)}(x)}{7!}.$$

Now graph $|D_7(0, x)x^{-7}|$, $0 \le x \le 1$, where

$$D_7(0, x) = \ln(1 + x) - \left(x - \frac{x^2}{2} + \frac{x^3}{3} - \frac{x^4}{4} + \frac{x^5}{5} - \frac{x^6}{6}\right),$$

and see how it compares with these bounding functions.

2.6   Emerging Doubts

Calculus derives its name from its use as a tool of calculation. At its most basic level, it is a collection of algebraic techniques that yield exact numerical answers to geometric problems. One does not have to know why it works to use it. But the question of *why* kept arising, partly because no one could satisfactorily answer it, partly because sometimes these techniques would fail.

Newton and his successors thought in terms of velocities and rates of change and talked of *fluxions*. For Leibniz and his school, the founding concept was the *differential*, a small increment that was not zero yet smaller than any positive quantity. Neither of these approaches is entirely satisfactory. George Berkeley (1685–1753) attacked both understandings in his classic treatise *The Analyst*, published in 1734. His point was that a belief in mechanistic principles of science that could explain everything was self-deception. Not even the calculus was sufficiently well understood that it could be employed without reliance on faith.

> By moments we are not to understand finite particles. These are said not to be moments, but quantities generated from moments, which last are only the nascent principles of finite quantities. It is said that the minutest errors are not to be neglected in mathematics: that the fluxions are celerities, not proportional to the finite increments, though ever so small; but only to the moments or nascent increments, whereof the proportion alone, and not the magnitude, is considered . . . It seems still more difficult to conceive the abstracted velocities of such nascent imperfect entities. But the velocities of the velocities, the second, third, fourth, and fifth velocities, &c, exceed, if I mistake not, all human understanding. The further the mind analyseth and pursueth these fugitive ideas the more it is lost and bewildered; the objects, at first fleeting and minute, soon vanishing out of sight. Certainly in any sense, a second or third fluxion seems an obscure mystery. The incipient celerity of an incipient celerity, the nascent augment of a nascent augment, *i.e.*, of a thing which hath no magnitude; take it in what light you please, the clear conception of it will, if I mistake not, be found impossible; whether it be so or no I appeal to the trial of every thinking reader. And if a second fluxion be inconceivable, what are we to think of third, fourth, fifth fluxions, and so on without end!
>
> The foreign mathematicians are supposed by some, even of our own, to proceed in a manner less accurate, perhaps, and geometrical, yet more intelligible. Instead of flowing quantities and their fluxions, they consider the variable finite quantities as increasing or diminishing by the continual addition or subduction of infinitely small quantities. Instead of the velocities wherewith increments are generated, they consider the increments or decrements themselves, which they call differences, and which are supposed to be infinitely small. The difference of a line is an infinitely little line; of a plane an infinitely little plane. They suppose finite quantities to consist of parts infinitely little, which by the angles they make one with another determine the curvity of the line. Now to conceive a quantity, or than any the least finite magnitude is, I confess, above my capacity. But to conceive a part of such infinitely small quantity that shall be still infinitely less than it, and consequently though multiplied infinitely shall never equal the minutest finite quantity is, I suspect, an infinite difficulty to any man whatsoever; and will be allowed such by those who candidly say what they think; provided they really think and reflect, and do not take things upon trust.

This is only a small piece of Berkeley's attack, but it illustrates the fundamental weakness of calculus which is hammered upon in the second paragraph: the need to use infinity without ever clearly defining what it means. The abuse of infinity has yielded rich rewards, but it *is* abuse. Berkeley recognizes this.

No one was prepared to abandon calculus, but the doubts that had been voiced were unsettling. Many mathematicians tried to answer the question of why it was so successful.

Berkeley himself suggested that there was a system of compensating errors underlying calculus. Jean Le Rond d'Alembert relied on the notion of limits. In 1784, the Berlin Academy offered a prize for a "clear and precise theory of what is called the infinite in mathematics." They were not entirely satisfied with any of the entrants, although the prize was awarded to the Swiss mathematician Simon Antoine Jean L'Huillier (1750–1840) who had adopted d'Alembert's limits.

To the reader who has seen the derivative and integral defined in terms of limits, it may seem that d'Alembert and L'Huillier got it right. This was not so clear to their contemporaries. In his article of 1754 on the *différentiel* for Diderot's *Encyclopédie*, d'Alembert speaks of the limit as that number that is approached "as closely as we please" by the slope of the approximating secant line. We still use this phrase to explain limits, but its meaning is not entirely clear. Mathematicians of the 18th century were not yet ready to embrace the full radicalism of the Archimedean understanding, that there is, strictly speaking, no such thing as an infinite summation or process.

Problems with Infinite Series

If the trouble had only lain in the definition of the derivative and integral, then it would not have received the attention that it did. Infinite series were also causing misgivings. Euler worked with divergent series and, as we saw in section 2.1, determined the value from the genesis of the series. He would assign the value $\sqrt{1 + 200/199}$ to the divergent series

$$1 + (1/2)\frac{200}{199} + \frac{(1/2)(1/2 - 1)}{2!}\left(\frac{200}{199}\right)^2 + \frac{(1/2)(1/2 - 1)(1/2 - 2)}{3!}\left(\frac{200}{199}\right)^3 + \cdots$$
$$(2.67)$$

because it arises from the Taylor series for $\sqrt{1 + x}$.

There is a difficulty with this point of view that was exposed by Johann Bernoulli's son Daniel (1700–1782) in 1772: different machinery can give rise to the same series with different values. The alternating series of 1's and -1's can arise when x is set equal to 1 in

$$1 - x + x^2 - x^3 + x^4 - x^5 + \cdots = \frac{1}{1 + x},$$

or

$$1 - x + x^3 - x^4 + x^6 - x^7 + \cdots = (1 - x)(1 + x^3 + x^6 + \cdots)$$
$$= \frac{1 - x}{1 - x^3}$$
$$= \frac{1}{1 + x + x^2},$$

or

$$1 - x^2 + x^3 - x^5 + x^6 + x^8 + \cdots = (1 - x^2)(1 + x^3 + x^6 + \cdots)$$
$$= \frac{1 - x^2}{1 - x^3}$$
$$= \frac{1 + x}{1 + x + x^2}.$$

The first gives a value of 1/2, the second 1/3, and the last 2/3. We can vary these exponents to get any rational number between 0 and 1. *The same divergent series can have different values depending upon the context in which it arises.* Many mathematicians found this to be a highly unsatisfactory state of affairs. On the other hand, to simply discard all divergent series is to lose those, like the error term in Stirling's formula, that are truly useful.

> For information on the error term in Stirling's formula, Go to **Appendix A.4, The Size of *n*!**.

The Vibrating String Problem

Fourier's 1807 paper on the propagation of heat was seen by Lagrange and the other members of the reviewing committee as another piece in a longstanding controversy within mathematics. This controversy had begun with the mathematical model of the vibrating string.

In 1747, d'Alembert published the differential equation governing the height y above position x at time t of a vibrating string:

$$\frac{\partial^2 y}{\partial x^2} = \frac{1}{c^2}\frac{\partial^2 y}{\partial t^2}, \tag{2.68}$$

where c depends on the length, tension, and mass of the string. To solve this equation, we need to know boundary conditions. If the ends of the string at $x = 0$ and l are fixed, then $y(0, t) = y(l, t) = 0$. We also need to know the original position of the string: $y(x, 0)$.

The situation is very similar to that of heat propagation. If $y(x)$ can be expressed as a linear combination of functions of the form $\sin(\pi x/l)$, $\sin(2\pi x/l)$, $\sin(3\pi x/l)$, ..., for example

$$y(x, 0) = 3\sin(\pi x/l) - 2\sin(3\pi x/l), \tag{2.69}$$

then the solution to equation (2.68) is

$$y(x, t) = 3\sin(\pi x/l)\,\cos(\pi ct/l) - 2\sin(3\pi x/l)\cos(3\pi ct/l). \tag{2.70}$$

It is worth noting that as a function of time, each piece of this solution is periodic. The first piece has period $2l/c$; the second has period $2l/3c$. That means that the first piece has a **frequency** of $c/2l$ vibrations per unit time; the second has frequency $3c/2l$. This explains the overtones or **harmonics** of a vibrating string.

Daniel Bernoulli suggested in 1753 that the vibrating string might be capable of infinitely many harmonics. The most general initial position should be an infinite sum of the form

$$y(x) = a_1\sin(\pi x/l) + a_2\sin(2\pi x/l) + a_3\sin(3\pi x/l) + \cdots. \tag{2.71}$$

Euler rejected this possibility. The reason for his rejection is illuminating. *The function in equation (2.71) is necessarily periodic with period 2l. Bernoulli's solution cannot handle an initial position that is not a periodic function of x.*

Euler seems particularly obtuse to the modern mathematician. We only need to describe the initial position between $x = 0$ and $x = l$. We do not care whether or not the function repeats itself outside this interval. But this misses the point of a basic misunderstanding that was widely shared in the eighteenth century.

For Euler and his contemporaries, a function was an expression: a polynomial, a trigonometric function, perhaps one of the more exotic series arising as a solution of a differential equation. As a function of x, it existed as an organic whole for all values of x. This is not to imply that it was necessarily well-defined for all values of x, but the values where it was not well-defined would be part of its intrinsic nature. Euler admitted that one could chop and splice functions. For example, one might want to consider $y = x^2$ for $x \le 1$ and $y = 2x - 1$ for $x \ge 1$. But these were two different functions that had been juxtaposed. To Euler, the shape of a function between 0 and l determined that function everywhere.

Lagrange built on this understanding when he asserted that every function has a power series representation and that the derivative of f at $x = a$ can be defined as the coefficient of $(x - a)$ in the expansion of $f(x)$ in powers of $(x - a)$. In other words, he used the Taylor series for f to define $f'(a)$, $f''(a)$, $f'''(a)$, ... As late as 1816, Charles Babbage (1792–1871), John Herschel (1792–1871), and George Peacock (1791–1858) would champion Lagrange's viewpoint. It implies that the values of a function and all of its derivatives at one point completely determine that function at every value of x.

Lagrange was asssuming that all functions, at least all functions worthy of study, possess infinitely many derivatives and so have power series expansions. We now know that that is far too limited a view. Lagrange's functions are now given a special designation. They are called **analytic functions**.

Definition: C^p and analytic functions

Given an interval I, a function with a continuous first derivative in I is said to belong to the class C^1. If the pth derivative exists and is continuous in I, the function belongs to the class C^p. If all derivatives exist, the function belongs to C^∞ and is called **analytic**.

The most revolutionary thing that Fourier accomplished in 1807 was to assert that both Daniel Bernoulli and Leonhard Euler were right. Any initial position can be expressed as an infinite sum of the form given in equation (2.71). Fourier showed how to compute the coefficients. But it is equally true that any function represented by such a trigonometric expansion is periodic. The implication is that the description of a function between 0 and l tells us nothing about the function outside this interval.

To Lagrange especially, but probably also to the other members of the committee that reviewed this manuscript, there had to be a flaw. The easiest way out was to assume a problem with the convergence of Fourier's series. In succeeding years, Fourier and others demonstrated that there was no problem with the convergence. This forced a critical reevaluation of what was meant by a function, an infinite series, a derivative. As each object of the structure that is calculus came under scrutiny, it was found to rest on uncertain foundations that needed to be examined and reconstructed. Above all, it was the notion of infinity that was in need of correction. It was Augustin Louis Cauchy (1789–1857) who

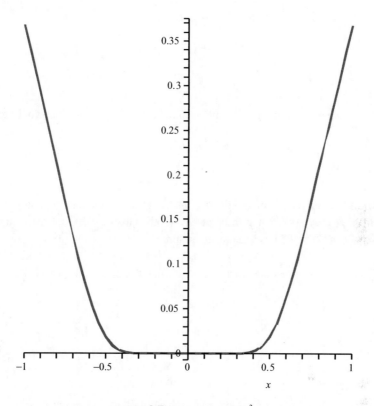

FIGURE 2.7. Plot of $y = e^{-1/x^2}$.

realized that the only true foundation was a return to the Archimedean understanding. In the 1820s, he set himself the task of reconstructing calculus upon this bedrock.

Cauchy's Counterexample

The death knell for Lagrange's definition of the derivative was sounded by Cauchy in 1821. He exhibited a counterexample to Lagrange's assertion that distinct functions have distinct power series (Figure 2.7):

$$f(x) = e^{-1/x^2}, \quad f(0) = 0.$$

All of the derivatives of $f(x)$ at $x = 0$ are equal to 0. At $x = 0$, this function has the same power series expansion as the constant function 0. The determination of the derivatives of $f(x)$ at $x = 0$ will be demonstrated in section 3.1.

Exercises

2.6.1. Rewrite the series

$$1 - x^2 + x^5 - x^7 + x^{10} - x^{12} + \cdots$$

as a rational function of x. Set $x = 1$. What value does this give for the series $1 - 1 + 1 - 1 + \cdots$?

2.6.2. Find a power series in x that would imply that

$$1 - 1 + 1 - 1 + \cdots = \frac{4}{7}$$

when x is set equal to 1.

2.6.3. Given *any* nonzero integers m and n, find a power series in x that would imply that

$$1 - 1 + 1 - 1 + \cdots = \frac{m}{n}$$

when x is set equal to 1.

2.6.4. All of the derivatives of e^x at $x = 0$ are equal to 1. Find another analytic function on $\mathbb{R}$, call it f, that is not equal to e^x at any x other than $x = 0$, but for which the nth derivative of f at 0, $f^{(n)}(0)$, is equal to 1 for all $n \geq 1$.

2.6.5. Find an analytic function on $\mathbb{R}$, call it g, such that $g(0) = 1 = g^{(n)}(0)$ for all $n \geq 1$ and $g(1) = 1$.

3

Differentiability and Continuity

In this chapter we shall compress roughly fifty years of struggle to understand differentiability and continuity. Our goal is to prove Theorem 2.1 on page 44, the Lagrange remainder theorem, but we cannot do this without first coming to grips with these two concepts. Our modern interpretations of differentiability and continuity were certainly in the air by the early 1800s. Fr. Bernhard Bolzano (1781–1848) in Prague described these concepts in terms that clearly conform to our present definitions although it is uncertain how much influence he had. Carl Gauss in private notebooks of 1814 showed considerable insight into these foundational questions, but he never published his results.

Credit for our current interpretation of differentiability and continuity usually goes to Augustin Louis Cauchy (1789–1857) and the books that he wrote in the 1820s to support the courses he was teaching, especially his *Cours d'analyse de l'École Royale Polytechnique* of 1821. Cauchy was born in the summer of the fall of the Bastille. Laplace and Lagrange were family friends who admired and encouraged this precocious child. His first job came in 1810 as a military engineer in Cherbourg preparing for the invasion of England. In the midst of it, he wrote and published mathematics. By the time he returned to Paris in 1813, he had made his reputation, amply confirmed in the succeeding years. His contemporaries described him as bigoted and cold, but he was loyal to his king and his church, and he was brilliant. He dominated French mathematics through the golden years of the 1820s. In 1830 he left France, following the last Bourbon king of France, Charles X, into exile.

It is a mistake to think that Cauchy got all of his mathematics right the first time. He was often very confused. We shall spend considerable time on the points that confused him, precisely because those are still difficulties for those who are first entering this subject. Nevertheless, Cauchy's writings were important and influential. He brought foundational questions to the forefront of mathematical research and discovered many of the definitions that would make progress possible.

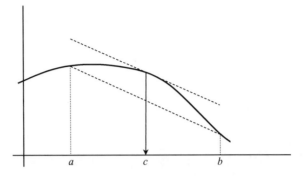

FIGURE 3.1. Illustration of the Mean Value Theorem.

It was not until the 1850s and 1860s that anything like our modern standards of rigor came into analysis. This was the result of efforts by people such as Karl Weierstrass (1815–1897) and Bernhard Riemann (1826–1866) who will be introduced in Chapter 5.

3.1 Differentiability

The Lagrange Remainder Theorem is a statement about the difference between the partial sum of the first n terms of the Taylor series and the target value of that series. We will prove it by induction on n, which means that we first need to prove the case $n = 1$:

$$f(x) = f(a) + f'(c)(x - a) \tag{3.1}$$

for some real number c strictly between a and x. Equation (3.1) is commonly known as the **Mean Value Theorem**.

Theorem 3.1 (Mean Value Theorem). *Given a function f that is differentiable at all points strictly between a and x and continuous at all points on the closed interval from a to x, there exists a real number c strictly between a and x such that*

$$\frac{f(x) - f(a)}{x - a} = f'(c). \tag{3.2}$$

This theorem says that the average rate of change of the function f over the interval from a to x is equal to the instantaneous rate of change at some point between a and x. If we look at what this means graphically (see Figure 3.1), we see that it makes sense. The slope of the secant line connecting $(a, f(a))$ to $(x, f(x))$ should be parallel to the tangent at some point between a and x. It makes sense, but how do we *prove* it? How do we know the right conditions under which this theorem is true? Could we weaken the assumption of either differentiability or continuity and still get this conclusion? Is this really all we need to assume?

Cauchy was the first mathematician to fully appreciate the importance of this theorem and to wrestle with its proof. He was not entirely successful. In the next section, we will follow his struggles because they are very informative of the complexities of working with derivatives. Right now we need to address a more basic issue: what do mean by the derivative of a function f at a?

The standard definition of the derivative given in first-year calculus is

$$f'(a) = \lim_{x \to a} \frac{f(x) - f(a)}{x - a}, \tag{3.3}$$

where this is understood to mean that $(f(x) - f(a)) / (x - a)$ is a pretty good approximation to $f'(a)$ that gets better as x gets closer to a. But what precisely do we mean by a limit? We can apply the Archimedean understanding to limits.

Definition: Archimedean understanding of limits

When we write any limit statement such as

$$\lim_{x \to a} f(x) = T,$$

what we actually mean is that if we take any number $M > T$, then we can force $f(x) < M$ by taking x sufficiently close to a. Similarly, if we take any $L < T$, then we can force $f(x) > L$ by taking x sufficiently close to (but not equal to) a.

When we apply this strict definition of the limit to the slope of the secant line, we get a precise definition of the derivative.

Definition: derivative of f at $x = a$

The **derivative** of f at a is that value, denoted $f'(a)$, such that for any $L < f'(a)$ and any $M > f'(a)$, we can force

$$L < \frac{f(x) - f(a)}{x - a} < M$$

simply by taking x sufficiently close to (but not equal to) a.

For example, if $f(x) = x^3$ and $a = 2$, then

$$\frac{x^3 - 8}{x - 2} = x^2 + 2x + 4.$$

Near $x = 2$, this is an increasing function. Our candidate for $f'(2)$ is 12. Let $L = 11.99$ and $M = 12.01$. If we take $1.999 < x < 2.001$, then

$$\frac{x^3 - 8}{x - 2} > \frac{1.999^3 - 8}{1.999 - 2} = 11.994001 > 11.99 = L,$$

$$\frac{x^3 - 8}{x - 2} < \frac{2.001^3 - 8}{2.001 - 2} = 12.006001 < 12.01 = M.$$

Note that this does not *prove* that the derivative is 12. We must show that there is an interval around $x = 2$ that works for *every* possible pair (L, M) for which $L < 12 < M$.

To accomplish this, it is useful to study the actual difference between the derivative and the average rate of change:

$$E(x, a) = f'(a) - \frac{f(x) - f(a)}{x - a}. \tag{3.4}$$

The function $E(x, a)$ is the discrepancy or error introduced when we use the average rate of change in place of the derivative, or vice-versa.

We now present notation that was introduced by Cauchy in 1823. Rather than working with L and M, he used $f'(a) - \epsilon$ and $f'(a) + \epsilon$ where ϵ is a positive amount. It is the same idea, and we do not lose anything by insisting that L and M must be the same distance from $f'(a)$.

Following Cauchy, we let $\delta > 0$ be the distance x is allowed to vary from a: $a - \delta < x < a + \delta$, $x \neq a$. It is important that $x \neq a$ because the slope of the secant line is not defined at $x = a$. These conditions are neatly summarized by $0 < |x - a| < \delta$.

Definition: Cauchy definition of derivative of f at $x = a$

The **derivative** of f at a is that value, denoted $f'(a)$, such that for any $\epsilon > 0$, we have a response $\delta > 0$ so that if $0 < |x - a| < \delta$, then this forces

$$|E(x, a)| = \left| f'(a) - \frac{f(x) - f(a)}{x - a} \right| < \epsilon.$$

In our example, $f(x) = x^3$ at $x = 2$, we have

$$E(x, 2) = 12 - (x^2 + 2x + 4) = -(x - 2)(x + 4).$$

If we are given any $\epsilon > 0$ and restrict our x to $2 - \delta < x < 2 + \delta$ where δ is the smaller of $\epsilon/10$ and 1, then

$$|E(x, s)| = |x - 2|\,|x + 4| < \delta(6 + \delta) = 6\delta + \delta^2 < 7\delta < \frac{7}{10}\epsilon.$$

We note that there is at most one number that can serve as the value of the derivative at a. To prove this, we assume that both α and β will work, $\alpha \neq \beta$. Let

$$E_1(x, a) = \alpha - \frac{f(x) - f(a)}{x - a}$$

and

$$E_2(x, a) = \beta - \frac{f(x) - f(a)}{x - a}.$$

The distance between α and β can be expressed in terms of these error functions:

$$|\beta - \alpha| = \left| \left(\beta - \frac{f(x) - f(a)}{x - a} \right) - \left(\alpha - \frac{f(x) - f(a)}{x - a} \right) \right|$$

$$= |E_2(x, a) - E_1(x, a)|$$

$$\leq |E_2(x, a)| + |E_1(x, a)|.$$

Since we can make the absolute value of each error as small as we want—certainly less than $|\beta - \alpha|/2$—this hands us a contradiction.

> **Web Resource:** To see how this error function plays a role in analyzing the effectiveness of the Newton–Raphson method for finding roots, go to **Newton–Raphson Method** .

Caution!

The definition just given is neither obvious nor easy to absorb. The reader encountering it for the first time should keep in mind that it is the fruit of two centuries of searching. It looks deceptively like the casual definition, for it compares $f'(a)$ to the average rate of change, $(f(x) - f(a))/(x - a)$. There may be a tendency to ignore the ϵ and δ and hope that they are not important. They are. This has proven to be the definition of the derivative that explains those grey areas where differentiation does not seem to be working the way we expect. Later in this section, we shall use it to explain why it is that sometimes you can differentiate an infinite series by differentiating each term, and sometimes you cannot.

The key to this definition is the emphasis on the error or difference between the derivative and the average rate of change. We have differentiability when this error can be made as small as desired by tightening up the distance between x and a. We are not just saying that we can make this error small. It is not enough to show that the absolute value of the error can be made less than 0.01 or 0.0001 or even 10^{-100}. We have to be able to get the error inside *any* specific positive bound.

ϵ and δ

I find it useful to think of ϵ and δ as a two-person game. You play ϵ; I play δ. The particular game is specified by a function, for example $f(x) = 3x^2 - 5x + 2$, and a point at which we want to check for differentiability, perhaps $x = 2$. If this function has a derivative at 2, the value of that derivative would have to be

$$f'(2) = 6 \cdot 2 - 5 = 7.$$

Our error term is

$$E(x, 2) = 7 - \frac{f(x) - f(2)}{x - 2} = 7 - \frac{3x^2 - 5x - 2}{x - 2} = 7 - (3x + 1) = -3x + 6.$$

You now get to choose any ϵ. The only constraint is that it must be positive. If you choose $\epsilon = 0.01$, then this is saying that you want to see the absolute value of the error, $|-3x + 6|$, less than 0.01. My challenge is to find a positive distance δ such that if x is within δ of 2, then the error will satisfy your constraint (see Figure 3.2). I could respond with $\delta = 0.001$. This means that $1.999 < x < 2.001$, and so

$$-0.003 < E(x, 2) < 0.003.$$

I have met your ϵ. (Note that I did not have to find the best possible δ, I only had to find some δ that would make $|E(x, 2)| < 0.01$.)

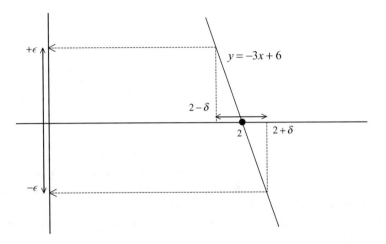

FIGURE 3.2. ϵ and δ when $E(x, 2) = -3x + 6$.

But I have not yet won. The game is not over. You are now permitted to reply with a smaller ϵ, maybe $\epsilon = 10^{-100}$. I counter with $\delta = 10^{-101}$, and we check that if $2 - 10^{-101} < x < 2 + 10^{-101}$, then

$$-3 \cdot 10^{-101} < E(x, 2) < 3 \cdot 10^{-101},$$

and so

$$|E(x, 2)| < 3 \cdot 10^{-101} < 10^{-100}.$$

At this point you realize that I always have a comeback. If you propose $\epsilon = 10^{-1000000}$, I can counter with $\delta = 10^{-1000001}$. When it is recognized that *every* positive ϵ can be countered, then I have won and we declare that $f(x) = 3x^2 - 5x + 2$ is differentiable at 2 and its derivative is 7.

On the other hand, if you ever succeed in stumping me, then the function is not differentiable at that point. Let us take the function defined by $f(x) = |x^2 - x/200|$ at $x = 0$ (see Figure 3.3). We shall try setting $f'(0) = 0$. If this is not right, then I am allowed to change it because the definition of differentiability only asks that there is some number $f'(0)$ for which I always have a comeback. The error function is

$$|E(x, 0)| = \left| 0 - \frac{|x^2 - x/200| - 0}{x - 0} \right| = |x - 0.005|.$$

We are ready to play.

Your first challenge is $\epsilon = 0.01$. I successfully counter with $\delta = 0.001$. If $-0.001 < x < 0.001$, then $|E(x, 0)| < 0.006 < 0.01$. You return with $\epsilon = 0.001$. I cannot reply. If I want

$$|x - 0.005| < 0.001,$$

then x must lie between 0.004 and 0.006. I cannot put x between those limits. I can only restrict how close x lies to 0. In desperation, I can try a different $f'(0)$. The error

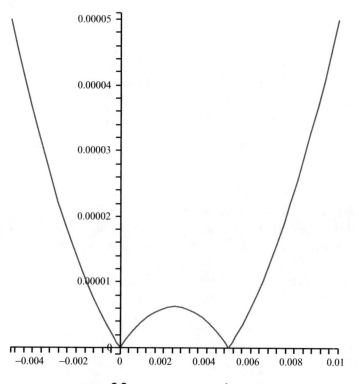

FIGURE 3.3. Graph of $f(x) = |x^2 - x/200|$.

function is

$$|E(x, 0)| = \left| f'(0) - \frac{|x^2 - 0.005x|}{x} \right|$$

$$= \begin{cases} |f'(0) - 0.005 + x|, & 0 < x < 0.005, \\ |f'(0) - x + 0.005|, & x < 0. \end{cases}$$

To make this error less than 0.001, I would have to find an $f'(0)$ for which $|f'(0) - 0.005|$ *and* $|f'(0) + 0.005|$ are each less than 0.001. No number $f'(0)$ is going to lie within $\epsilon = 0.001$ of both 0.005 and -0.005. I have lost. This function is not differentiable at $x = 0$.

Derivatives of Sums

The definition of the derivative that we have given is pointless unless we can show that it tells us something about the derivative that we did not already know. One problem that we can begin to tackle is that of determining when we are allowed to differentiate an infinite series by differentiating each term.

We recall Fourier's series,

$$F(x) = \frac{4}{\pi} \left[\cos\left(\frac{\pi x}{2}\right) - \frac{1}{3} \cos\left(\frac{3\pi x}{2}\right) + \frac{1}{5} \cos\left(\frac{5\pi x}{2}\right) - \frac{1}{7} \cos\left(\frac{7\pi x}{2}\right) + \cdots \right],$$

which is equal to 1 for $-1 < x < 1$. Anywhere between -1 and 1, the derivative of this series is 0,

$$F'(x) = 0, \quad -1 < x < 1.$$

However, if we try to differentiate each term, we obtain the series

$$G(x) = -2 \left[\sin\left(\frac{\pi x}{2}\right) - \sin\left(\frac{3\pi x}{2}\right) + \sin\left(\frac{5\pi x}{2}\right) - \sin\left(\frac{7\pi x}{2}\right) + \cdots \right].$$

For $-1 < x < 1$, this series does not converge unless $x = 0$. For example, at $x = 0.5$ we have

$$G(0.5) = -\sqrt{2}\,[1 - 1 - 1 + 1 + 1 - 1 - 1 + 1 + \cdots].$$

The conclusion is that G is not the derivative of F.

To understand the difference between differentiating a finite sum and an infinite sum, we shall begin by proving that if f and g are both differentiable at $x = a$, then so is $f + g$ and the derivative of this sum at $x = a$ is $f'(a) + g'(a)$.

Since f and g are both differentiable at a, we know that we have two error functions satisfying

$$f'(a) = \frac{f(x) - f(a)}{x - a} + E_1(x, a), \tag{3.5}$$

$$g'(a) = \frac{g(x) - g(a)}{x - a} + E_2(x, a), \tag{3.6}$$

where $|E_1|$ and $|E_2|$ can each be made as small as we wish by taking x sufficiently close to a. We now define the error function for $f + g$:

$$\begin{aligned} E(x, a) &= f'(a) + g'(a) - \frac{f(x) + g(x) - f(a) - g(a)}{x - a} \\ &= f'(a) - \frac{f(x) - f(a)}{x - a} + g'(a) - \frac{g(x) - g(a)}{x - a} \\ &= E_1(x, a) + E_2(x, a). \end{aligned} \tag{3.7}$$

The error of the sum is the sum of the errors.

We have to show that we shall always win the ϵ–δ game for the error function $E(x, a)$. We know that it is winnable for E_1 and E_2 separately. Someone serves us an $\epsilon > 0$. We give half of it to E_1 and half to E_2. In other words, we find a distance δ_1 that guarantees that

$$|E_1(x, a)| < \frac{\epsilon}{2},$$

and another distance δ_2 that guarantees that

$$|E_2(x, a)| < \frac{\epsilon}{2}.$$

We choose whichever distance is smaller and call it δ. If any restriction on the distance works, then any tighter restriction will also work. If $0 < |x - a| < \delta$, then both of the individual errors are less than $\epsilon/2$. It follows that

$$|E(x, a)| = |E_1(x, a) + E_2(x, a)| \le |E_1(x, a)| + |E_2(x, a)| < \epsilon. \tag{3.8}$$

We always have a comeback. The sum $f + g$ is differentiable at $x = a$ and the derivative is $f'(a) + g'(a)$.

What Goes Wrong with Infinite Series?

If we want to prove the differentiability of the sum of three functions, the same argument shows us that the error function for the sum is the sum of the three error functions, and we allot to each of them a third of our error bound ϵ. With a sum of four functions, each error gets a quarter of ϵ. How do we allocate the error bound if we have an infinite sum?

It can be done. Let us assume that we have

$$F(x) = f_1(x) + f_2(x) + f_3(x) + \cdots,$$

and that each function in the summation is differentiable at $x = a$:

$$f_n'(a) = \frac{f_n(x) - f_n(a)}{x - a} + E_n(x, a), \tag{3.9}$$

where we can make $|E_n(x, a)|$ arbitrarily small by taking x sufficiently close to a. Given a total error bound of ϵ, we can allocate $\epsilon/2$ to the first error, $\epsilon/4$ to the second, $\epsilon/8$ to the third, $\epsilon/16$ to the fourth, and so on. Each function has a response: $\delta_1, \delta_2, \delta_3, \ldots$. Our difficulty arises when we try to find a single positive δ that is less than or equal to each of these. If there were finitely many δs we could do it. But there are infinitely many, and there is no guarantee that they do not get arbitrarily close to 0. In other words, there may be no positive number that is less than or equal to all of our δs.

As we shall see, another way of looking at this is to sum both sides of equation (3.9):

$$\sum_{n=1}^{\infty} f_n'(a) = \sum_{n=1}^{\infty} \frac{f_n(x) - f_n(a)}{x - a} + \sum_{n=1}^{\infty} E_n(x, a)$$

$$= \frac{F(x) - F(a)}{x - a} + \sum_{n=1}^{\infty} E_n(x, a). \tag{3.10}$$

The derivative of $F = \sum_{n=1}^{\infty} f_n$ at a is $\sum_{n=1}^{\infty} f_n'(a)$ if and only if

$$E(x, a) = \sum_{n=1}^{\infty} E_n(x, a)$$

can be made arbitrarily close to 0 by taking x sufficiently close to a. Sometimes it can, and sometimes it cannot. An infinite sum of very small numbers might be very small or very large. When we learn more about infinite series, we shall see some useful criteria for determining when the derivative of an infinite series is the infinite series of the derivatives.

Strange Derivatives: $x^2 \sin(x^{-2})$ and e^{-1/x^2}

Our ϵ–δ definition of the derivative enables us to demonstrate the existence of a derivative in certain cases where a simple application of the techniques of differentiation would suggest that no derivative exists. One example is the function defined by (see Figure 3.4)

$$f(x) = x^2 \sin(x^{-2}), \ x \neq 0, \qquad f(0) = 0. \tag{3.11}$$

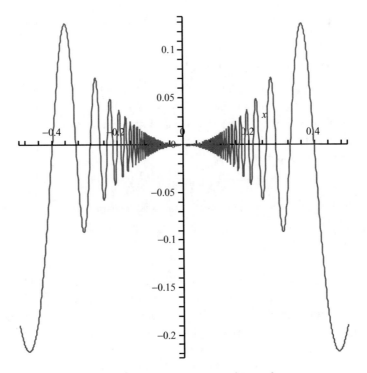

FIGURE 3.4. Graph of $f(x) = x^2 \sin(x^{-2})$.

When x is not 0, we can use our standard algorithms to find the derivative:

$$f'(x) = 2x \, \sin(x^{-2}) - 2x^{-1} \cos(x^{-2}), \quad x \neq 0.$$

(Figure 3.5). Since x cannot be set equal to 0 in the expression on the right, there is a common misconception that $f'(0)$ does not exist. But if we look at the definition of the derivative, we see that if we try setting $f'(0) = 0$, then this is the correct value provided

$$|E(x, 0)| = \left| 0 - \frac{x^2 \sin(x^{-2}) - 0}{x - 0} \right| = \left| -x \sin(x^{-2}) \right| \tag{3.12}$$

can be made less than any specified ϵ by restricting x to within some δ of 0. Since $|\sin(x^{-2})| \leq 1$, we see that

$$|E(x, 0)| \leq |x|, \tag{3.13}$$

and therefore given a bound $\epsilon > 0$, we can always reply with $\delta = \epsilon$. If $|x| < \delta = \epsilon$, then $|E(x, 0)| \leq |x| < \epsilon$. There is a response that forces $|E(x, 0)|$ to be less than ϵ, and therefore the derivative of f at $x = 0$ is zero.

Our second example is the function from section 2.6:

$$g(x) = e^{-1/x^2}, \ x \neq 0, \qquad g(0) = 0. \tag{3.14}$$

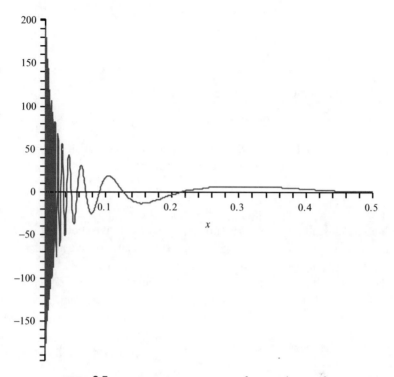

FIGURE 3.5. Graph of $f'(x) = 2x \sin(x^{-2}) - 2x^{-1} \cos(x^{-2})$.

We claimed that not only is $g'(0) = 0$, but *all* derivatives of g at $x = 0$ are zero: $g^{(n)}(0) = 0$. When $x \neq 0$, the derivatives of g are given by

$$g'(x) = 2x^{-3}e^{-1/x^2},$$
$$g''(x) = (4x^{-6} - 6x^{-4})e^{-1/x^2},$$
$$g'''(x) = (8x^{-9} - 36x^{-7} + 24x^{-5})e^{-1/x^2},$$
$$\vdots$$

In general, $g^{(n)}(x)$ will be a polynomial in x^{-1} multiplied by e^{-1/x^2} (see exercise 3.1.21). We write this as

$$g^{(n)}(x) = P_n(x^{-1})e^{-1/x^2}, \quad x \neq 0. \tag{3.15}$$

We first prove that $g'(0) = 0$. Our error function is given by

$$|E_1(x, 0)| = \left| 0 - \frac{e^{-1/x^2} - 0}{x - 0} \right| = \left| x^{-1}e^{-1/x^2} \right|. \tag{3.16}$$

It follows from equation (2.57) on page 44 that if z is positive, then e^z is strictly larger than $1 + z$. Therefore, for $x \neq 0$:

$$x^2 e^{1/x^2} > x^2 \left(1 + \frac{1}{x^2} \right) = x^2 + 1 > 1, \tag{3.17}$$

and so

$$|x| > \frac{1}{|x|\,e^{1/x^2}} = \left|x^{-1}e^{-1/x^2}\right|. \qquad (3.18)$$

Given an error bound ϵ, we can respond with $\delta = \epsilon$. If $|x| < \epsilon$, then

$$|E_1(x, 0)| = \left|x^{-1}e^{-1/x^2}\right| < |x| < \epsilon.$$

We now move to the higher derivatives of g. If we have shown that $g^{(n)}(0) = 0$, then to prove that $g^{(n+1)}(0)$ is also 0, we need to demonstrate that we can force

$$E_{n+1}(x, 0) = 0 - \frac{P_n(x^{-1})\,e^{-1/x^2} - 0}{x - 0} = x^{-1}P_n(x^{-1})\,e^{-1/x^2}$$

to be arbitrarily close to 0 by keeping x sufficiently close to 0. We know that $x^{-1}P_n(x^{-1})$ is a finite sum of powers of x^{-1}. It is enough to show that for any positive integer k, $x^{-k}e^{-1/x^2}$ can be forced arbitrarily close to 0 by keeping x sufficiently close to 0.

We choose an integer j so that $k + 1 - 2j \leq -1$ and then restrict our choices of δ to $\delta \leq 1/j!$. It follows that $|x| < \delta$ implies that $|x|^{-1} > j!$. We know that when z is positive, e^z is strictly larger than $1 + z + z^2/2! + \cdots + z^j/j!$. Since $|x| < 1/j! < 1$, we have that

$$|x|^{k+1}e^{1/x^2} > |x|^{k+1}\left(1 + \frac{1}{x^2} + \frac{1}{2x^4} + \cdots + \frac{1}{j!\,x^{2j}}\right)$$

$$> \frac{|x|^{k+1-2j}}{j!} \geq \frac{|x|^{-1}}{j!} > 1. \qquad (3.19)$$

It follows that

$$|x| > \left|x^{-k}e^{-1/x^2}\right|. \qquad (3.20)$$

Given the challenge ϵ, we respond with δ equal to either ϵ or $1/j!$, whichever is smaller.

We have proven that **every derivative of** g **at** $x = 0$ **is zero**, and therefore the Taylor series for $g(x) = e^{-1/x^2}$, expanded about $a = 0$, is the same as the Taylor series for the constant function 0.

Exercises

3.1.1. You are the teacher in a calculus class. You give a quiz in which one of the questions is, "Find the derivative of $f(x) = x^2$ at $x = 3$." One of your students writes

$$f(3) = 3^2 = 9, \qquad \frac{d}{dx}9 = \boxed{0}$$

Write a response to this student.

3.1.2. Find the derivatives (where they exist) of the following functions. The function denoted by $\lfloor x \rfloor$ sends x to the greatest integer less than or equal to x.

a. $f(x) = x\,|x|$, $\quad x \in \mathbb{R}$
b. $f(x) = \sqrt{|x|}$, $\quad x \in \mathbb{R}$

c. $f(x) = \lfloor x \rfloor \sin^2(\pi x), \quad x \in \mathbb{R}$

d. $f(x) = (x - \lfloor x \rfloor) \sin^2(\pi x), \quad x \in \mathbb{R}$

e. $f(x) = \ln|x|, \quad x \in \mathbb{R} \setminus \{0\}$

f. $f(x) = \arccos(1/|x|), \quad |x| > 1$

3.1.3. Find the derivatives of the following functions.

a. $f(x) = \log_x 2, \quad x > 0, \ x \neq 1$

b. $f(x) = \log_x (\cos x), \quad 0 < x < \pi/2, \ x \neq 1$

3.1.4. Find the derivatives (where they exist) of the following functions. The signum function, sgn x, is $+1$ if $x > 0$, -1 if $x < 0$, and 0 if $x = 0$.

a. $f(x) = \begin{cases} \arctan x, & \text{if } |x| \leq 1 \\ (\pi/4)\text{sgn } x + (x-1)/2, & \text{if } |x| > 1 \end{cases}$

b. $f(x) = \begin{cases} x^2 e^{-x^2}, & \text{if } |x| \leq 1 \\ 1/e, & \text{if } |x| > 1 \end{cases}$

c. $f(x) = \begin{cases} \arctan(1/|x|), & \text{if } x \neq 0 \\ \pi/2, & \text{if } x = 0 \end{cases}$

3.1.5. Show that the function given by

$$f(x) = \begin{cases} x^2 |\cos(\pi/x)|, & \text{if } x \neq 0 \\ 0, & \text{if } x = 0 \end{cases}$$

is not differentiable at $x_n = 2/(2n+1)$, $n \in \mathbb{Z}$, but is differentiable at 0 which is the limit of this sequence.

3.1.6. Determine the constants a, b, c, and d so that f is differentiable for all real values of x.

a. $f(x) = \begin{cases} 4x & \text{if } x \leq 0 \\ ax^2 + bx + c & \text{if } 0 < x < 1 \\ 3 - 2x & \text{if } x \geq 1 \end{cases}$

b. $f(x) = \begin{cases} ax + b & \text{if } x \leq 0 \\ cx^2 + dx & \text{if } 0 < x < 1 \\ 1 - \frac{1}{x} & \text{if } x \geq 1 \end{cases}$

c. $f(x) = \begin{cases} ax + b & \text{if } x \leq 1 \\ ax^2 + c & \text{if } 1 < x < 2 \\ \frac{dx^2 + 1}{x} & \text{if } x \geq 2 \end{cases}$

3.1.7. Let $f(x) = x^2$, $f'(a) = 2a$. Find the error $E(x, a)$ in equation (3.4) in terms of x and a. How close must x be to a if $|E(x, a)|$ is to be less than 0.01, less than 0.0001?

3.1.8. Let $f(x) = x^3$. Find the error $E(x, 1)$ in equation (3.4) as a function of x. Graph $E(x, 1)$. How close must x be to 1 if $|E(x, 1)|$ is to be less than 0.01, less than 0.0001?

3.1.9. Let $f(x) = x^3$. Find the error $E(x, 10)$ in equation (3.4) as a function of x. Graph $E(x, 10)$. How close must x be to 10 if $|E(x, 10)|$ is to be less than 0.01, less than 0.0001?

3.1.10. Let $f(x) = \sin x$. Find the error $E(x, \pi/2)$ in equation (3.4) as a function of x. Graph $E(x, \pi/2)$. Find a δ to respond with if you are given $\epsilon = 0.1, 0.0001, 10^{-100}$.

3.1.11. Use the definition of differentiability to prove that $f(x) = |x|$ is not differentiable at $x = 0$, by finding an ϵ for which there is no δ response. Explain your answer.

3.1.12. Graph the function $f(x) = x \sin(1/x)$ ($f(0) = 0$) for $-2 \le x \le 2$. Prove that $f(x)$ is not differentiable at $x = 0$, by finding an ϵ for which there is no δ response. Explain your answer.

3.1.13. Graph the function $f(x) = x^2 \sin(1/x)$ ($f(0) = 0$) for $-2 \le x \le 2$. Use the definition of differentiability to prove that $f(x)$ *is* differentiable at $x = 0$. Show that this derivative cannot be obtained by differentiating $x^2 \sin(1/x)$ and then setting $x = 0$.

3.1.14. Prove that if f is continuous at a and $\lim_{x \to a} f'(x)$ exists, then so does $f'(a)$ and they must be equal.

Web Resource: For help with finding and then presenting proofs, go to **How to find and write a proof**.

3.1.15. Assume that f and g are differentiable at a. Find

a. $\displaystyle \lim_{x \to a} \frac{xf(a) - af(x)}{x - a}$,

b. $\displaystyle \lim_{x \to a} \frac{f(x)g(a) - f(a)g(x)}{x - a}$.

3.1.16. Let f be differentiable at $x = 0$, $f'(0) \ne 0$. Find

$$\lim_{x \to 0} \frac{f(x)e^x - f(0)}{f(x)\cos x - f(0)}.$$

3.1.17. Let f be differentiable at a with $f(a) > 0$. Find

$$\lim_{x \to a} \left(\frac{f(x)}{f(a)} \right)^{1/(\ln x - \ln a)}.$$

3.1.18. Let f be differentiable at $x = 0$, $f(0) = 0$, and let k be any positive integer. Find the value of

$$\lim_{x \to 0} \frac{1}{x} \left(f(x) + f\left(\frac{x}{2}\right) + f\left(\frac{x}{3}\right) + \cdots + f\left(\frac{x}{k}\right) \right).$$

3.1.19. Let f be differentiable at a and let $\{x_n\}$ and $\{z_n\}$ be two sequences converging to a and such that $x_n \neq a$, $z_n \neq a$, $x_n \neq z_n$ for $n \in \mathbb{N}$. Give an example of f for which

$$\lim_{n \to \infty} \frac{f(x_n) - f(z_n)}{x_n - z_n}$$

a. is equal to $f'(a)$,
b. does not exist or exists but is different from $f'(a)$.

3.1.20. Let f be differentiable at a and let $\{x_n\}$ and $\{z_n\}$ be two sequences converging to a such that $x_n < a < z_n$ for $n \in \mathbb{N}$. Prove that

$$\lim_{n \to \infty} \frac{f(x_n) - f(z_n)}{x_n - z_n} = f'(a).$$

3.1.21. Consider the polynomials $P_n(x)$ defined by equation 3.15 on page 67. We have seen that

$$P_0(x) = 1,$$
$$P_1(x) = 2x^3,$$
$$P_2(x) = 4x^6 - 6x^4,$$
$$P_3(x) = 8x^9 - 36x^7 + 24x^5.$$

Find $P_4(x)$. Show that

$$P_{n+1}(x) = 2x^3 P_n(x) - x^2 P'_n(x),$$

and therefore $P_n(x)$ is a polynomial of degree $3n$.

3.1.22. Sketch the graph of the function

$$f(x) = \begin{cases} e^{1/(x^2-1)} & \text{if } |x| < 1, \\ 0 & \text{if } |x| \geq 1. \end{cases}$$

Show that all derivatives of f exist at $x = \pm 1$.

3.2 Cauchy and the Mean Value Theorems

Cauchy knew that he needed to prove Lagrange's form of Taylor's theorem with its explicit remainder term:

$$f(x) = f(0) + f'(0)x + \frac{f''(0)}{2!}x^2 + \frac{f'''(0)}{3!}x^3 + \cdots + \frac{f^{(k-1)}(0)}{(k-1)!}x^{k-1} + R_k(x), \quad (3.21)$$

where $R_k(x) = f^k(c)x^k/k!$ for some constant c between 0 and x. If we replace x by $x - a$ and then shift our function by a, we obtain the equivalent but seemingly more general equation:

$$f(x) = f(a) + f'(a)(x - a) + \frac{f''(a)}{2!}(x - a)^2 + \frac{f'''(a)}{3!}(x - a)^3$$

$$+ \cdots + \frac{f^{k-1}(a)}{(k-1)!}(x - a)^{k-1} + R_k(x), \quad (3.22)$$

where $R_k(x) = f^k(c)(x-a)^k/k!$ for some constant c between a and x. He began the proof of equation (3.22) with the very simplest case, $k = 1$, the **mean value theorem**, Theorem 3.1 on page 58.

Equation (3.2) simply asserts that the average rate of change of f between $x = a$ and $x = b$ is equal to the instantaneous rate of change at some point between a and b (see Figure 3.1). Geometrically, this seems obvious. But to actually verify it and thus understand when it is or is not valid requires a great deal of insight.

None of the early proofs of the mean value theorem was without flaws. The proof that is found in most textbooks today is a very slick approach that was discovered by Ossian Bonnet (1819–1892) and first published in Joseph Alfred Serret's calculus text of 1868, *Cours de Calcul Différentiel et Intégral*. It avoids all of the difficulties that we shall be encountering as we attempt to follow Cauchy. We shall postpone Bonnet's proof until after we have discussed continuity and its consequences in section 3.4.

Cauchy's First Proof of the Mean Value Theorem

The following proof of the mean value theorem, equation (3.2), was given by Cauchy in his *Résumé des Leçons données a l'École Royale Polytechnique sur le calcul infinitésimal* of 1823. We shall run through it and then analyze the difficulties.

We first assume that f is differentiable at every point in the interval $[a, b]$. The definition of the derivative given in the last section guarantees that for any given positive error ϵ, we can find a distance δ such that if $|h| < \delta$, then

$$f'(x) - \epsilon < \frac{f(x+h) - f(x)}{h} < f'(x) + \epsilon. \tag{3.23}$$

We partition the interval from a to b into n steps of size $h < \delta$, $x_{i+1} = x_i + h$:

We note that $nh = b - a$.

Applying the inequality (3.23) to each of these intervals, we obtain

$$f'(x_0) - \epsilon < \frac{f(x_1) - f(x_0)}{h} < f'(x_0) + \epsilon,$$

$$f'(x_1) - \epsilon < \frac{f(x_2) - f(x_1)}{h} < f'(x_1) + \epsilon,$$

$$f'(x_2) - \epsilon < \frac{f(x_3) - f(x_2)}{h} < f'(x_2) + \epsilon,$$

$$\vdots$$

$$f'(x_{n-1}) - \epsilon < \frac{f(x_n) - f(x_{n-1})}{h} < f'(x_{n-1}) + \epsilon. \tag{3.24}$$

Let A be the minimal value of $f'(x)$ over the interval $[a, b]$ and B the maximal value. We can replace the left side of each of the lower inequalities by the same bound, $A - \epsilon$, and the right side of the upper inequalities by $B + \epsilon$. Adding the central terms, we see that

$$n(A - \epsilon) < \frac{f(x_1) - f(x_0)}{h} + \frac{f(x_2) - f(x_1)}{h} + \cdots + \frac{f(x_n) - f(x_{n-1})}{h} < n(B + \epsilon),$$

(3.25)

$$n(A - \epsilon) < \frac{f(x_n) - f(x_0)}{h} < n(B + \epsilon). \tag{3.26}$$

We divide by n and recall that $nh = b - a$, $x_n = b$, $x_0 = a$:

$$A - \epsilon < \frac{f(b) - f(a)}{b - a} < B + \epsilon. \tag{3.27}$$

This is true for *any* positive ϵ, no matter how small, and so we must have that

$$A \leq \frac{f(b) - f(a)}{b - a} \leq B. \tag{3.28}$$

Cauchy now assumes that f' is continuous and he invokes the intermediate value property of continuous functions. This result will need to be proven, a fact that Cauchy recognized.

Definition: intermediate value property

A function f is said to have the **intermediate value property** on the interval $[a, b]$ if given any two points $x_1, x_2 \in [a, b]$ and any number N satisfying

$$f(x_1) < N < f(x_2),$$

then there is at least one value c between x_1 and x_2 for which $f(c) = N$.

If $f'(x_1) = A$ and $f'(x_2) = B$ and $N = [f(b) - f(a)]/[b - a]$ lies between A and B (see Figure 3.6), and if $f'(x)$ is continuous, then $f'(x)$ must cross the line $y = [f(b) - f(a)]/[b - a]$ somewhere between x_1 and x_2. We let c denote the value of x where this happens:

$$\frac{f(b) - f(a)}{b - a} = f'(c).$$

Q.E.D.

The initials **Q.E.D.** stand for *Quod Erat Demonstrandum*, a Latin translation of the phrase $o\pi\epsilon\rho \ \epsilon\delta\epsilon\iota \ \delta\epsilon\iota\xi\alpha\iota$ with which Euclid concluded most of his proofs. It means "what was to be proved" and signifies that the proof has been concluded.

The Problems with this Proof

This is an ingenious proof that demonstrates a profound understanding of the derivative, but it would not pass muster today. We can prove something stronger than Cauchy's theorem. Cauchy required that f be differentiable at every point in the interval $[a, b]$. This is necessary if we want to get all of the inequalities of (3.24). As we shall see,

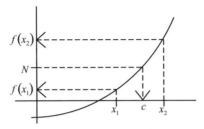

FIGURE 3.6. Intermediate value property in Cauchy's proof of the mean value theorem.

Bonnet's proof works under the slightly weaker assumption that we have differentiability on the open interval (a, b) and continuity on the closed interval $[a, b]$. For example, Cauchy's proof would not apply to the function $x \sin(1/x)$ on $[0, 1]$ (see exercise 3.1.12 of section 3.1 and exercise 3.2.1 of this section). Bonnet's would. Furthermore, Cauchy only proves that c lies somewhere in the interval $[a, b]$. Bonnet's proof shows that it must lie strictly between a and b.

But these are quibbles. The question is not whether we can improve on Cauchy's statement of the mean value theorem, but whether his proof is valid. There are two questionable assertions in his proof. The first is that given the error bound ϵ we can find a δ that works over the entire interval $[a, b]$. It is true that at x_0 we can find a δ so that if $|x_1 - x_0| < \delta$, then

$$f'(x_0) - \epsilon < \frac{f(x_1) - f(x_0)}{x_1 - x_0} < f'(x_0) + \epsilon,$$

but the inequalities of (3.24) assume that the same δ can be used at x_1 and x_2 and $x_3 \ldots$ all the way to x_{n-1}. This is not always so (see exercise 3.2.2 for a counterexample).

The second assumption is that f' actually has an upper bound B and a lower bound A over the interval $[a, b]$ and that it achieves these bounds. That is to say, we can find c_1, $c_2 \in [a, b]$ such that

$$f'(c_1) \leq f'(x) \leq f'(c_2)$$

for all $x \in [a, b]$. In fact, such bounds do not always exist. Consider the function defined by

$$f(x) = x^2 \sin(x^{-2}), \quad x \neq 0; \qquad f(0) = 0. \tag{3.29}$$

We saw this function in the last section (pages 65–66) where it was demonstrated that the derivative exists at every x in $[-1, 1]$, but the derivative is not bounded over this interval.

Cauchy could have avoided these problems by assuming that the derivative of f is continuous. When he does invoke continuity to pass from the double inequality (3.28) to the statement that $[f(b) - f(a)]/[b - a] = f'(c)$ for some $c \in [a, b]$, it is no longer needed. While the derivative is not always continuous, we shall see that it does always possess the intermediate value property. Part of the confusion that will have to be straightened out in section 3.3 is the difference between continuity and the intermediate value property.

Cauchy's Second Attempt

Cauchy gave another proof of the mean value theorem in an appendix to *Résumé des Leçons*. He actually proves a far more powerful result. We shall state it in the form in which we will eventually prove it.

Theorem 3.2 (Generalized Mean Value Theorem). *If f and F are both continuous at every point of* $[a, b]$ *and differentiable at every point in the open interval* (a, b) *and if* F' *is never zero in this interval, then*

$$\frac{f(b) - f(a)}{F(b) - F(a)} = \frac{f'(c)}{F'(c)} \tag{3.30}$$

for at least one point c, $a < c < b$.

We note that if $F(x) = x$, then this becomes the ordinary mean value theorem.

We begin by defining $g(x) = f(x) - f(a)$ and $G(x) = F(x) - F(a)$ so that $g(a) = 0$, $g'(x) = f'(x)$, $G(a) = 0$, and $G'(x) = F'(x)$. We consider the function defined by

$$\frac{f'(x)}{F'(x)} = \frac{g'(x)}{G'(x)},$$

and let A be its minimal value over $[a, b]$, B its maximal value:

$$A \leq \frac{g'(x)}{G'(x)} \leq B. \tag{3.31}$$

This implies that

$$g'(x) - A\,G'(x) \geq 0, \tag{3.32}$$
$$g'(x) - B\,G'(x) \leq 0, \tag{3.33}$$

and so $g - A\,G$ is an increasing function which is equal to $g(a) - A\,G(a) = 0$ at $x = a$, and $g - B\,G$ is a decreasing function which is equal to $g(a) - B\,G(a) = 0$ at $x = a$ (Figure 3.7). It follows that

$$g(x) - A\,G(x) \geq 0, \tag{3.34}$$
$$g(x) - B\,G(x) \leq 0, \tag{3.35}$$

for all x, $a \leq x \leq b$, and therefore

$$A \leq \frac{g(b)}{G(b)} \leq B. \tag{3.36}$$

We now assume that g'/G' is a continuous function and so takes on every value between its minimum A and its maximum B. For some $c \in [a, b]$,

$$\frac{g(b)}{G(b)} = \frac{g'(c)}{G'(c)} = \frac{f'(c)}{F'(c)}. \tag{3.37}$$

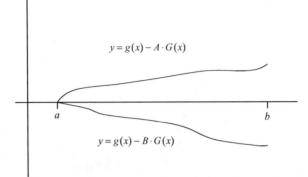

FIGURE 3.7. Graphs of $g(x) - A\,G(x)$ and $g(x) - B\,G(x)$.

When we substitute $f(b) - f(a)$ for $g(b)$ and $F(b) - F(a)$ for $G(b)$, we obtain the desired equation (3.30).

Q.E.D.

The principal difficulty with this approach is the use of the result that if $h(a) = 0$ and $h'(x) \geq 0$ for $a \leq x \leq b$, then $h(b) \geq 0$. This may seem obvious from our geometric understanding of the derivative: a positive derivative means that the function is increasing. But how do we *prove* this statement without relying on geometric intuition? If you look in any current calculus text, you will see that the proof uses the mean value theorem:

$$\frac{h(b) - h(a)}{b - a} = h'(c) \geq 0,$$

and so

$$h(b) - h(a) \geq 0.$$

This sends us into a circular argument: the mean value theorem is true because positive derivative implies increasing, and positive derivative implies increasing because of the mean value theorem.

Cauchy does give an independent proof that a positive derivative implies an increasing function. He first points out that when x is sufficiently close to a, $f'(a)$ will be close to $[f(x) - f(a)]/[x - a]$. If $f'(a)$ is positive we can, by keeping x close to a, force $[f(x) - f(a)]/[x - a]$ to be positive. If x is larger than a, then $f(x)$ will have to be larger than $f(a)$. So far, so good.

The problem arises when Cauchy tries to use this fact to conclude that if $f'(x)$ is positive for all $x \in [a, b]$, then $f(b)$ is strictly larger than $f(a)$. To quote Cauchy, "If one increases the variable x by insensible degrees from the first limit to the second, the function y will grow with it as long as it has a finite derivative with positive value." The difficulty is subtle but important. How large are these "insensible degrees"? Can we use them to get from a to b?

There is a way of proving that a positive derivative implies an increasing function without using the mean value theorem, but it requires a deeper understanding of continuity. This proof is described in exercises 3.4.12 and 3.4.13 at the end of section 3.4.

Our present state of affairs with respect to the mean value theorem is far from satisfactory. We still have not seen an acceptable proof. The key to such a proof is a better understanding of continuity, and for this we shall have to wait until the next section.

Exercises

3.2.1. Where does Cauchy's proof of the mean value theorem break down if we try to apply it to the function defined by $f(x) = x \sin(1/x)$ $(f(0) = 0)$ over the interval $[0, 1]$. Note: the mean value theorem *does* apply to this function, but Cauchy's approach cannot be used to establish this fact.

3.2.2. The purpose of this exercise is to demonstrate that even though the function defined by

$$f(x) = x^2 \sin(x^{-2}), \quad x \neq 0; \qquad f(0) = 0,$$

is differentiable at all points of $[0, 1]$, if we are given a bound $0 < \epsilon < 1$ for the error function

$$|E(x, a)| = \left| f'(a) - \frac{f(x) - f(a)}{x - a} \right|,$$

there is no response δ that works for all values of $a \in [0, 1]$.

a. Prove that f is differentiable at $x = 0$ and that $f'(0) = 0$. Given a bound ϵ, find a response δ (as a function of ϵ) such that $|x| < \delta$ guarantees that $|E(x, 0)| < \epsilon$.

b. Graph $f(x)$ and $f'(x)$ over the interval $[0, 1]$.

c. Given $\delta > 0$, show that we can always find a and x such that $0 < x < a < \delta$ *and* both a^{-2} and x^{-2} are even multiples of π.

d. Given the values of a and x that were found in the previous part of this problem, show that

$$|E(x, a)| = 2a^{-1} > 2\delta^{-1}.$$

e. Complete the proof that there is no single value of δ that will serve as a response to the bound ϵ at every value of $a \in [0, 1]$.

3.2.3. Using the function f given in exercise 3.2.2, let $a = 1/\sqrt{2\pi}$. Graph $E(x, a)$. How close must x be to a if we are to keep $|E(x, a)|$ within the error bound $\epsilon = 0.1$?, $\epsilon = 0.01$?

3.2.4. Repeat exercise 3.2.3 with $a = 1/\sqrt{200\pi}$.

3.2.5. Find another function that is differentiable at every point in $[0, 1]$ but whose derivative is not bounded on $[0, 1]$.

3.2.6. Find a function that is *not* continuous on a closed interval $[a, b]$ but which *does* have the intermediate value property on this interval.

3.2.7. Explain why a function f that satisfies the intermediate value property on the interval $[a, b]$ and satisfies $f(a) < 0 < f(b)$ or $f(a) > 0 > f(b)$ must have at least one root between a and b.

3.2.8. Graph the function defined by

$$f(x) = \frac{1-x}{|x|}, \quad x \neq 0; \qquad f(0) = 1.$$

Explain how you know that it does not have the intermediate value property on the interval $[-2, 2]$.

3.2.9. Use the generalized mean value theorem to prove that if f is twice differentiable, then there is some number c between x_0 and $x_0 + 2\Delta x$ such that

$$\frac{f(x_0 + 2\Delta x) - 2f(x_0 + \Delta x) + f(x_0)}{\Delta x^2} = f''(c).$$

3.2.10. Explain why the generalized mean value theorem implies that

$$\lim_{\Delta x \to 0} \frac{f(x_0 + 3\Delta x) - 3f(x_0 + 2\Delta x) + 3f(x_0 + \Delta x) - f(x_0)}{\Delta x^3} = f'''(x_0).$$

3.3 Continuity

Continuity is such an obvious geometric phenomenon that only slowly did it dawn on mathematicians that it needed a precise definition. Well into the 19th century it was simply viewed as a descriptive term for curves that are unbroken. The preeminent calculus text of that era was S. F. Lacroix's *Traité élémentaire de calcul différentiel et de calcul intégral*. It was first published in 1802. The sixth edition appeared in 1858. Unchanged throughout these editions was its definition of continuity: "*By the law of continuity is meant that which is observed in the description of lines by motion, and according to which the consecutive points of the same line succeed each other without any interval.*"

This intuitive notion of continuity is useless when one tries to prove anything. The first appearance of the modern definition of continuity was published by Bernhard Bolzano in 1817 in the Proceedings of the Prague Scientific Society under the title *Rein analytischer Beweis des Lehrsatzes dass zwischen je zwey [sic] Werthen, die ein entgegengesetztes Resultat gewaehren, wenigstens eine reele Wurzel der Gleichung liege*. This roughly translates as *Purely analytic proof of the theorem that between any two values that yield results of opposite sign there will be at least one real root of the equation*. The title says it all. Bolzano was proving that any continuous function has the intermediate value property.

Bolzano's article raises an important point. If he has to prove that continuity implies the intermediate value property, then he is not using the intermediate value property to define continuity. Why not? Such a definition would agree with the intuitive notion of continuity. If a function is defined at every point on the interval $[a, b]$, then to say that it has the intermediate value property is equivalent to saying that it has no jumps or breaks.

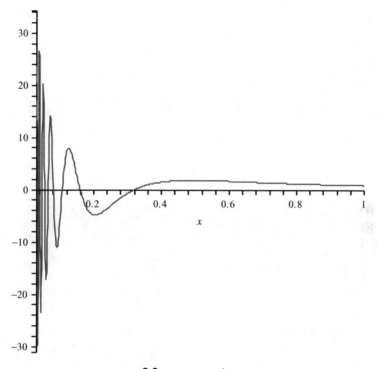

FIGURE 3.8. Graph of $x^{-1}\sin(1/x)$.

There are several problems with choosing this definition of continuity. A function that has the intermediate value property on $[0, 1]$ is not necessarily bounded on that interval. An example is the function defined by

$$f(x) = x^{-1}\sin(1/x), \quad x \neq 0; \quad \cdot \quad f(0) = 0, \tag{3.38}$$

(see Figure 3.8).

Another problem with using the intermediate value property to define continuity is that two functions can have it while their sum does not:

$$f(x) = \sin^2(1/x), \quad x \neq 0; \quad f(0) = 0,$$
$$g(x) = \cos^2(1/x), \quad x \neq 0; \quad g(0) = 0,$$
$$f(x) + g(x) = \sin^2(1/x) + \cos^2(1/x) = 1, \quad x \neq 0; \quad f(0) + g(0) = 0$$

(see Figure 3.9). This is not damning, but it is unsettling.

Perhaps the most important aspect of continuity that the intermediate value property lacks, and the one that may have suggested the modern definition, is that if f is continuous in a neighborhood of a and if there is a small error in the input so that instead of evaluating f at a we evaluate it at something very close to a, then we want the output to be very close to $f(a)$ (see Figure 3.10). The function defined by

$$f(x) = \sin(1/x), \quad x \neq 0; \quad f(0) = 0,$$

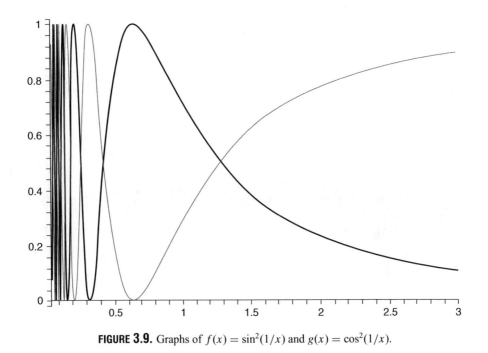

FIGURE 3.9. Graphs of $f(x) = \sin^2(1/x)$ and $g(x) = \cos^2(1/x)$.

satisfies the intermediate value property no matter how we define $f(0)$, provided that $-1 \leq f(0) \leq 1$ (see Figure 3.11), but at $a = 0$ any allowance for error in the input will result in an output that could be any number from -1 to 1. We want to be able to control the variation in the output by setting a tolerance on the input.

Definition of Continuity

We require that if f is continuous in a neighborhood of a and if x is close to a, then $f(x)$ must be close to $f(a)$. More precisely, we want to be able to force $f(x)$ to be arbitrarily close to $f(a)$ by controlling the distance between x and a. This is the definition of continuity

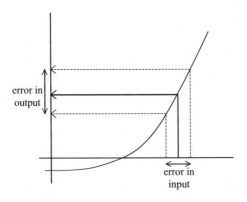

FIGURE 3.10. The effect of a small error in input.

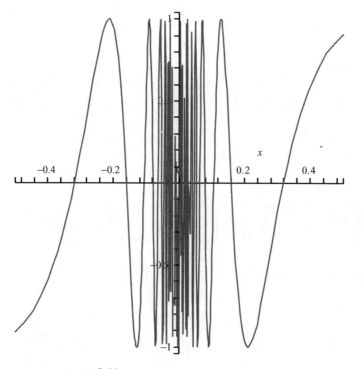

FIGURE 3.11. Graph of $f(x) = \sin(1/x)$, $x \neq 0$.

that Bolzano stated in 1817 and that Cauchy proposed in his *Cours d'analyse* of 1821. To make it as precise as possible, Cauchy used the Archimedean understanding.

Definition: continuity

We say that f is **continuous at** a if given any positive error bound ϵ, we can always reply with a tolerance δ such that if x is within δ of a, then $f(x)$ is within ϵ of $f(a)$:

$$|x - a| < \delta \quad \text{implies that} \quad |f(x) - f(a)| < \epsilon.$$

To say that f is **continuous on an interval** I means that it is continuous at every point a in the interval I.

Neither $\sin(1/x)$ nor $x^{-1}\sin(1/x)$ satisfy this definition of continuity at $x = 0$. Neither of these functions are forced to take values close to 0 simply by restricting x to be close to 0.

This definition does it all for us: It implies the intermediate value property, and it implies that a continuous function of $[a, b]$ is bounded and achieves its bounds on that interval. Continuity is preserved when we add two continuous functions or multiply them or even take compositions of continuous functions. All of the difficulties that we encountered in Cauchy's first proof of the mean value theorem evaporate if we add the assumption that f' is continuous. Even better, our analysis of the properties of continuous functions will eventually lead us to Bonnet's proof of the mean value theorem in which we can weaken Cauchy's assumptions.

Strange Examples

This definition does have its own idiosyncracies that run counter to our intuitive notion of continuity. The emphasis is not on what happens over an interval but rather at what happens near a specific point. The result is that it is possible for a function to be continuous at one and only one point as the following example demonstrates. The examples given in this section are elaborations of a basic idea that first occurred to Dirichlet: to define a function one way over the rationals and in a different manner over the irrationals.

Define a function f by

$$f(x) = \begin{cases} x, & \text{if } x \text{ is rational,} \\ 0, & \text{if } x \text{ is irrational.} \end{cases} \tag{3.39}$$

We observe that this function is continuous at $x = 0$. If we are given a bound $\epsilon > 0$, we can always reply with $\delta = \epsilon$:

$$|x - 0| < \epsilon \quad \text{implies that} \quad |f(x) - f(0)| < \epsilon$$

because $|f(x) - f(0)|$ is always either $|x - 0| = |x|$ (when x is rational) or $|0 - 0| = 0$ (when x is irrational).

If $a \neq 0$, then f is not continuous at a because we have no reply to any bound $\epsilon < |a|$. If a is rational, then no matter how small we choose δ there is an irrational x within distance δ of a:

$$|x - a| < \delta \quad \text{but} \quad |f(x) - f(a)| = |0 - a| = |a| > \epsilon.$$

If a is irrational, then no matter how small we choose δ there is a rational x within distance δ of a and with an absolute value slightly larger than $|a|$:

$$|x - a| < \delta \quad \text{but} \quad |f(x) - f(a)| = |x - 0| > |a| > \epsilon.$$

This function is continuous at $x = 0$ and *only* at $x = 0$.

An example that is even stranger is the following function which is not continuous at any rational number but which *is* continuous at every irrational number. In other words, the points where this function is continuous form a discontinuous set. When we write a nonzero rational number as p/q, we shall choose p and $q > 0$ to be the unique pair of integers with no common factor. Let g be defined by

$$g(x) = \begin{cases} 1, & \text{if } x = 0, \\ 1/q, & \text{if } x = p/q \text{ is rational,} \\ 0, & \text{if } x \text{ is irrational.} \end{cases} \tag{3.40}$$

If $a = 0$, there is no response to any $\epsilon < 1$. If $a = p/q$ is rational, then we cannot respond to a bound of $\epsilon < 1/q$. In both cases, this is because within any distance δ of a, we can always find an irrational number x for which $|g(x) - g(a)| = g(a) > \epsilon$.

On the other hand, if a is irrational and we are given a bound ϵ, then the change in g is bounded by ϵ, $|g(x) - g(a)| = g(x) < \epsilon$, provided that whenever $x = p/q$ is rational, $1/q$

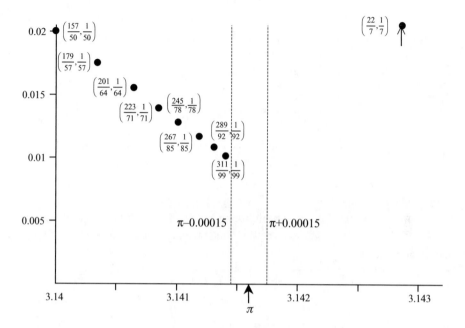

FIGURE 3.12. Rational numbers between 3.14 and 3.143 with denominators ≤ 100.

is less than ϵ. Equivalently, we want to choose a distance δ so that if $x = p/q$ is a rational number within δ of a, then $q > \epsilon^{-1}$. We locate the rational numbers within distance 1 of a for which the denominator is less than or equal to ϵ^{-1}. The critical observation is that there are at most finitely many of them. For example, if $a = \pi$ and $\epsilon = 0.01$, then we only need to exclude those fractions p/q with

$$\pi - 1 < \frac{p}{q} < \pi + 1, \quad \text{and} \quad q \leq 100.$$

We mark their positions on the interval $(\pi - 1, \pi + 1)$ and choose our response δ to be less than the distance between π and the closest of these unacceptable rational numbers. The closest is $311/99 = 3.141414\ldots$ which is just over 0.00017 from π (see Figure 3.12). If we respond with $\delta = 0.00015$, then none of the fractions inside the interval $(\pi - 0.00015, \pi + 0.00015)$ has a denominator less than or equal to 100:

$$|x - \pi| < 0.00015 \quad \text{implies that} \quad |g(x) - g(\pi)| < 0.01.$$

> **Web Resource:** To discover how continued fractions can be used to find these approximations to π with very small denominators, go to **Continued Fractions**.

An Equivalent Definition of Continuity

If we look back at the definition of continuity on page 81 and compare it with the definition of limit on page 59, we see that f is continuous at $x = a$ if and only if

$$\lim_{x \to a} f(x) = f(a). \tag{3.41}$$

This says that a function is continuous at $x = a$ if and only if we can force $f(x)$ to be as close as we wish to $f(a)$ simply by restricting the distance between x and a, excluding the value $x = a$. In particular, continuity at $x = a$ implies that if $(x_1, x_2, x_3, \ldots)$ is *any* sequence that converges to a, then

$$\lim_{k \to \infty} f(x_k) = f\left(\lim_{k \to \infty} x_k\right). \tag{3.42}$$

Usually, this approach is not helpful when we are trying to prove continuity; there are too many possible sequences to check. But if we know that a function is continuous, then this characterization of continuity can be very useful (see the proof of Theorem 3.3). And there are times when we want to prove that a function is discontinuous (not continuous) at a given value, a. A common method of accomplishing this is to find a sequence of values of x, $(x_1, x_2, x_3, \ldots)$, that converges to a, but for which the sequence $(f(x_1), f(x_2), f(x_3), \ldots)$ does *not* converge to $f(a)$,

$$\lim_{k \to \infty} f(x_k) \neq f(a).$$

If this can be done, then f cannot be continuous at $x = a$.

An example is provided by $f(x) = \sin(1/x)$, $x \neq 0$, $f(0) = 0$. The sequence $(1/\pi, 2/3\pi, 2/5\pi, 2/7\pi, \ldots)$ converges to 0, but the sequence

$$\left(f\left(\frac{2}{\pi}\right), f\left(\frac{2}{3\pi}\right), f\left(\frac{2}{5\pi}\right), f\left(\frac{2}{7\pi}\right), \ldots\right) = (1, -1, 1, -1, \ldots)$$

does not converge.

The Intermediate Value Theorem

The key to proving that any continuous function satisfies the intermediate value property is the nested interval principle. It was stated on page 32. It is repeated here for convenience:

Definition: nested interval principle

Given an increasing sequence, $x_1 \leq x_2 \leq x_3 \leq \cdots$, and a decreasing sequence, $y_1 \geq y_2 \geq y_3 \geq \cdots$, such that y_n is always larger than x_n but the difference between y_n and x_n can be made arbitrarily small by taking n sufficiently large, there is *exactly* one real number that is greater than or equal to every x_n and less than or equal to every y_n.

As was mentioned in section 2.4, this principle is taken as an axiom or unproven assumption. Both Bolzano and Cauchy used it without proof. When in the later 19th century mathematicians began to realize that it might be in need of justification, they saw that it depends upon the definition of the real numbers. In 1872, Richard Dedekind (1831–1916), Georg Cantor (1845–1918), Charles Méray (1835–1911), and Heinrich Heine (1821–1881) each gave a different definition of the real numbers that would imply this principle. It is referred to as the **Bolzano–Weierstrass theorem** when it is proven as a consequence of carefully stated properties of the real numbers. The name acknowledges the first two mathematicians to recognize the need to state it explicitly.

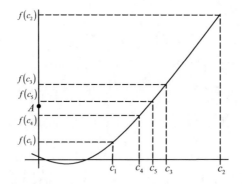

FIGURE 3.13. Proof of intermediate value theorem.

We have been searching for bedrock, a solid and unequivocal foundation on which to construct analysis. One of the lessons of the twentieth century has been that this search can be continued forever. To define the real numbers we require a careful definition of the rationals. This in turn rests on a precise description of the integers which is impossible without a clear understanding of sets and cardinality. At this point, the very principles of logic need underpinning. The solution is to draw a line somewhere and state that this is what we shall assume, this is where we shall begin. Not everyone will agree that the nested interval principle is the right place to draw that line, but it has the advantage of being simple and yet sufficient for all we want to prove. Here is where we shall begin to build the theorems of analysis.

Theorem 3.3 (Intermediate Value Theorem). *If f is continuous on the interval $[a, b]$, then f has the intermediate value property on this interval.*

Proof: We assume that f is continuous on the interval $[a, b]$. We want to show that if c_1 and c_2 are any two points on this interval and if A is any number strictly between $f(c_1)$ and $f(c_2)$, then there is at least one value c between c_1 and c_2 for which $f(c) = A$. The trick is to shrink the interval in which c is located, show that we can make this interval arbitrarily small, and then invoke the nested interval principle to justify the claim that there is something left in our net when we are done.

We can assume that $c_1 < c_2$. We begin to define the sequences for the nested interval principle by setting $x_1 = c_1$ and $y_1 = c_2$. We split the difference between these endpoints, call it

$$c_3 = \frac{x_1 + y_1}{2}.$$

If $f(c_3) = A$, then we are done. We have found our c. If not, then $f(c_3)$ is either on the same side of A as $f(x_1)$ or it is on the same side of A as $f(y_1)$ (Figure 3.13). We are in one of two possibilities:

1. If $f(x_1)$ and $f(c_3)$ are on opposite sides of A, then we define $x_2 = x_1$ and $y_2 = c_3$.
2. If $f(c_3)$ and $f(y_1)$ are on opposite sides of A, then we define $x_2 = c_3$ and $y_2 = y_1$.

In either case, the result is that

$$x_1 \leq x_2 < y_2 \leq y_1,$$
$$y_2 - x_2 = (y_1 - x_1)/2,$$

and $f(x_2)$, $f(y_2)$ lie on opposite sides of A. We have cut in half the size of the interval where c must lie.

We repeat what we have just done. We find the midpoint of our last interval:

$$c_4 = \frac{x_2 + y_2}{2}.$$

If $f(c_4) = A$, then we are done. Otherwise, we are in one of two situations:

1. If $f(x_2)$ and $f(c_4)$ are on opposite sides of A, then we define $x_3 = x_2$ and $y_3 = c_4$.
2. If $f(c_4)$ and $f(y_2)$ are on opposite sides of A, then we define $x_3 = c_4$ and $y_3 = y_2$.

In either case, the result is that

$$x_1 \leq x_2 \leq x_3 < y_3 \leq y_2 \leq y_1,$$
$$y_3 - x_3 = (y_2 - x_2)/2 = (y_1 - x_1)/4,$$

and $f(x_3)$, $f(y_3)$ lie on opposite sides of A.

We can keep on doing this as long as we please. Once we have found x_k and y_k, we find the midpoint:

$$c_{k+2} = \frac{x_k + y_k}{2}.$$

If $f(c_{k+2}) = A$, then we are done. Otherwise, we have that either

1. $f(x_k)$ and $f(c_{k+2})$ are on opposite sides of A in which case we define $x_{k+1} = x_k$ and $y_{k+1} = c_{k+2}$, or
2. $f(c_{k+2})$ and $f(y_k)$ are on opposite sides of A in which case we define $x_{k+1} = c_{k+2}$ and $y_{k+1} = y_k$.

In either case, the result is that

$$x_1 \leq x_2 \leq \cdots \leq x_k \leq x_{k+1} < y_{k+1} \leq y_k \leq \cdots \leq y_2 \leq y_1,$$
$$y_{k+1} - x_{k+1} = (y_k - x_k)/2 = \cdots = (y_2 - x_2)/2^{k-1} = (y_1 - x_1)/2^k,$$

and $f(x_{k+1})$, $f(y_{k+1})$ lie on opposite sides of A and can be forced as close as we wish to A by taking k sufficiently large. This is the Archimedean definition of limit:

$$\lim_{k \to \infty} f(x_k) = \lim_{k \to \infty} f(y_k) = A. \tag{3.43}$$

Our sequences $x_1 \leq x_2 \leq \cdots$ and $y_1 \geq y_2 \geq \cdots$ satisfy the conditions of the nested interval principle and so there is a number c that lies in all of these intervals. Again by the Archimedean definition of limit, we see that

$$\lim_{k \to \infty} x_k = \lim_{k \to \infty} y_k = c. \tag{3.44}$$

Since f is continuous at $x = c$, we know that

$$\lim_{k \to \infty} f(x_k) = f(c). \tag{3.45}$$

Since this limit is also equal to A and any sequence has at most one limit, we have proved that $f(c) = A$.

Q.E.D.

The Modified Converse to the Intermediate Value Theorem

As we have seen by example, the intermediate value property is not enough to imply continuity. The converse of the intermediate value theorem is not true. But a very reasonable question to pose is whether we can find a broad class of functions for which the intermediate value property is equivalent to continuity. One such class consists of the functions that are **piecewise monotonic** on any finite interval.

Definition: monotonic

A function is **monotonic** on $[a, b]$ if it is **increasing** on this interval,

$$a \le x_1 < x_2 \le b \quad \text{implies that} \quad f(x_1) \le f(x_2),$$

or if it is **decreasing** on this interval,

$$a \le x_1 < x_2 \le b \quad \text{implies that} \quad f(x_1) \ge f(x_2).$$

A function is **piecewise monotonic** on $[a, b]$ if we can find a partition of the interval into a finite number of subintervals

$$a = x_1 < x_2 < \cdots < x_{n-1} < x_n = b$$

for which the function is monotonic on each open subinterval (x_i, x_{i+1}).

The key word is *finite*. It excludes all of the strange functions we have encountered so far. They all jumped or oscillated infinitely often within our interval.

Theorem 3.4 (Modified Converse to IVT). *If f is piecewise monotonic and satisfies the intermediate value property on the interval $[a, b]$, then f is continuous at every point c in (a, b).*

Proof: We shall assume that c lies inside one of the intervals (x_i, x_{i+1}) on which f is monotonic. The proof that f is also continuous at the ends of these intervals is similar and is left as an exercise. For convenience, we assume that f is increasing on (x_i, x_{i+1}). If not, then we replace f with $-f$. Since f is increasing and $x_i < c < x_{i+1}$, it follows that $f(x_i) \le f(c) \le f(x_{i+1})$.

We are given an error bound ϵ. Our challenge is to show that we always have a response δ so that keeping x within δ of c guarantees that $f(x)$ will be within ϵ of $f(c)$. We begin by finding two numbers, c_1 and c_2, that satisfy

$$x_i \le c_1 < c < c_2 \le x_{i+1}$$

and

$$f(c) - \epsilon < f(c_1) \le f(c) \le f(c_2) < f(c) + \epsilon.$$

If $f(c) - \epsilon < f(x_i) \le f(c)$, then we let $c_1 = x_i$. Otherwise, we have that

$$f(x_i) \le f(c) - \epsilon < f(c).$$

The intermediate value property promises us a c_1 between x_i and c for which $f(c) - \epsilon < f(c_1) < f(c)$. For example, we could choose c_1 to be a value for which $f(c_1) = f(c) - \epsilon/2$. In either case, we have that $x_i \le c_1 < c$ and

$$f(c) - \epsilon < f(c_1) \le f(c).$$

We find c_2 similarly. If $f(x_{i+1}) < f(c) + \epsilon$, then $c_2 = x_{i+1}$. Otherwise, we choose c_2 so that $c < c_2 < x_{i+1}$ and $f(c) \le f(c_2) < f(c) + \epsilon$. In either case, we have that $c < c_2 \le x_{i+1}$ and

$$f(c) \le f(c_2) < f(c) + \epsilon.$$

Now that we have found c_1 and c_2, we choose δ to be the smaller of the distances $c - c_1 > 0$ and $c_2 - c > 0$ so that if x is within δ of c then

$$c_1 < x < c_2.$$

Since f is increasing on $[c_1, c_2]$, we can conclude that

$$f(c) - \epsilon < f(c_1) \le f(x) \le f(c_2) < f(c) + \epsilon.$$

Q.E.D.

Most functions that you are likely to encounter are piecewise monotonic. It should come as a relief that in this case our two definitions of continuity are interchangeable. When we reach Dirichlet's proof of the validity of the Fourier series expansion, we shall see that piecewise monotonicity is a critical assumption.

Sums, Products, Reciprocals, and Compositions

Combinations of continuous functions using addition, multiplication, division, or composition yield continuous functions. The proofs follow directly from the definition of continuity.

We begin by assuming that f and g are continuous at $x = c$. In order to show that $f + g$ is also continuous at c, we have to demonstrate that if someone gives us a bound $\epsilon > 0$, then we can find a response δ so that if we keep x within distance δ of c, $|x - c| < \delta$, then we are guaranteed that

$$\left| [f(x) + g(x)] - [f(c) + g(c)] \right| < \epsilon.$$

We split our error bound, giving half to f and half to g. The continuity of f and g at c promises us responses δ_1 and δ_2 such that

$$|x - c| < \delta_1 \quad \text{implies that} \quad |f(x) - f(c)| < \epsilon/2$$

and

$$|x - c| < \delta_2 \quad \text{implies that} \quad |g(x) - g(c)| < \epsilon/2.$$

We choose δ to be the smaller of these two responses. When $|x - c| < \delta$ we have that

$$|\,[f(x) + g(x)] - [f(c) + g(c)]\,| \leq |f(x) - f(c)| + |g(x) - g(c)|$$
$$< \epsilon/2 + \epsilon/2 \quad = \quad \epsilon.$$

The product fg is a little trickier. We again begin with the assumption that both f and g are continuous at $x = c$. Before deciding how to divide our assigned bound ϵ we observe that

$$|f(x)\,g(x) - f(c)\,g(c)| = |f(x)\,g(x) - f(c)\,g(x) + f(c)\,g(x) - f(c)\,g(c)|$$
$$\leq |f(x) - f(c)|\,|g(x)| + |f(c)|\,|g(x) - g(c)|. \qquad (3.46)$$

We want each of these two pieces to be less than $\epsilon/2$. If $f(c) = 0$, then the second piece gives us no problem. If $f(c)$ is not zero, then we need to have $|g(x) - g(c)|$ bounded by $(\epsilon/2)\,|f(c)|$. Let δ_1 be the response:

$$|x - c| < \delta_1 \quad \text{implies that} \quad |g(x) - g(c)| < \frac{\epsilon}{2|f(c)|}.$$

The first piece is slightly more problematic. Since c is fixed, $f(c)$ is a constant. We are now faced with a multiplier, $|g(x)|$, that can take on different values. Our first task is to use the continuity of g to pin down $|g(x)|$. We choose a δ_2 so that $|x - c| < \delta_2$ guarantees that $|g(x) - g(c)| < 1$. This implies that

$$|g(x)| < 1 + |g(c)|.$$

We find a δ_3 for which

$$|x - c| < \delta_3 \quad \text{implies that} \quad |f(x) - f(c)| < \frac{\epsilon}{2\,(1 + |g(c)|)}.$$

In either case, choosing a δ that is less than or equal to both δ_2 and δ_3 gives us the desired bound:

$$|f(x) - f(c)|\,|g(x)| < \frac{\epsilon}{2\,(1 + |g(c)|)}\,(1 + |g(c)|) = \frac{\epsilon}{2}.$$

If we choose our final response δ to be the smallest of δ_1, δ_2, and δ_3, then

$$|f(x)\,g(x) - f(c)\,g(c)| \leq |f(x) - f(c)|\,|g(x)| + |f(c)|\,|g(x) - g(c)| < \frac{\epsilon}{2} + \frac{\epsilon}{2}.$$

Reciprocals require a similar finesse. If f is continuous at $x = c$ and if $f(c) \neq 0$, then we need to find a δ that will force

$$\left| \frac{1}{f(x)} - \frac{1}{f(c)} \right| = \frac{|f(c) - f(x)|}{|f(x)|\,|f(c)|}$$

to be less than any prespecified bound ϵ. We need an upper bound on $1/|f(x)|$ which means finding a lower bound on $|f(x)|$. We use the continuity of f to find δ_1 guaranteeing that

$$|f(x) - f(c)| < \frac{|f(c)|}{2},$$

and therefore

$$|f(x)| > \frac{|f(c)|}{2}.$$

We now have the bound

$$\frac{|f(c) - f(x)|}{|f(x)| \, |f(c)|} < |f(x) - f(c)| \frac{2}{|f(c)|^2}.$$

We again use the continuity of f at c to find δ_2 for which

$$|x - c| < \delta_2 \quad \text{implies that} \quad |f(x) - f(c)| < \frac{\epsilon \, |f(c)|^2}{2}.$$

Choosing δ to be the smaller of δ_1 and δ_2, we see that

$$\left| \frac{1}{f(x)} - \frac{1}{f(c)} \right| < \frac{\epsilon \, |f(c)|^2}{2} \cdot \frac{2}{|f(c)|^2} = \epsilon.$$

The easiest has been saved for last: compositions of continuous functions. If $g(x)$ is continuous at c and $f(y)$ is continuous at $g(c)$, then given a bound ϵ we first feed it to f and find the response δ_1:

$$|y - g(c)| < \delta_1 \quad \text{implies that} \quad |f(y) - f(g(c))| < \epsilon.$$

To get $|g(x) - g(c)| < \delta_1$, we feed δ_1 to g, getting a response δ_2:

$$|x - c| < \delta_2 \text{ implies that } |g(x) - g(c)| < \delta_1$$
$$\text{implies that } |f(g(x)) - f(g(c))| < \epsilon.$$

Differentiability Implies Continuity

We conclude this section with the observation that differentiability at $x = c$ implies continuity at $x = c$. Once it was realized that continuity was a significant property that actually needed verification, it was seen that we could not have differentiability without continuity. The converse remained an enigma for many years. It is possible to have a continuous function that fails to be differentiable at a single point or even at several discrete points. The function $f(x) = |x|$ is the simplest example. It is continuous at 0. Given a bound ϵ, one can always reply with $\delta = \epsilon$:

$$|x - 0| < \delta = \epsilon \quad \text{implies that} \quad |x| - 0 < \epsilon.$$

On the other hand, if we look at the error term in the definition of differentiability:

$$E(x, 0) = f'(0) - \frac{|x| - 0}{x - 0} = f'(0) - \left\{ \begin{array}{ll} 1, & x > 0 \\ -1, & x < 0 \end{array} \right\},$$

we see that there is no value that we can assign to $f'(0)$: If it is close to 1 then it will be far from -1, and if it is close to -1 then it will be far from 1.

How nondifferentiable can a continuous function be? In particular, can we find a function that is continuous at every point in some interval $[a, b]$ but that is not differentiable at any point in this interval? To most people's surprise, the answer to this question is *yes*. Bolzano found an example in the early 1830s, although it was not published until almost a century later.

In 1872, Weierstrass shocked the mathematical community with his example,

$$f(x) = \sum_{n=0}^{\infty} b^n \cos(a^n \pi x), \tag{3.47}$$

where a is an odd integer, $0 < b < 1$, and $ab > 1 + 3\pi/2$. What is so astonishing is that this is a reasonable Fourier series. For example, if $a = 13$ and $b = 1/2$, then this is the series

$$\cos(\pi x) + \frac{1}{2}\cos(13\pi x) + \frac{1}{4}\cos(169\pi x) + \frac{1}{8}\cos(2197\pi x) + \cdots .$$

This example and others will be explained in section 6.4. To verify that this function is continuous but not differentiable at any value of x, we shall first need to study properties of infinite series in more detail. For now, we shall content ourselves with the verification that differentiability requires continuity.

> **Theorem 3.5 (Differentiable ⟹ Continuous).** *If f is differentiable at $x = c$, then f is continuous at $x = c$.*

Proof: From the definition of differentiability, we know that there is a value $f'(c)$ for which the error term

$$E(x, c) = f'(c) - \frac{f(x) - f(c)}{x - c}$$

can be made as small as we want by restricting x to be sufficiently close to c. We solve this equation for $f(x) - f(c)$:

$$f(x) - f(c) = (x - c)f'(c) - (x - c)E(x, c),$$
$$|f(x) - f(c)| \leq |x - c|\,|f'(c)| + |x - c|\,|E(x, c)|.$$

Given a bound ϵ, we give half of it to each of the terms on the right side of this inequality. If $f'(c) = 0$, then the first term is zero. If not, then to make $|x - c|\,|f'(c)|$ less than $\epsilon/2$ we need to have

$$|x - c| < \frac{\epsilon}{2|f'(c)|}.$$

We can make $|E(x, c)|$ as small as we want. We find a δ_1 so that $|x - c| < \delta_1$ implies that $|E(x, c)| < \epsilon/2$. The second term will be the right size as long as $|x - c|$ is less than 1.

We choose δ to be the smallest of $\epsilon/2|f'(c)|$, δ_1, and 1 and we get the desired bound:

$$|f(x) - f(c)| \leq |x - c|\,|f'(c)| + |x - c|\,|E(x, c)|$$
$$< \frac{\epsilon}{2|f'(c)|}\,|f'(c)| + 1 \cdot \frac{\epsilon}{2} \ = \ \epsilon.$$

Q.E.D.

In exercise 3.3.34 you will get a chance to prove this theorem with a weaker hypothesis, only using one-sided derivatives.

Definition: one-sided limits and derivatives

The **limit from the right**, $\lim_{x \to a^+} f(x)$, is the target value T with the property that for any $\epsilon > 0$, there is a response δ so that if $a < x < a + \delta$, then $|f(x) - T| < \epsilon$. Similarly, the **limit from the left** implies this inequality when $a - \delta < x < a$. The **one-sided derivatives** are defined by

$$f'_+(a) = \lim_{x \to a^+} \frac{f(x) - f(a)}{x - a}, \qquad f'_-(a) = \lim_{x \to a^-} \frac{f(x) - f(a)}{x - a}.$$

Exercises

The symbol (**M&M**) indicates that *Maple* and *Mathematica* codes for this problem are available in the **Web Resources** at **www.macalester.edu/aratra**.

3.3.1. Prove that the function defined by

$$f(x) = \sin(1/x), \quad x \neq 0; \qquad f(0) = 0$$

is not continuous at $x = 0$ by finding an ϵ for which there is no reply.

3.3.2. Prove that the function defined by

$$f(x) = \begin{cases} 1, & x \text{ rational} \\ 0, & x \text{ irrational} \end{cases}$$

is not continuous at any x.

3.3.3. For the function g given in equation (3.40) on page 82, find a response δ that will work at $a = \sqrt{2}$ when $\epsilon = 0.2$. What are the rational numbers x in the interval $(\sqrt{2} - 1, \sqrt{2} + 1)$ for which $|g(x) - g(\sqrt{2})| \geq 0.2$?

3.3.4. At what values of x is the function f continuous? Justify your answer.

$$f(x) = \begin{cases} 0, & \text{if } x \text{ is irrational} \\ \sin x, & \text{if } x \text{ is rational} \end{cases}$$

3.3.5. At what values of x is the function f continuous? Justify your answer.

$$f(x) = \begin{cases} x^2 - 1, & \text{if } x \text{ is irrational} \\ 0, & \text{if } x \text{ is rational} \end{cases}$$

3.3.6. At what values of x is the function f continuous? Justify your answer.

$$f(x) = \begin{cases} x, & \text{if } x \text{ is irrational or } x = 0 \\ qx/(q+1), & \text{if } x = p/q, \text{ where } p \text{ and } q \text{ are relatively prime integers,} \\ & q > 0 \end{cases}$$

3.3.7. Prove that if f is continuous on $[a, b]$, then $|f|$ is continuous on $[a, b]$. Show by an example that the converse is not true.

3.3.8. Let f be a continuous function from $[0, 1]$ to $[0, 1]$. Show that there must be an x in $[0, 1]$ for which $f(x) = x$.

3.3.9. Let f and g be continuous functions on $[0, 1]$ such that $f(0) < g(0)$ and $f(1) > g(1)$. Show that there must be an x in $(0, 1)$ for which $f(x) = g(x)$.

3.3.10. Prove that the equation $(1 - x) \cos x = \sin x$ has at least one solution in $(0, 1)$.

3.3.11. Let f be a continuous function on $[0, 2]$ such that $f(0) = f(2)$. Show that there must be values a and b in $[0, 2]$ such that

$$a - b = 1 \quad \text{and} \quad f(a) = f(b).$$

3.3.12. Let f be a continuous function on $[0, 2]$. Show that there must be values a and b in $[0, 2]$ such that

$$a - b = 1 \quad \text{and} \quad f(a) - f(b) = \frac{f(2) - f(0)}{2}.$$

3.3.13. Let f be a continuous function on $[0, n]$ such that $f(0) = f(n)$, where $n \in \mathbb{N}$. Show that there must be values a and b in $[0, n]$ such that

$$a - b = 1 \quad \text{and} \quad f(a) = f(b).$$

3.3.14. Let $f(x) = \lfloor x^2 \rfloor \sin(\pi x)$ for all real x. Study the continuity of f. ($\lfloor a \rfloor$ is the *floor* of a, the greatest integer less than or equal to a.)

3.3.15. Let

$$f(x) = \lfloor x \rfloor + (x - \lfloor x \rfloor)^{\lfloor x \rfloor}, \quad \text{for } x \geq 1/2.$$

Show that f is continuous. Show that f is strictly increasing on $[1, \infty)$.

3.3.16. Find an x between 0 and 0.1 for which $\sin(1/x) = 1$. Find such an x for which $\sin(1/x) = -1$. Find x's between 0 and 0.001 for which $\sin(1/x) = 1, = -1$. Find x's between 0 and 10^{-100} for which $\sin(1/x) = 1, = -1$.

3.3.17. Find a $\delta > 0$ such that $0 < h \leq \delta$ guarantees that

$$|\sin(x + h) - \sin x| < 0.1.$$

3.3.18. Find a $\delta > 0$ such that $0 < h \leq \delta$ guarantees that

$$|(x + h)^2 - x^2| < 0.1,$$

when $0 \leq x \leq 1$.

3.3.19. If we do not restrict the size of x, can we find a $\delta > 0$ that does not depend on x and for which $0 < h \leq \delta$ guarantees that

$$|(x+h)^2 - x^2| < 0.1?$$

Explain why or why not.

3.3.20. Find a $\delta > 0$ such that $0 < h \leq \delta$ guarantees that

$$|e^{x+h} - e^x| < 0.1,$$

when $0 \leq x \leq 1$. Do we need to restrict the size of x?

3.3.21. Give an example of a function other than $f(x) = |x|$ that is continuous for all real x but that is not differentiable for at least one value of x.

3.3.22. Prove that

$$|\ln x - \ln c| < \begin{cases} |x-c|/c, & \text{if } x > c, \\ |x-c|/x, & \text{if } x < c. \end{cases} \tag{3.48}$$

3.3.23. Use the inequality in exercise 3.3.22 to find a positive number $\delta > 0$ such that $|h| \leq \delta$ implies that

$$|\ln(x+h) - \ln(x)| < 0.1$$

for all $x \geq 1$.

3.3.24. Does it seem strange to you that a function can be continuous at *exactly* one point? Find another function that is continuous at exactly one point.

3.3.25. **(M&M)** Graph the functions defined by

$$f_n(x) = \frac{\ln(x+2) - x^{2n}(\sin x)}{1 + x^{2n}},$$

for $n = 2$, 5, 10, and 20 over the interval $[0, \pi/2]$. Describe what you see. Find the approximate location of the root.

3.3.26. The functions f_n of exercise 3.3.25 are all continuous. Graph the function defined by

$$f(x) = \lim_{n \to \infty} f_n(x)$$

and prove that it is *not* continuous on the interval $[0, \pi/2]$. What is the value of $f(1)$? The conclusion is that the limit of a family of continuous functions might not be continuous.

3.3.27. Let $f(x) = \sin(1/x)$ when $x \neq 0$. Prove that if we choose *any* value from the interval $[-1, 1]$ to assign to $f(0)$, then f will have the intermediate value property.

3.3.28. Consider the function that takes the tenths digit in the decimal expansion of x and replaces it with a 1. For example, $f(2.57) = 2.17$, $f(3) = 3.1$, $f(\pi) = 3.14159\ldots = \pi$. Where is this function continuous? Where is this function not continuous? Justify your assertions.

3.3.29. Consider the function that takes the digits in the decimal expansion of $x \in (0, 1)$ and inserts 0's between them so that $0.a_1a_2a_3\ldots$ becomes $0.0a_10a_20a_3\ldots$. Is there any x in this interval that has a finite decimal expansion and for which this function is continuous?

3.3.30. Using the same function as in exercise 3.3.29, is there any $x \in (0, 1)$ that has an infinite decimal expansion and for which this function is not continuous?

3.3.31. Prove the intermediate value theorem with the weaker assumption that f is continuous on (a, b), continuous from the right at $x = a$ ($\lim_{x \to a^+} f(x) = f(a)$), and continuous from the left at $x = b$ ($\lim_{x \to b^-} f(x) = f(b)$).

3.3.32. Prove that if f has the intermediate value property on the interval $[a, c]$ and if it is monotonic on (a, c) then $\lim_{x \to c^-} f(x) = f(c)$. This completes the proof of Theorem 3.4 and shows that we can add the conclusion $\lim_{x \to a^+} f(x) = f(a)$ and $\lim_{x \to b^-} f(x) = f(b)$.

3.3.33. Prove that if f and g are both continuous at $x = c$ and if $g(c) \neq 0$, then f/g must be continuous at $x = c$.

3.3.34. Show that if the one-sided derivatives $f'_-(a)$ and $f'_+(a)$ exist, then f is continuous at a.

3.4 Consequences of Continuity

Continuity is a powerful concept. There is much that we can conclude about any continuous function. In this section, we shall pursue some of these consequences and investigate the even richer rewards that accrue when differentiability is also brought in to play.

Theorem 3.6 (Continuous $\Rightarrow$ Bounded). *If f is continuous on the interval $[a, b]$, then there exist finite bounds A and B such that*

$$A \leq f(x) \leq B$$

for all $x \in [a, b]$.

Before proving this theorem, we note that we really do need all of the conditions. If f only satisfies the intermediate value property, then it could be the function defined by

$$f(x) = x^{-1} \sin(1/x), \quad x \neq 0; \qquad f(0) = 0$$

which is not bounded on $[0, 1]$. If the endpoints of the interval are not included, then we could have a continuous function such as $f(x) = 1/x$ which is not bounded on $(0, 1)$.

Proof: We assume that f is *not* bounded and show that this implies at least one point $c \in [a, b]$ where f is not continuous. Again, we use the nested interval principle to find the point c. Let $x_1 = a$ and $y_1 = b$ and let c_1 be the midpoint of this interval:

$$c_1 = \frac{x_1 + y_1}{2}.$$

If we consider the two intervals $[x_1, c_1]$ and $[c_1, y_1]$, our function must be unbounded on at least one of them (if it were bounded on both, then the greater of the upper bounds would work for both intervals, the lesser of the lower bounds would serve as lower bound for both intervals). We choose one of these intervals on which f is unbounded and define x_2, y_2 to be the endpoints of this shorter interval:

$$x_1 \leq x_2 \; < \; y_2 \leq y_1,$$

$$y_2 - x_2 = (y_1 - x_1)/2.$$

We repeat this operation, setting $c_2 = (x_2 + y_2)/2$ and choosing a shorter interval on which f is still unbounded. Continuing in this manner, we obtain a sequence of nested intervals of arbitrarily short length,

$$x_1 \leq x_2 \leq \cdots \leq x_k \; < \; y_k \leq \cdots \leq y_2 \leq y_1,$$

$$y_{k+1} - x_{k+1} = (y_k - x_k)/2 = \cdots = (y_2 - x_2)/2^{k-1} = (y_1 - x_1)/2^k,$$

each with the property that f is unbounded on $[x_k, y_k]$. We let c be the point in all of these intervals that is promised to us by the nested interval principle.

All that is left is to prove that f is not continuous at c. We play the ϵ–δ game with an interesting twist: *no matter which ϵ is chosen, there is no δ with which we can respond.* To see this, let ϵ be any positive number, and let us claim that a certain δ will work. Our opponent points out that there is a k for which $y_k - x_k < \delta$ and so any point in $[x_k, y_k]$ is less than distance δ from c. We are also reminded that f is unbounded on $[x_k, y_k]$ which means that there is at least one x in this interval for which $f(x) > f(c) + \epsilon$ or $f(x) < f(c) - \epsilon$ (otherwise we could use $f(c) - \epsilon$ and $f(c) + \epsilon$ as our bounds). But then the distance from $f(x)$ to $f(c)$ is larger than ϵ.

Q.E.D.

Least Upper and Greatest Lower Bounds

If we want to patch up Cauchy's first proof of the mean value theorem by assuming that the derivative is continuous on $[a, b]$, it is not enough to prove that a continuous function on a closed interval is bounded, it must actually *achieve* the best possible bounds. That is to say, if f is continuous on $[a, b]$ then we must be able to find c_1 and c_2 in $[a, b]$ for which

$$f(c_1) \leq f(x) \leq f(c_2)$$

for all $x \in [a, b]$. The theorem we have just proved only promises us that bounds exist. It says nothing about how close these bounds come to the actual values of the function.

What we are usually interested in are the *best possible* bounds. In the case of $f(x) = x^3$ on $[-2, 3]$, these are -8 and 27. Respectively, these are called the greatest lower bound and the least upper bound. The **greatest lower bound** is a lower bound with the property that any larger number is not a lower bound. Similarly, the **least upper bound** is an upper bound with the property that any smaller number is not an upper bound. Before we can ask whether or not f achieves these best possible bounds, we must know whether they always exist. The precise definition is similar to the Archimedean understanding of a limit.

Definition: least upper, greatest lower bounds

Given a set $\mathcal{S}$, the **least upper bound** or **supremum** of $\mathcal{S}$, denoted sup $\mathcal{S}$, is the number G with the property that for any numbers $L < G$ and $M > G$, there is at least one element of $\mathcal{S}$ that is strictly larger than L and at least one upper bound for $\mathcal{S}$ that is strictly smaller than M. The **greatest lower bound** or **infimum** of $\mathcal{S}$, denoted inf $\mathcal{S}$, is the negative of the least upper bound of $-\mathcal{S} = \{-s \mid s \in \mathcal{S}\}$.

It may seem obvious that every bounded set has greatest lower and least upper bounds, but this is a subtle point that would not be recognized as a potential difficulty until the latter half of the 19th century. We have the machinery at hand for tackling it, and so we shall proceed. To convince you that there is something worth investigating, we consider the sequence

$$
\begin{aligned}
x_1 &= 1 - 1/3 \\
&= 2/3, \\
x_2 &= 1 - 1/3 + 1/5 - 1/7 \\
&= 76/105, \\
x_3 &= 1 - 1/3 + 1/5 - 1/7 + 1/9 - 1/11 \\
&= 2578/3465, \\
x_4 &= 1 - 1/3 + 1/5 - 1/7 + 1/9 - 1/11 + 1/13 - 1/15 \\
&= 33976/45045, \ \ldots .
\end{aligned}
$$

As we know from our earlier work on series, these numbers are increasing and approaching $\pi/4$. If we define

$$
\mathcal{S} = \{x_k \mid k \geq 1\},
$$

then $\pi/4$ is the least upper bound for this set. But what happens if we restrict our attention to *rational* numbers? Every element of $\mathcal{S}$ is rational, but $\pi/4$ is not. In the set of rational numbers, we cannot call on $\pi/4$ to serve as our least upper bound. We are required to choose a rational number. We can certainly find a rational number that is an upper bound. The number 1 will do, but it is not a least upper bound: 83/105 would be better. Still better would be 2722/3465. If we restrict our sights to the rational numbers, then we can always find a better upper bound, there is no best. That is because the best, $\pi/4$, is outside the domain in which we are searching. No matter how close to $\pi/4$ we choose our rational number, there is always another rational number that is a little bit closer.

The problem is that the set of rational numbers has holes in it: precisely those irrational numbers like $\pi/4$. What characterizes the real numbers is that they include all of the rationals plus what is needed to plug the holes. This property of the real numbers is implicit in the nested interval principle.

> **Theorem 3.7 (Upper Bound $\Rightarrow$ Least Upper Bound).** *In the real numbers, every set that has an upper bound also has a least upper bound and every set that has a lower bound also has a greatest lower bound.*

Proof: Since the greatest lower bound can be defined in terms of the least upper bound, it is enough to prove the existence of the least upper bound.

We assume that S has an upper bound (and therefore is not empty) and construct our sequences for the nested interval principle as follows: let x_1 be a number that is *not* an upper bound of S (choose some $x \in S$ and then choose any $x_1 < x$) and let y_1 be an upper bound for S. We let c_1 be the midpoint of $[x_1, y_1]$, $c_1 = (x_1 + y_1)/2$. If c_1 is an upper bound, then we set $x_2 = x_1$ and $y_2 = c_1$. If c_1 is not an upper bound, then we set $x_2 = c_1$ and $y_2 = y_1$. In either case, x_2 is not an upper bound and y_2 is an upper bound for S,

$$x_1 \leq x_2 \ < \ y_2 \leq y_1,$$

and

$$y_2 - x_2 = (y_1 - x_1)/2.$$

This can be repeated as often as we like. Once we have found x_k and y_k, we split the difference: $c_k = (x_k + y_k)/2$. If c_k is an upper bound, then $x_{k+1} = x_k$ and $y_{k+1} = c_k$. If c_k is not an upper bound, then $x_{k+1} = c_k$ and $y_{k+1} = y_k$. In either case, x_{k+1} is not an upper bound and y_{k+1} is an upper bound for S,

$$x_1 \leq x_2 \leq \cdots \leq x_k \leq x_{k+1} \ < \ y_{k+1} \leq y_k \leq \cdots \leq y_2 \leq y_1,$$

and

$$y_{k+1} - x_{k+1} = (y_k - x_k)/2 = \cdots = (y_2 - x_2)/2^{k-1} = (y_1 - x_1)/2^k.$$

We claim that the c that lies in all of these intervals is the least upper bound. If we take any $L < c$, then we can find an $x_k > L$ and so there is an $x \in S$ with $L < x_k < x$. If we take any $M > c$, then we can find a $y_k < M$, and y_k is an upper bound for S.

$$\text{Q.E.D.}$$

We could have taken the existence of least upper bounds as an axiom of the real numbers. As you are asked to prove exercise 3.4.11, the statement "every set with an upper bound has a least upper bound" implies the nested interval principle.

Achieving the Bounds

> **Theorem 3.8 (Continuous $\Rightarrow$ Bounds Achieved).** *If f is continuous on $[a, b]$, then it achieves its greatest lower bound and its least upper bound. Equivalently, there exist $k_1, k_2 \in [a, b]$ such that*
>
> $$f(k_1) \leq f(x) \leq f(k_2)$$
>
> *for all $x \in [a, b]$.*

Proof: We shall only prove the existence of k_2. The proof for k_1 follows by substituting $-f$ for f.

As we saw in Theorem 3.6, the set $\{f(x) \mid a \leq x \leq b\}$ has an upper bound. Theorem 3.7 then promises us a least upper bound; call it A. Our problem is to show that there is some $c \in [a, b]$ for which $f(c) = A$. By now you should expect that we use the nested interval principle to find our candidate for c. We start by defining $x_1 = a$, $y_1 = b$, and $c_1 = (x_1 + y_1)/2$. Since A is an upper bound for $f(x)$ over the entire interval $[a, b]$, it is also an upper bound for $f(x)$ over each of the shorter intervals $[x_1, c_1]$ and $[c_1, y_1]$. It must be the least upper bound for $f(x)$ over at least one of these subintervals, because if something smaller worked for both subintervals, then A would not be the least upper bound over $[a, b]$. If A is the least upper bound over $[x_1, c_1]$, then we define $x_2 = x_1$ and $y_2 = c_1$. If not, then A is the least upper bound over $[c_1, y_1]$ and we define $x_2 = c_1$ and $y_2 = y_1$.

We continue in this manner, each time chopping our interval in half and choosing a half on which A is still the least upper bound. We get our sequences

$$x_1 \leq x_2 \leq \cdots \leq x_k \ < \ y_k \leq \cdots \leq y_2 \leq y_1,$$

$$y_k - x_k = (y_{k-1} - x_{k-1})/2 = \cdots = (y_2 - x_2)/2^{k-1} = (y_1 - x_1)/2^k.$$

Let c be the point that is in all of these intervals.

Since A is an upper bound for $f(x)$ over $[a, b]$, we know that $f(c)$ is less than or equal to A. If $f(c)$ is strictly less than A, then we choose an $\epsilon < A - f(c)$ and use the continuity of f at c to find a δ such that $|x - c| < \delta$ guarantees that $|f(x) - f(c)| < \epsilon$. This in turn implies that

$$f(c) - \epsilon < f(x) < f(c) + \epsilon < A.$$

We now choose our k so that $y_k - x_k < \delta$. It follows that every point in $[x_k, y_k]$ is at most distance δ from c. The quantity $f(c) + \epsilon$—which is less than A—is an upper bound for $f(x)$ over $[x_k, y_k]$. This contradicts the fact that A is the least upper bound for $f(x)$ over the interval $[x_k, y_k]$. Our assumption that $f(c)$ is strictly less than A cannot be valid, and so $f(c) = A$.

Q.E.D.

Fermat's Theorem on Extrema

As was mentioned in section 3.2, the best proof of the mean value theorem is slick, but it is neither direct nor obvious. It is the result of knowing enough about continuous and differentiable functions that someone, eventually, observed a better route. Here we begin to look at the consequences of differentiability, starting with an observation that had been made by Pierre de Fermat in 1637 or 1638, and in a less precise form by Johann Kepler in 1615, well before Newton or Leibniz were born (1642 and 1646, respectively). It is the observation that one finds the extrema (maxima or minima) of a function where the derivative is zero. This observation was a principal impetus behind the search for the algorithms of differentiation.

> **Theorem 3.9 (Fermat's Theorem on Extrema).** *If f has an extremum at a point* $c \in (a, b)$ *[*$f(c) \geq f(x)$ *for all* $x \in (a, b)$ *or* $f(c) \leq f(x)$ *for all* $x \in (a, b)$*] and if f is differentiable at every point in* (a, b)*, then* $f'(c) = 0$.

Proof: We shall actually prove that if $f'(c) \neq 0$, then we can find $x_1, x_2 \in (a, b)$ for which

$$f(x_1) < f(c) < f(x_2).$$

It follows that if $f'(c) \neq 0$, then we do not have an extremum at $x = c$. This is logically equivalent to what we want to prove.

Without loss of generality, we can assume that $f'(c) > 0$ (if it is less than zero, then we replace f by $-f$). It should be evident that we want x_1 to be a little less than c and x_2 to be a little more, but both have to be very close to c. How close? Here is where we use the definition of differentiability.

We let $E(x, c)$ be the error introduced when the derivative is replaced by the average rate of change:

$$E(x, c) = f'(c) - \frac{f(x) - f(c)}{x - c}.$$

If the absolute value of the error is smaller than $|f'(c)|$, then $[f(x) - f(c)]/[x - c]$ will have the same sign as $f'(c)$. Since we have assumed that $f'(c)$ is positive, we have

$$\frac{f(x) - f(c)}{x - c} > 0.$$

When x is less than c, $f(x)$ will have to be less than $f(c)$. When x is larger than c, $f(x)$ will have to be larger than $f(c)$. The solution is therefore to find a δ for which

$$|x - c| < \delta \quad \text{implies that} \quad |E(x, c)| < |f'(c)|.$$

The definition of differentiability promises us such a δ. We choose x_1 and x_2 so that

$$c - \delta < x_1 < c < x_2 < c + \delta.$$

Q.E.D.

Rolle's Theorem

There is a special case of the mean value theorem that was noted by Michel Rolle (1652–1719) in 1691 and was periodically resurrected over the succeeding years. At the time, it seemed so obvious that no one bothered to prove it. In fact, it is equivalent to the mean value theorem. Once we have proved it, we shall be almost there.

> **Theorem 3.10 (Rolle's Theorem).** *Let f be a function that is continuous on* $[a, b]$ *and differentiable on* (a, b) *and for which* $f(a) = f(b) = 0$. *There exists at least one* c, $a < c < b$, *for which* $f'(c) = 0$.

Proof: Since our function is continuous on $[a, b]$, Theorem 3.8 promises us that it must achieve its maximal and minimal values somewhere on this interval. At least one of these

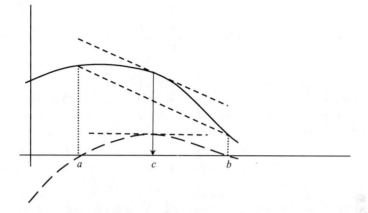

FIGURE 3.14. Proof of the mean value theorem.

extrema must occur at some x strictly between a and b. The only possible counterexample would be a function with both extrema at the endpoints, but then 0 is the largest value of the function and 0 is the smallest value of the function. The function would be identically 0 and so have an extremum at *every* point in $[a, b]$. Let $c \in (a, b)$ be a point at which f has an extremum. By Theorem 3.9, $f'(c) = 0$.

Q.E.D.

We note that this is a special case of the mean value theorem because the average rate of change over this interval is

$$\frac{f(b) - f(a)}{b - a} = \frac{0 - 0}{b - a} = 0.$$

It is equivalent to the mean value theorem because we can find an auxiliary function that enables us to reduce the mean value theorem to this case.

Mean Value Theorem

We are now ready to prove the mean value theorem, Theorem 3.1 on page 58, that if f is continuous on $[a, b]$ and differentiable on (a, b), then there is at least one $c, a < c < b$, for which

$$f'(c) = \frac{f(b) - f(a)}{b - a}.$$

Proof: (Mean Value Theorem) We subtract from our function the straight line passing through $[a, f(a)]$ and $[b, f(b)]$ (Figure 3.14). The result is a new function that is continuous on $[a, b]$, differentiable on (a, b) and for which $g(a) = g(b) = 0$:

$$g(x) = f(x) - \frac{f(b) - f(a)}{b - a}(x - a) - f(a).$$

We apply Rolle's theorem to g:

$$0 = g'(c) = f'(c) - \frac{f(b) - f(a)}{b - a},$$

and therefore

$$f'(c) = \frac{f(b) - f(a)}{b - a}.$$

Q.E.D.

As was pointed out in section 3.2, this proof was discovered by Ossian Bonnet and published in Serret's calculus text of 1868. It should be noted that we have weakened the assumptions that Cauchy made. Our function does not need to be differentiable at the endpoints, and we certainly do not need the derivative to be continuous. There is no reason why we cannot have an unbounded derivative, for example the function defined by

$$f(x) = x \sin(1/x), \quad x \neq 0; \qquad f(0) = 0.$$

This is continuous on $[0, 1]$, and it is differentiable on $(0, 1)$, but it is not differentiable on $[0, 1]$. The mean value theorem that we have just proven assures us that for every positive x there is a point c, $0 < c < x$, for which

$$\frac{x \sin(1/x) - 0}{x - 0} = \sin(1/x) = f'(c) = \sin(1/c) - c^{-1} \cos(1/c). \qquad (3.49)$$

Exercises

The symbol (**M&M**) indicates that *Maple* and *Mathematica* codes for this problem are available in the **Web Resources** at **www.macalester.edu/aratra**.

3.4.1. Give an example of a function that exists and is bounded for all x in the interval $[0, 1]$ but which never achieves either its least upper bound or its greatest lower bound over this interval.

3.4.2. Give an example of a bounded, continuous function which does not achieve its least upper bound. Notice that the domain was not specified.

3.4.3. Give an example of a function whose derivative vanishes at $x = 1$, $f'(1) = 0$, but which does not have an extremum at $x = 1$.

3.4.4. Prove that if A is not zero and $|A - B|$ is less than $|A|$, then A and B must have the same sign.

3.4.5. Prove that if a function is continuous at every point except $x = c$ and is so discontinuous at $x = c$ that there is no response δ for *any* error bound ϵ, then the function must have a vertical asymptote at $x = c$. Part of proving this statement is coming up with a careful definition of a vertical asymptote.

3.4.6. Find the greatest lower bound and the least upper bound of each of the following sets.

a. the interval $(0, 3)$

b. $\{1, 1/2, 1/4, 1/8, \ldots\}$

c. $\{1, 1 + 1/2, 1 + 1/2 + 1/4, 1 + 1/2 + 1/4 + 1/8, \ldots\}$

d. $\{2/1, (2 \cdot 2)/(1 \cdot 3), (2 \cdot 2 \cdot 4)/(1 \cdot 3 \cdot 3), (2 \cdot 2 \cdot 4 \cdot 4)/(1 \cdot 3 \cdot 3 \cdot 5), \ldots\}$ (See equation (2.18) on page 24.)

e. $\{0.2, 0.22, 0.222, \ldots\}$

f. the set of decimal fractions between 0 and 1 whose only digits are 0's and 1's

g. $\{(n + 1)^2/2^n \mid n \in \mathbb{N}\}$

h. $\{(m + n)^2/2^{mn} \mid m, n \in \mathbb{N}, m < 2n\}$

i. $\{m/n \mid m, n \in \mathbb{N}\}$

j. $\{\sqrt{n} - \lfloor \sqrt{n} \rfloor \mid n \in \mathbb{N}\}$

k. $\{x \mid x^2 + x + 1 > 0\}$

l. $\{x + x^{-1} \mid x > 0\}$

m. $\{2^x + 2^{1/x} \mid x > 0\}$

n. $\{m/n + 4n/m \mid m, n \in \mathbb{N}\}$

o. $\{mn/(4m^2 + n^2) \mid m \in \mathbb{Z}, n \in \mathbb{N}\}$

p. $\{m/(m + n) \mid m, n \in \mathbb{N}\}$

q. $\{m/(|m| + n) \mid m \in \mathbb{Z}, n \in \mathbb{N}\}$

r. $\{mn/(1 + m + n) \mid m, n \in \mathbb{N}\}$

3.4.7. Prove that for any set S, the negative of the least upper bound of $-S$ is a lower bound for S, and there is no lower bound for S that is larger.

3.4.8. Modify the proof of Theorem 3.6 to prove that if f is continuous on (a, b), $\lim_{x \to a^+} f(x) = f(a)$, and $\lim_{x \to b^-} f(x) = f(b)$, then there exist finite bounds A and B such that $A \le f(x) \le B$ for all $x \in [a, b]$.

3.4.9. Modify the proof of Theorem 3.8 to prove that if f is continuous on (a, b), $\lim_{x \to a^+} f(x) = f(a)$, and $\lim_{x \to b^-} f(x) = f(b)$, then there exist $k_1, k_2 \in [a, b]$ such that $f(k_1) \le f(x) \le f(k_2)$ for all $x \in [a, b]$.

3.4.10. For the mean value theorem (Theorem 3.1) and the generalized mean value theorem (Theorem 3.2), explain how the proofs need to be modified in order to weaken the hypotheses so that instead of continuity at every point on a closed interval we only need continuity on the open interval and continuity from one side at each of the endpoints.

3.4.11. Prove that if "every set with an upper bound has a least upper bound," then the nested interval principle holds.

3.4.12. Use the existence of a least upper bound for any bounded set to prove that if $g'(x) > 0$ for all $x \in [a, b]$, then g is increasing over $[a, b]$ ($a \le x_1 < x_2 \le b$ implies that $g(x_1) < g(x_2)$).

3.4.13. Use the result from exercise 3.4.12 to prove that if $f'(x) \ge 0$ for all $x \in [a, b]$ and if $a \le x_1 < x_2 \le b$, then $f(x_1) \le f(x_2)$.

3.4.14. Prove that

$$y = \frac{f(b) - f(a)}{b - a}(x - a) + f(a)$$

is the equation of the straight line through $[a, f(a)]$ and $[b, f(b)]$.

3.4.15. **(M&M)** In equation 3.49 on page 102, we showed that for any positive x, the mean value theorem implies that there exists a value c, $0 < c < x$, for which

$$\sin(1/x) = \sin(1/c) - c^{-1}\cos(1/c).$$

Find (to ten-digit accuracy) such a c for each of the following values of x: 1, 1/3, and 0.01.

3.4.16. Using exercise 3.4.15, we can define a function g for which $g(0) = 0$ and if $x > 0$ then $g(x) = c$ where c is the largest number less than x for which

$$\sin(1/x) = \sin(1/c) - c^{-1}\cos(1/c).$$

Prove that g does not have the intermediate value property on $[0, 1]$.

3.4.17. Explain why the conclusion of the mean value theorem still holds if we only assume that f is continuous and differentiable on the open interval (a, b) and that $\lim_{x \to a^+} f(x) = f(a)$, $\lim_{x \to b^-} f(x) = f(b)$.

3.4.18. Define

$$f(x) = x\sin(\ln x), \quad x > 0; \qquad f(0) = 0.$$

Show that f on the interval $[0, 1]$ satisfies the conditions of the mean value theorem given in exercise 3.4.17. Prove that if x is positive, then there exists a c, $0 < c < x$, for which

$$\sin(\ln c) + \cos(\ln c) = \sin(\ln x).$$

3.4.19. Using exercise 3.4.18, we can define a function h for which $h(0) = 0$ and if $x > 0$ then $h(x) = c$ where c is the largest number less than x for which

$$\sin(\ln c) + \cos(\ln c) = \sin(\ln x).$$

Prove that h does not have the intermediate value property on $[0, 1]$.

3.4.20. Prove that if f is differentiable on $[a, b]$ and if f' is piecewise monotonic on $[a, b]$, then f' is continuous on $[a, b]$.

3.4.21. **(M&M)** Graph the function

$$f(x) = \frac{x}{1 + x\sin(1/x)}, \quad x \neq 0; \qquad f(0) = 0,$$

over the interval $[0, 1]$. Prove that it is differentiable and piecewise monotonic on the interval $[0, 1]$. Is the derivative f' continuous on $[0, 1]$? Discuss this result in light of exercise 3.4.20.

3.4.22. Let $P(x)$ be any polynomial of degree at least 2, all of whose roots are real and distinct. Prove that all of the roots of $P'(x)$ must be real. What happens if some of the roots of P are multiple roots?

3.4.23. Prove that if f is defined on (a, b), f achieves its maximum value at $c \in (a, b)$, and the one-sided derivatives $f'_-(c)$ and $f'_+(c)$ exist, then $f'_+(c) \leq 0 \leq f'_-(c)$.

3.4.24. Prove that if f is continuous on $[a, b]$, $f(a) = f(b)$, and the one-sided derivative f'_- exists for all $x \in (a, b)$, then

$$\inf\{f'_-(x) \mid x \in (a, b)\} \leq 0 \leq \sup\{f'_-(x) \mid x \in (a, b)\}.$$

3.4.25. Prove that if f is continuous on $[a, b]$ and the one-sided derivative f'_- exists for all $x \in (a, b)$, then

$$\inf\{f'_-(x) \mid x \in (a, b)\} \leq \frac{f(b) - f(a)}{b - a} \leq \sup\{f'_-(x) \mid x \in (a, b)\}.$$

3.4.26. Prove that if f'_- exists and is continuous for all $x \in (a, b)$, then f is differentiable on (a, b) and $f'(x) = f'_-(x)$ for all $x \in (a, b)$.

3.4.27. Does there exist a function f on $(1, 2)$ such that $f'_-(x) = x$ and $f'_+(x) = 2x$ for all $x \in (1, 2)$?

3.4.28. Let f be differentiable on $[a, b]$ such that $f(a) = 0 = f(b)$ and $f'(a) > 0$, $f'(b) > 0$. Prove that there is at least one $c \in (a, b)$ for which $f(c) = 0$ and $f'(c) \leq 0$.

3.4.29. Suppose that f is continuous on $[a, \infty)$ and $\lim_{x \to \infty} f(x)$ is finite. Show that f is bounded on $[a, \infty)$.

3.4.30. Prove that if f is continuous on a closed interval $[a, b]$, differentiable on the open interval (a, b), and $f(a) = f(b) = 0$, then for any real number α there is an $x \in (a, b)$ such that

$$\alpha f(x) + f'(x) = 0.$$

3.5 Consequences of the Mean Value Theorem

Cauchy used the generalized mean value theorem to prove Lagrange's form of Taylor's theorem, so we begin by proving Theorem 3.2.

Proof: (Generalized Mean Value Theorem) We are given that f and F are continuous in $[a, b]$ and differentiable in (a, b) and that $F'(x)$ is never zero for $a < x < b$. We define a new function g by

$$g(x) = F(x)[f(b) - f(a)] - f(x)[F(b) - F(a)].$$

The function g is also continuous on $[a, b]$ and differentiable on (a, b) and

$$g(a) = F(a)f(b) - f(a)F(b) = g(b).$$

By the mean value theorem, there is a c strictly between a and b for which

$$g'(c) = F'(c)[f(b) - f(a)] - f'(c)[F(b) - F(a)] = 0. \tag{3.50}$$

Since $F'(x)$ is never zero for $a < x < b$, $F(b)$ cannot equal $F(a)$ (why not?, exercise 3.5.1) and $F'(c) \neq 0$. We can rewrite equation 3.50 as

$$\frac{f(b) - f(a)}{F(b) - F(a)} = \frac{f'(c)}{F'(c)}. \tag{3.51}$$

Q.E.D.

Finally, we are ready to prove Theorem 2.1 from page 44.

Proof: (Lagrange Remainder Theorem) We assume that the first k derivatives of f exist in some neighborhood of $x = a$. We define F to be the difference between f and the truncated Taylor series:

$$F(x) = f(x) - f(a) - f(a)(x - a)$$
$$- \frac{f''(a)}{2!}(x - a)^2 - \cdots - \frac{f^{(k-1)}(a)}{(k-1)!}(x - a)^{k-1}. \tag{3.52}$$

We observe that

$$F(a) = F'(a) = F''(a) = \cdots = F^{(k-1)}(a) = 0, \tag{3.53}$$

and

$$F^{(k)}(x) = f^{(k)}(x). \tag{3.54}$$

We consider the fraction $F(x)$ divided by $(x - a)^k$. Since both expressions are 0 when $x = a$, we can subtract $F(a)$ from the numerator and $(a - a)^k$ from the denominator and then apply the generalized mean value theorem. There must be some x_1 between x and a for which

$$\frac{F(x)}{(x - a)^k} = \frac{F(x) - F(a)}{(x - a)^k - (a - a)^k} = \frac{F'(x_1)}{k(x_1 - a)^{k-1}}. \tag{3.55}$$

We apply the generalized mean value theorem to this function of x_1:

$$\frac{F'(x_1)}{k(x_1 - a)^{k-1}} = \frac{F'(x_1) - F'(a)}{k(x_1 - a)^{k-1} - k(a - a)^{k-1}} = \frac{F''(x_2)}{k(k - 1)(x_2 - a)^{k-2}}, \tag{3.56}$$

for some x_2 between x_1 and a. We continue in this manner:

$$\frac{F(x)}{(x-a)^k} = \frac{F'(x_1)}{k(x_1-a)^{k-1}} = \frac{F''(x_2)}{k(k-1)(x_2-a)^{k-2}} = \cdots$$

$$= \frac{F^{(k-1)}(x_{k-1})}{k(k-1)\cdots 2(x_{k-1}-a)} = \frac{F^{(k)}(c)}{k!}, \tag{3.57}$$

where $a < c < x_{k-1} < x_{k-2} < \cdots < x_2 < x_1 < x$. Since $F^{(k)}(c) = f^{(k)}(c)$, we can rewrite this last equation as

$$F(x) = \frac{f^{(k)}(c)}{k!}(x-a)^k. \tag{3.58}$$

Q.E.D.

Cauchy realized that there was another way of expressing this error.

Theorem 3.11 (Cauchy's Remainder Theorem). *Given a function f for which all derivatives exist at $x = a$, let $D_n(a, x)$ denote the difference between the nth partial sum of the Taylor series for f expanded about $x = a$ and the target value $f(x)$,*

$$D_n(a, x) = f(x) - \left(f(a) + f'(a)(x-a) + \frac{f''(a)}{2!}(x-a)^2 \right.$$

$$\left. + \cdots + \frac{f^{(n-1)}(a)}{(n-1)!}(x-a)^{n-1} \right). \tag{3.59}$$

There is at least one real number c strictly between a and x for which

$$D_n(a, x) = \frac{f^{(n)}(c)}{(n-1)!}(x-c)^{n-1}(x-a). \tag{3.60}$$

Proof: We look at the difference between $f(x)$ and the truncated series not as a function of x but as a function of a:

$$\phi(a) = f(x) - f(a) - f'(a)(x-a) - \frac{f''(a)}{2!}(x-a)^2 - \cdots - \frac{f^{(k-1)}(a)}{(k-1)!}(x-a)^{k-1}. \tag{3.61}$$

We note that $\phi(x) = 0$. Taking the derivative with respect to a, we see that

$$\phi'(a) = 0 - f'(a) - \left(f''(a)(x-a) - f'(a) \right)$$

$$- \left(\frac{f'''(a)}{2!}(x-a)^2 - 2\frac{f''(a)}{2!}(x-a) \right) - \cdots$$

$$- \left(\frac{f^{(k)}(a)}{(k-1)!}(x-a)^{k-1} - (k-1)\frac{f^{(k-1)}(a)}{(k-1)!}(x-a)^{k-2} \right)$$

$$= -\frac{f^{(k)}(a)}{(k-1)!}(x-a)^{k-1}. \tag{3.62}$$

We now use the mean value theorem just once:

$$\frac{\phi(a)}{a-x} = \frac{\phi(a) - \phi(x)}{a-x} = \phi'(c) = -\frac{f^{(k)}(c)}{(k-1)!}(x-c)^{k-1}, \tag{3.63}$$

for some c between a and x. In this case, the remainder is

$$\phi(a) = \frac{f^{(k)}(c)}{(k-1)!}(x-c)^{k-1}(x-a). \tag{3.64}$$

Q.E.D.

Comparing Remainders

A good series for illustrating the distinction between these two expressions for the remainder is the logarithmic series:

$$\ln(1+x) = x - \frac{x^2}{2} + \frac{x^3}{3} - \frac{x^4}{4} + \cdots + (-1)^k \frac{x^{k-1}}{k-1} + R_k(x).$$

The Lagrange form of the remainder is

$$R_k(x) = \frac{f^{(k)}(c)}{k!} x^k = (-1)^{k-1} \frac{x^k}{k(1+c)^k}. \tag{3.65}$$

The Cauchy form is

$$R_k(x) = \frac{f^{(k)}(c)}{(k-1)!}(x-c)^{k-1}x = (-1)^{k+1} \frac{x(x-c)^{k-1}}{(1+c)^k}. \tag{3.66}$$

In each case, c is some constant (different in each case) lying between 0 and x.

If $x = 2/3$, then $0 < c < 2/3$ and the absolute values of the respective remainders are

$$\frac{(2/3)^k}{k(1+c)^k} < \frac{2^k}{k \cdot 3^k}$$

(the Lagrange remainder is maximized when $c = 0$), and

$$\frac{(2/3)(2/3 - c)^{k-1}}{(1+c)^k} < \frac{2^k}{3^k}$$

(the Cauchy remainder is also maximized when $c = 0$). We see that the Lagrange form gives a tighter bound.

If $x = -2/3$, then $-2/3 < c < 0$ and the absolute values of the respective remainders are

$$\frac{(2/3)^k}{k(1+c)^k} < \frac{2^k}{k}$$

(the Lagrange remainder is maximized when $c = -2/3$), and

$$\frac{(2/3)(2/3 + c)^{k-1}}{(1+c)^k} < \frac{2^k}{3^k}$$

(the Cauchy remainder is maximized when $c = 0$). We see that the Cauchy form gives a tighter bound in this case. In fact, the Cauchy bound approaches zero as k goes to infinity while the Lagrange bound diverges to infinity.

L'Hospital's Rule

It is a familiar story that the Marquis de L'Hospital (1661–1704) stole what has come to be known as L'Hospital's rule from Johann Bernoulli. It needs to be tempered with the observation that while the result is almost certainly Bernoulli's, L'Hospital was a respectable mathematician who had paid for the privilege of publishing Bernoulli's results under his own name.

> To learn more about the Marquis de l'Hospital, his role in the early development of calculus, our uncertainty over how to spell his name, and why we do not pronounce the "s" in his name, go to **The Marquis de l'Hospital**.

We work with the Archimedean definition of limit given on page 59. We also need to be careful about what we mean by an infinite limit and a limit at infinity.

Definition: infinite limit and limit at infinity

The statement

$$\lim_{x \to a} f(x) = \infty$$

means that for any real number L, we can force $f(x) > L$ by restricting x to be sufficiently close to a. That is to say, there is a $\delta > 0$ so that $|x - a| < \delta$ implies that $f(x) > L$. When we write

$$\lim_{x \to \infty} f(x) = T,$$

we mean that for any positive ϵ, we can force $f(x)$ to be within ϵ of T by taking x sufficiently large. In other words, there is an N so that $x > N$ implies that $|f(x) - T| < \epsilon$.

Theorem 3.12 (L'Hospital's Rule: 0/0). *If f and F are both differentiable inside an open interval that contains a, if*

$$\lim_{x \to a} f(x) = 0 = \lim_{x \to a} F(x),$$

if $F'(x) \neq 0$ for all x in this open interval, and if $\lim_{x \to a} f'(x)/F'(x)$ exists, then

$$\lim_{x \to a} \frac{f(x)}{F(x)} = \lim_{x \to a} \frac{f'(x)}{F'(x)}. \tag{3.67}$$

Note that this theorem has a lot of hypotheses. They are all important. As you work through this proof, identify the places where the hypotheses are used. In the exercises, you will show that each of these hypotheses is necessary by finding examples where the conclusion does not hold when one of the hypotheses is removed. You will also be asked to prove that this theorem remains true if a is replaced by ∞.

Proof: Since f and F are differentiable in this interval, they are continuous, and therefore $f(a) = 0 = F(a)$. The generalized mean value theorem tells us that

$$\frac{f(x)}{F(x)} = \frac{f(x) - f(a)}{F(x) - F(a)} = \frac{f'(c)}{F'(c)},$$

for some c between a and x. Let L be the limit of $\lim_{x \to a} f'(x)/F'(x)$. Given any $\epsilon > 0$, there is a response δ so that if $|c - a| < \delta$, then

$$\left| \frac{f'(c)}{F'(c)} - L \right| < \epsilon.$$

If $\delta > |x - a|$, then we also have that $\delta > |c - a|$ and so

$$\left| \frac{f(x)}{F(x)} - L \right| = \left| \frac{f'(c)}{F'(c)} - L \right| < \epsilon.$$

Q.E.D.

Note that there is nothing in this proof that requires that we work with values of x on both sides of a or that either f or F is differentiable at a. In particular, L'Hospital's rule works equally well with one-sided limits (see exercise 3.5.6).

Theorem 3.13 (L'Hospital's Rule: ∞/∞). *If f and F are both differentiable at every point except $x = a$ in an open interval that contains a, if*

$$\lim_{x \to a} |F(x)| = \infty,$$

if $F'(x) \neq 0$ for all x in this open interval, and if $\lim_{x \to a} f'(x)/F'(x)$ exists, then

$$\lim_{x \to a} \frac{f(x)}{F(x)} = \lim_{x \to a} \frac{f'(x)}{F'(x)}. \tag{3.68}$$

While we do not insist that $\lim_{x \to a} |f(x)| = \infty$, that is the only interesting case of this theorem. Otherwise, the limit of $f(x)/F(x)$ is 0 or does not exist (see exercise 3.5.5).

Proof: Here we need a great deal more finesse. We shall assume that L, the limit of $f'(x)/F'(x)$, is finite. The proof can be modified to handle an infinite limit.

We are given an error bound ϵ. We must find a response δ so that if x is within δ of a, then $f(x)/F(x)$ will lie within ϵ of L. We begin by observing that if we take two values inside our open interval and both on the same side of a, call them x and s, then

$$\frac{f(x) - f(s)}{F(x) - F(s)} = \frac{f'(c)}{F'(c)}, \tag{3.69}$$

for some c between x and s. We choose x and s so that x lies between a and s and so that s is close enough to a to guarantee that $|L - f'(c)/F'(c)| < \epsilon/2$ for any c between s and *a*. The generalized mean value theorem implies that for any choice of x between s and a, we have

$$L - \frac{\epsilon}{2} < \frac{f(x) - f(s)}{F(x) - F(s)} = \frac{f'(c)}{F'(c)} < L + \frac{\epsilon}{2}. \qquad (3.70)$$

We fix our value for s. Since $\lim_{x \to a} |F(x)| = \infty$, there is a δ_1 for which $|x - a| < \delta_1$ implies that $|F(x)| > |F(s)|$ and therefore

$$1 - \frac{F(s)}{F(x)} \geq 1 - \left| \frac{F(s)}{F(x)} \right| > 0.$$

Multiplying equation (3.70) by

$$1 - \frac{F(s)}{F(x)} = \frac{F(x) - F(s)}{F(x)}$$

gives us

$$\left(L - \frac{\epsilon}{2} \right) \left(1 - \frac{F(s)}{F(x)} \right) < \frac{f(x)}{F(x)} - \frac{f(s)}{F(x)} < \left(L + \frac{\epsilon}{2} \right) \left(1 - \frac{F(s)}{F(x)} \right). \qquad (3.71)$$

This is equivalent to

$$L - \frac{\epsilon}{2} - \frac{F(s)[L - \epsilon/2] - f(s)}{F(x)} < \frac{f(x)}{F(x)} < L + \frac{\epsilon}{2} - \frac{F(s)[L + \epsilon/2] - f(s)}{F(x)}. \qquad (3.72)$$

Since s, L, and ϵ are fixed, we can find a δ_2 so that $|x - a| < \delta_2$ implies that

$$\left| \frac{F(s)[L - \epsilon/2] - f(s)}{F(x)} \right| < \frac{\epsilon}{2}, \qquad (3.73)$$

$$\left| \frac{F(s)[L + \epsilon/2] - f(s)}{F(x)} \right| < \frac{\epsilon}{2}. \qquad (3.74)$$

Choose δ to be the smaller of δ_1 and δ_2. Equations (3.72–3.74) imply that if $|x - a| < \delta$ then

$$L - \epsilon < \frac{f(x)}{F(x)} < L + \epsilon.$$

Q.E.D.

Intermediate Value Property for Derivatives

In exercise 3.1.14 of section 3.1, you were asked to prove that if $\lim_{x \to a} f'(x)$ exists, then so does $f'(a)$, and they must be equal. This implies that where the limit exists, the derivative must be continuous. Gaston Darboux (1842–1917) was the first to observe that even more is true. Even if a derivative is not continuous, it must have the intermediate value property. By Theorem 3.4, the modified converse of the intermediate value theorem, if a derivative is not continuous then it cannot be piecewise monotonic. All examples of discontinuous derivatives are similar to the derivative of $x^2 \sin(x^{-1})$ which exists but is not continuous at $x = 0$ because the derivative oscillates infinitely often in any neighborhood of 0. Our proof is based on one discovered by Lars Olsen.

> **Theorem 3.14 (Darboux's Theorem).** *If f is differentiable on $[a, b]$, then f' has the intermediate value property on $[a, b]$.*

Proof: We define a new function g:

$$g(x) = \begin{cases} f'(a), & x = a, \\ \dfrac{f(2x - a) - f(a)}{2x - 2a}, & a < x \le (a + b)/2, \\ \dfrac{f(b) - f(2x - b)}{2b - 2x}, & (a + b)/2 \le x < b, \\ f'(b), & x = b. \end{cases}$$

The function g is continuous on $[a, b]$ (see exercise 3.5.19). Given any T between $f'(a)$ and $f'(b)$, the intermediate value theorem promises us an x in $[a, b]$ at which $g(x) = T$. For all x in (a, b), $g(x)$ is equal to

$$\frac{f(t) - f(s)}{t - s}$$

for some pair s and t with $a \le s < t \le b$. By the mean value theorem, every value of g is a value of the derivative of f at some point in $[a, b]$.

Q.E.D.

Note that we could have weakened the hypotheses and only assumed that f is differentiable on (a, b) and that $f'_+(a)$ and $f'_-(b)$ exist. The conclusion then applies to the function defined as f' on (a, b), $f'_+(a)$ at $x = a$, and $f'_-(b)$ at $x = b$.

Exercises

3.5.1. Prove that if F is continuous on $[a, b]$ and differentiable on (a, b) and if $F'(x)$ is not zero for any x strictly between a and b, then $F(b) \ne F(a)$.

3.5.2. Show that the approximation formula

$$\sqrt{1 + x} \approx 1 + \frac{1}{2}x - \frac{1}{8}x^2$$

gives $\sqrt{1 + x}$ with an error not greater than $|x|^3/2$, if $|x| < 1/2$.

3.5.3. For $x > -1$, $x \ne 0$, show that

$$(1 + x)^\alpha > 1 + \alpha x, \quad \text{if } \alpha > 1 \text{ or } \alpha < 0,$$
$$(1 + x)^\alpha < 1 + \alpha x \quad \text{of } 0 < \alpha < 1.$$

3.5.4. Show that each of the following equations has exactly one real root.

a. $x^{13} + 7x^3 - 5 = 0$

b. $3^x + 4^x = 5^x$

3.5.5. Prove that under the hypothesis of Theorem 3.13, if $\lim_{x \to a} |f(x)| \neq \infty$, then

$$\lim_{x \to a} \frac{f(x)}{F(x)} = 0 \quad \text{or does not exist.}$$

3.5.6. Show that the 0/0 and ∞/∞ forms of L'Hospital's rule also work for one-sided limits. That is to say, explain how to modify the given proofs so that if f and F are both differentiable in an open interval whose left-hand endpoint is a, if

$$f(a) = \lim_{x \to a^+} f(x) = 0 = \lim_{x \to a^+} F(x) = F(a),$$

or

$$\lim_{x \to a^+} F(x) = \infty,$$

if $F'(x) \neq 0$ for all x in this open interval, and if $\lim_{x \to a^+} f'(x)/F'(x)$ exists, then

$$\lim_{x \to a^+} \frac{f(x)}{F(x)} = \lim_{x \to a^+} \frac{f'(x)}{F'(x)}. \tag{3.75}$$

3.5.7. Explain what is wrong with the following application of L'Hospital's rule:

To evaluate $\lim_{x \to 0}(3x^2 - 1)/(x - 1)$, apply l'Hospital's rule:

$$\lim_{x \to 0} \frac{3x^2 - 1}{x - 1} = \lim_{x \to 0} \frac{6x}{1} = 0.$$

From the original function, however, it can be seen that as x approaches zero, the function approaches 1.

3.5.8. Explain what is wrong with the following application of L'Hospital's rule:

Let $f(x) = x^2 \sin(1/x)$, $F(x) = x$. Each of these functions approaches 0 as x approaches 0, so by L'Hospitals' rule

$$\lim_{x \to 0} \frac{f(x)}{F(x)} = \lim_{x \to 0} \frac{f'(x)}{F'(x)} = \frac{2x \sin(1/x) - \cos(1/x)}{1},$$

which does not exist.

3.5.9. This exercise pursues a more subtle misapplication of the ∞/∞ form of L'Hospital's rule for limits from the right (see exercise 3.5.6). We begin with the functions

$$f(x) = \cos(x^{-1}) \sin(x^{-1}) + x^{-1},$$

$$F(x) = \left(\cos(x^{-1}) \sin(x^{-1}) + x^{-1} \right) e^{\sin(x^{-1})}.$$

a. Show that $\lim_{x \to 0^+} f(x) = \infty = \lim_{x \to 0^+} F(x)$.

b. Show that the ratio of the derivatives of f and F simplifies to

$$\frac{f'(x)}{F'(x)} = \frac{2x \cos(x^{-1}) e^{-\sin(x^{-1})}}{2x \cos(x^{-1}) + x \cos(x^{-1}) \sin(x^{-1}) + 1},$$

and that this approaches 0 as x approaches 0 from the right.

c. Show that $f(x)/F(x)$ simplifies to

$$\frac{f(x)}{F(x)} = e^{-\sin(x^{-1})}$$

which oscillates between e and e^{-1} as x approaches 0 and thus does not have a limit.

d. Which hypothesis of L'Hospital's rule is violated by these functions?

e. Where was that hypothesis used in the proof? Identify the point at which the proof breaks down for these functions.

3.5.10. Modify the proof of the ∞/∞ form of L'Hospital's rule to prove that if f and F are differentiable at every x in some neighborhood of a, if F' is never zero in this neighborhood, if $\lim_{x\to a} |F(x)| = \infty$, and if

$$\lim_{x\to a} \frac{f'(x)}{F'(x)} = \infty,$$

then

$$\lim_{x\to a} \frac{f(x)}{F(x)} = \infty.$$

3.5.11. Use L'Hospital's rule to prove that

$$\lim_{x\to 0} \frac{e^{-1/x^2}}{x} = 0. \tag{3.76}$$

Use this to prove that if $f(x) = e^{-1/x^2}$ when $x \neq 0$ and $f(0) = 0$, then $f'(0) = 0$.

3.5.12. Prove by induction that for any positive integer n,

$$\lim_{x\to 0} \frac{e^{-1/x^2}}{x^n} = 0. \tag{3.77}$$

3.5.13. Compare the remainder terms of Lagrange and Cauchy for the truncated Taylor series for $f(x) = e^x$ when $x = 2$, expanded around $a = 0$. Which remainder gives a tighter bound on the error?

3.5.14. Prove that over the interval $[0, 2/3]$ with $k \geq 1$ both

$$\frac{(2/3)^k}{k(1+c)^k} \quad \text{and} \quad \frac{(2/3)(2/3 - c)^{k-1}}{(1+c)^k}$$

are maximized at $c = 0$.

3.5.15. Prove that over the interval $[-2/3, 0]$ with $k \geq 3$,

$$\frac{(2/3)^k}{k(1+c)^k}$$

is maximized at $c = -2/3$ while

$$\frac{(2/3)(2/3 + c)^{k-1}}{(1 + c)^k}$$

is maximized at $c = 0$.

3.5.16. Graph the function $y = x^{1/x}$, $x > 0$. Approximately where does it achieve its maximum? Use L'Hospital's rule to prove that

$$\lim_{x \to \infty} \ln(x^{1/x}) = 0.$$

It follows that

$$\lim_{x \to \infty} x^{1/x} = 1.$$

3.5.17. Let f and g be functions with continuous second derivatives on $[0, 1]$ such that $g'(x) \neq 0$ for $x \in (0, 1)$ and $f'(0)g''(0) - f''(0)g'(0) \neq 0$. Define a function θ for $x \in (0, 1)$ so that $\theta(x)$ is one of the values that satisfies the generalized mean value theorem,

$$\frac{f(x) - f(0)}{g(x) - g(0)} = \frac{f'(\theta(x))}{g'(\theta(x))}.$$

Show that

$$\lim_{x \to 0^+} \frac{\theta(x)}{x} = \frac{1}{2}.$$

3.5.18. Use L'Hospital's rule to evaluate the following limits.

a. $\displaystyle\lim_{x \to 1} \frac{\arctan\left(\frac{x^2-1}{x^2+1}\right)}{x - 1}$

b. $\displaystyle\lim_{x \to +\infty} x\left(\left(1 + \frac{1}{x}\right)^x - e\right)$

c. $\displaystyle\lim_{x \to 5}(6 - x)^{1/(x-5)}$

d. $\displaystyle\lim_{x \to 0^+} \left(\frac{\sin x}{x}\right)^{1/x}$

e. $\displaystyle\lim_{x \to 0^+} \left(\frac{\sin x}{x}\right)^{1/x^2}$

3.5.19. Prove that the function g defined in the proof of Darboux's theorem is continuous. L'Hospital's rule for one-sided limits looks tempting, but it assumes that the derivative of f is continuous on that side. To be safe, use the Cauchy definition of the derivative and find the δ response that forces $|E(2x - a, a)| < \epsilon$.

3.5.20. Find another function h that can be used to prove Darboux's theorem. What makes the function g work is that for $a < x < b$ it is equal to $(f(t) - f(s))/(t - s)$ where s and t are continuous functions of x, $a \leq s < t \leq b$, and

$$\lim_{x \to a} \frac{f(t) - f(s)}{t - s} = f'(a), \qquad \lim_{x \to b} \frac{f(t) - f(s)}{t - s} = f'(b).$$

4

The Convergence of Infinite Series

We have seen that when we talk about an infinite series, we are really talking about the sequence of partial sums. The definitions of infinite series and of convergence on pages 12 and 18 are stated in terms of the partial sums. This is the approach that will enable us to handle any infinite process. Thus the question "What is the value of 0.99999...?" is not well-posed until we clarify what we mean by such an infinite string of 9's. Our interpretation will be the limit of the sequence of finite strings of 9's: 0.9, 0.99, 0.999, 0.9999, 0.99999, ... If we combine this with the Archimedean understanding of such a limit: the number T such that for any $L < T$ and any $M > T$, all of the finite strings from some point on will lie strictly between L and M, then the meaning and value of 0.99999... are totally unambiguous. The value is 1.

Rather than using L and M, we shall follow the same procedure as we did in the last chapter and choose symmetric bounds. We choose an $\epsilon > 0$ and then use $L = T - \epsilon$ and $M = T + \epsilon$. In terms of ϵ, the definition of convergence of an infinite series is as follows.

Definition: convergence of an infinite series

An infinite series **converges** if there is a value T with the property that for each $\epsilon > 0$ there is a response N so that all of the partial sums with at least N terms lie strictly within the open interval $(T - \epsilon, T + \epsilon)$.

This chapter is devoted to answering two basic questions:

- How do we know if a particular infinite series converges?
- If we know that a particular infinite series converges, how do we find its value?

Neither question is easy, and there are no universal procedures for finding an answer. In some sense, the second question is meaningless. We know that

$$1 - \frac{1}{2} + \frac{1}{3} - \frac{1}{4} + \cdots$$

has the value $\ln 2$, but what do we mean by $\ln 2$? My calculator tells me that $\ln 2$ is .6931471806, which I know is wrong because the natural logarithm of 2 is not a rational number. Those ten digits give me an approximation. We have just finished seeing that a convergent series is a sequence of approximations that can be used to obtain any degree of accuracy we desire. It may require many terms, but the infinite series carries within itself a better approximation to $\ln 2$ than the ten digit decimal. We might as well call this number $1 - 1/2 + 1/3 - 1/4 + \cdots$.

This is a bit ingenuous. It is nice to know that the sequence of partial sums approaches a number which, when exponentiated, yields precisely 2. Recognizing the value of a convergent series as a number we have seen in another context can be very useful. But we need to be alert to the fact that asking for the precise value of a convergent series is not always meaningful. There may be no better way of expressing that value than as the limit of the partial sums of the series. We return to the first question. How do we know if a series converges?

4.1 The Basic Tests of Convergence

A highly unreliable method of deciding convergence is to actually calculate the partial sum of the first hundred or thousand or million terms. If you know something more than these first terms, then these calculations may give you some useful indications, but the first million summands in and of themselves tell you nothing about the next million summands, nor the million after them. It is even less true that as soon as the partial sums start agreeing to within the accuracy of your calculations, you have found the value of the series.

Stirling's Series

Stirling's formula (page 45) says that $n!$ is well approximated by $(n/e)^n \sqrt{2\pi n}$. One of the ways of making explicit what we mean by "well approximated" is that the logarithms of these two functions of n differ by an amount that approaches 0 as n increases:

$$\ln(n!) = n \ln n - n + \frac{1}{2} \ln(2\pi n) + E(n), \qquad \lim_{n \to \infty} E(n) = 0. \qquad (4.1)$$

There is an explicit series for $E(n)$ given in terms of the Bernoulli numbers, rational numbers that were first discovered by Jacob Bernoulli as an aid to calculating sums of powers,

$$1^k + 2^k + 3^k + \cdots + n^k, \qquad k = 1, 2, 3, \ldots .$$

> To learn Bernoulli's formula for the sum of consecutive integers raised to any fixed positive integer power, go to **Appendix A.2, Bernoulli's Numbers**.

Table 4.1. Values of S_n to ten-digit accuracy.

k	S_k
1	0.008333333333
2	0.008330555556
3	0.008330563492
4	0.008330563433
5	0.008330563433
6	0.008330563433
7	0.008330563433
8	0.008330563433
9	0.008330563433
10	0.008330563433
⋮	⋮

The easiest way to define these numbers is in terms of a power series expansion:

$$\frac{x}{e^x - 1} + \frac{x}{2} = 1 + \sum_{k=1}^{\infty} B_{2k} \frac{x^{2k}}{(2k)!}. \tag{4.2}$$

The first few values are

$$B_2 = \frac{1}{6}, \quad B_4 = \frac{-1}{30}, \quad B_6 = \frac{1}{42}, \quad B_8 = \frac{-1}{30}, \quad B_{10} = \frac{5}{66}, \quad B_{12} = \frac{-691}{2730}.$$

The series expansion of the error term $E(n) = \ln(n!) - (n \ln n - n + \ln(2\pi n)/2)$ is

$$\frac{B_2}{1 \cdot 2 \cdot n} + \frac{B_4}{3 \cdot 4 \cdot n^3} + \frac{B_6}{5 \cdot 6 \cdot n^5} + \cdots + \frac{B_{2k}}{(2k-1) \cdot 2k \cdot n^{2k-1}} + \cdots. \tag{4.3}$$

Does this series converge?

Web Resource: To explore the convergence of this error term for different values of n, go to **Stirling's formula**. More information on Stirling's formula including its derivation can be found in **Appendix A.4, The Size of $n!$**.

We let $n = 10$ and start calculating the partial sums:

$$S_k = \frac{B_2}{1 \cdot 2 \cdot 10} + \frac{B_4}{3 \cdot 4 \cdot 10^3} + \cdots + \frac{B_{2k}}{(2k-1) \cdot 2k \cdot 10^{2k-1}}.$$

It looks as if this series converges and that it converges quite rapidly. The values in Table 4.1. are given with ten-digit accuracy.

This is pretty good. The true value of $\ln(10!) - 10 \ln 10 + 10 - \ln(20\pi)/2$ to ten digits is 0.008330563433. It appears that this series converges to the true value of the error.

But a little after $k = 70$, something starts to go wrong (see Table 4.2.).

Table 4.2. Values of S_n to
ten-digit accuracy.

k	S_k
⋮	⋮
70	0.008330563433
71	0.008330563433
72	0.008330563432
73	0.008330563436
74	0.008330563418
75	0.008330563514
76	0.008330562971
77	0.008330566127
78	0.008330547295
79	0.008330662638
80	0.008329937885
81	0.008334608215
82	0.008303752990
83	0.008512682811
84	0.007063134389
85	0.017364593510
86	−0.05760318347
87	0.5009177478
88	−3.757762841
89	29.46731813
90	−235.6875347
⋮	⋮

In fact, this series does *not* converge. Even for $A = 0.008330563433$ and an error bound of $\epsilon = 0.01$, there is no N that we can use as a reply. For any k above 85, the partial sums all differ from A by more than 0.01.

A Preview of Abel's Test

On the other hand, it can take a convergent series a very long time before it closes in on its value. As we shall see at the end of this chapter,

$$\sum_{k=2}^{\infty} \frac{\sin(k/100)}{\ln k}$$

is a convergent series, but if we look at the partial sums:

$$S_n = \sum_{k=2}^{n} \frac{\sin(k/100)}{\ln k},$$

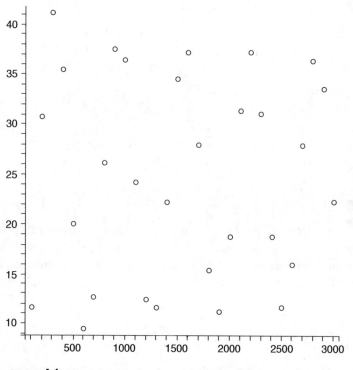

FIGURE 4.1. Plots of points $(n, S(n))$ where $S_n = \sum_{k=2}^{n} \sin(k/100)/\ln k$.

we see that at least as far as $n = 3000$ they do not seem to be settling down. Figure 4.1 is a plot of the values of the partial sums at the multiples of 100 from 100 to 3000. Among the partial sums are $S_{100} = 11.6084$, $S_{200} = 30.7754$, $S_{300} = 41.1982, \ldots, S_{1300} = 11.5691$, $S_{1400} = 22.2942, \ldots, S_{2200} = 37.2332$, $S_{2300} = 31.1325, \ldots, S_{2900} = 33.6201$, and $S_{3000} = 22.3079$.

It is also not enough to ask if the summands are approaching zero. The numbers 1, 1/2, 1/3, 1/4, 1/5, ... approach 0, but

$$1 + \frac{1}{2} + \frac{1}{3} + \frac{1}{4} + \cdots$$

is the harmonic series which we know does not converge. A common explanation is to say that these summands do not go to zero "fast enough," but we must be more careful than this. After all,

$$1 - \frac{1}{2} + \frac{1}{3} - \frac{1}{4} + \cdots = \ln 2$$

does converge and its summands have exactly the same absolute values as those in the harmonic series.

When the Summands Do Not Approach 0

In section 2.5 we saw d'Alembert's analysis of the binomial series and his proof that the summands do not approach zero when $|x| > 1$. He concluded that the series cannot converge. The justification for his conclusion is given in the next theorem.

> **Theorem 4.1 (The Divergence Theorem).** *Let $a_1 + a_2 + a_3 + \cdots$ be an infinite series. If this series converges, then the summands approach zero. More precisely, if this series converges and we are given any positive error bound ϵ, then there is a positive integer N for which all summands beyond the Nth have absolute value less than ϵ:*
>
> $$n \geq N \quad \text{implies that} \quad |a_n| < \epsilon.$$

Before we prove this theorem, I want to emphasize what it is does and what it does not say. The **converse** of any theorem reverses the direction of implication. The **inverse** states that the negation of the hypothesis implies the negation of the conclusion. The **contrapositive** is that the negation of the conclusion implies the negation of the hypothesis. For Theorem 4.1 these are

converse: "If the summands approach zero, then the series converges." We know that this is false.

inverse: "If the series diverges, then the summands do not approach zero." The harmonic series also contradicts this statement.

contrapositive: "If the summands do not approach zero, then the series diverges." This is logically equivalent to Theorem 4.1. It is the reason we call this the divergence theorem. We shall use it both ways. It can provide a fast and easy way of seeing that a series must diverge, but it also tells us something very useful about the summands whenever we know that our series converges.

Note that the inverse is the contrapositive of the converse, so these two statements are logically equivalent to each other. One is true if and only if the other is also. Whenever you see a theorem, it is always worth asking whether the converse is also true. If you think it might not always hold, can you think of an example for which it does not hold?

Proof: From the definition of convergence, we know that there is a number T for which we can always win the ϵ–N game on the partial sums. The nth summand is the difference between the nth partial sum and the one just before it:

$$\begin{aligned}
|a_n| &= |(a_1 + a_2 + \cdots + a_n) - (a_1 + a_2 + \cdots + a_{n-1})| \\
&= |(a_1 + a_2 + \cdots + a_n) - T + T - (a_1 + a_2 + \cdots + a_{n-1})| \\
&\leq |(a_1 + a_2 + \cdots + a_n) - T| + |T - (a_1 + a_2 + \cdots + a_{n-1})|.
\end{aligned} \quad (4.4)$$

We assign half of our bound to each of these differences. We find an N so that if $m \geq N$, then

$$|(a_1 + a_2 + \cdots + a_m) - T| < \epsilon/2.$$

As long as n is at least $N + 1$, we have that

$$|a_n| \leq |(a_1 + a_2 + \cdots + a_n) - T| + |T - (a_1 + a_2 + \cdots + a_{n-1})|$$
$$< \epsilon/2 + \epsilon/2 = \epsilon. \tag{4.5}$$

Q.E.D.

The Cauchy Criterion

It was Cauchy in his 1821 *Cours d'analyse* who presented the first systematic treatment of the question of convergence of infinite series. He began by facing the question: how can we determine whether the partial sums are approaching a value T when we do not know the value of T? The answer is known as the Cauchy criterion.

> **Theorem 4.2 (The Cauchy Criterion).** *Let $a_1 + a_2 + a_3 + \cdots$ be an infinite series whose partial sums are denoted by $S_n = a_1 + a_2 + \cdots + a_n$. This series converges if and only if the partial sums can be brought arbitrarily close together by taking the subscripts sufficiently large. Specifically, it converges if and only if for any positive error bound ϵ, we can always find a subscript N such that for any pair of partial sums beyond the Nth ($m, n \geq N$), we have*
>
> $$|S_m - S_n| < \epsilon. \tag{4.6}$$

In Cauchy's own words, "It is necessary and it suffices that, for infinitely large values of the number n, the sums $S_n, S_{n+1}, S_{n+2}, \ldots$ differ from the limit S, and in consequence among each other, by infinitely small quantities." This was not stated as a theorem by Cauchy, but merely as an observation. He did prove that if the series converges, then for every $\epsilon > 0$ there is a response N for which equation (4.6) must hold whenever m and n are at least as large as N. He stated the converse but did not prove it. As we shall see, this is the difficult part of the proof. It is also the heart of the theorem because it gives us a means for proving that a series converges even when we have no idea of the value to which it converges.

We say that a series is **Cauchy** if its partial sums can be forced arbitrarily close together by taking sufficiently many terms. Theorem 4.2 can be stated succinctly as: an infinite series converges if and only if it is Cauchy.

> **Definition: Cauchy sequence and series**
>
> An infinite sequence $\{S_1, S_2, S_3, \ldots\}$ is **Cauchy** if for any positive error bound ϵ, we can always find a subscript N such that $N \leq m < n$ implies that $|S_m - S_n| < \epsilon$. A series is **Cauchy** if the sequence of its partial sums is a Cauchy sequence.

Proof: We will work with the sequence of partial sums and prove that a sequence converges if and only if it is Cauchy. We start with the easy direction. If our sequence converges, then it has a value T, and we can force our terms to be arbitrarily close to T. Noting that

$$|S_m - S_n| = |S_m - T + T - S_n| \leq |S_m - T| + |T - S_n|,$$

we split our error bound in half and find an N such that $n \geq N$ implies that

$$|S_n - T| < \epsilon/2.$$

If m and n are both at least N, then

$$|S_m - S_n| \leq |S_m - T| + |T - S_n| < \epsilon/2 + \epsilon/2 = \epsilon.$$

The converse is harder. We need to show that there is a value T to which the sequence converges. We are going to use Theorem 3.7 which states that every set with an upper bound has a least upper bound. As we saw, this implies that any set with a lower bound has a greatest lower bound.

Start with the set of all terms of the sequence. The fact that the sequence is Cauchy guarantees that this set is bounded because we can find an n that is the response to $\epsilon = 1$. All of the terms from the nth on sit inside the interval $(S_n - 1, S_n + 1)$. We are left with $\{S_1, S_2, \ldots, S_{n-1}\}$, but any finite set is bounded. The entire sequence must be bounded. By Theorem 3.7, this set has a greatest lower bound; call it L_1.

While L_1 might be our target value, it also might not. Consider the sequence of partial sums of the alternating harmonic series: $1 - 1/2 + 1/3 - 1/4 + \cdots$. For this series, $L_1 = 1 - 1/2 = 1/2$. If we throw out the first two partial sums and consider the greatest lower bound of $\{S_3, S_4, \ldots\}$, then the greatest lower bound is $1 - 1/2 + 1/3 - 1/4 = 7/12$. That is still not the target value, but it is getting closer. We continue throwing away those partial sums with only a few terms. In general, we let L_k denote the greatest lower bound of the set $\{S_k, S_{k+1}, S_{k+2}, \ldots\}$. Notice that as we throw away terms, the greatest lower bound can only increase: $L_1 \leq L_2 \leq L_3 \leq \cdots$. These L_k are bounded by any upper bound on our sequence, and so they have a least upper bound. Let M be this least upper bound of the L_k. I claim that this is the target value for the series.

To prove that M is the target value, we have to demonstrate that given any $\epsilon > 0$, there is a response N so that all of the terms from the Nth term on lie inside the open interval $(M - \epsilon, M + \epsilon)$. By the definition of a least upper bound (page 97), we know that there is at least one L_k that is larger than $M - \epsilon$, call it L_K. Since L_K is the greatest lower bound of all terms starting with the Kth, all terms starting with the Kth are strictly greater than $M - \epsilon$.

We have used the fact that this sequence is Cauchy to conclude that it must be bounded, but not every bounded sequence also converges (consider $\{1, -1, 1, -1, \ldots\}$). We now need to use the full power of being Cauchy to find an N for which all terms starting with the Nth are strictly less than $M + \epsilon$.

We choose an $N \geq K$ such that $m, n \geq N$ implies that $|S_n - S_m| < \epsilon/2$. Since L_N is the greatest lower bound among all $S_n, n \geq N$, we can find an $m \geq N$ so that S_m is within $\epsilon/2$ of L_N. It follows that for any $n \geq N$,

$$M - \epsilon < L_K \leq L_N \leq S_n < S_m + \frac{\epsilon}{2} < L_N + \epsilon \leq M + \epsilon. \tag{4.7}$$

Q.E.D.

Completeness

We have shown that the nested interval principle implies that every set with an upper bound has a least upper bound, and we have shown that if every set with an upper bound has a least upper bound, then every Cauchy series converges. We will now complete the cycle by showing that if every Cauchy series converges, then the nested interval principle must hold. This does not prove the nested interval principle. What it shows is that these three statements are equivalent. They are different ways of looking at the same basic property of the real numbers, a property that is called **completeness**.

Theorem 4.3 (Completeness). *The following three properties of the real numbers are equivalent:*

- *The nested interval principle,*
- *Every set with an upper bound has a least upper bound,*
- *Every Cauchy sequence converges.*

Definition: completeness

A set of numbers is called **complete** if it has any of the three equivalent properties listed in Theorem 4.3. In particular, the set of all real numbers is complete. The set of all rational numbers is not complete.

Proof: We only have to prove that if every Cauchy sequence converges, then the nested interval principle holds. Let $x_1 \leq x_2 \leq x_3 \leq \cdots$ be the left-hand endpoints of our nested intervals. We first observe that this sequence is Cauchy: Given any $\epsilon > 0$, we can find an interval $[x_k, y_k]$ of length less than ϵ. All of the x_n with $n \geq k$ lie inside this interval, and so any two of them differ by at most ϵ.

Let T be the limit of this sequence. Since the x_n form an increasing sequence, T must be greater than or equal to every x_n. We only need to show that T is less than or equal to every y_n. What would happen if we could find a $y_k < T$? Since T is the limit of the x_n, we could find an x_n that is larger than y_k. This cannot happen because our intervals are nested. Our limit T lies inside all of the intervals.

Q.E.D.

Absolute Convergence

One of the consequences of the Cauchy criterion is the fact that if the sum of the absolute values of the terms in a series converges, then the original series must converge.

Definition: absolute convergence

If $|a_1| + |a_2| + |a_3| + \cdots$ converges, then we say that $a_1 + a_2 + a_3 + \cdots$ **converges absolutely**.

Corollary 4.4 (Absolute Convergence Theorem). *Given a series* $a_1 + a_2 + a_3 + \cdots$, *if* $|a_1| + |a_2| + |a_3| + \cdots$ *converges, then so does* $a_1 + a_2 + a_3 + \cdots$.

Proof: Let $T_n = |a_1| + |a_2| + \cdots + |a_n|$ be the partial sum of the absolute values and let $S_n = a_1 + a_2 + \cdots + a_n$ be the partial sum of the original series. Given any positive error ϵ, we know we can find an N such that for any $m, n \geq N$, $|T_n - T_m| < \epsilon$. We now show that the same response, N, will work for the series without the absolute values. We can assume that $m \leq n$, and therefore

$$
\begin{aligned}
|S_n - S_m| &= |a_{m+1} + a_{m+2} + \cdots + a_n| \\
&\leq |a_{m+1}| + |a_{m+2}| + \cdots + |a_n| \\
&= |T_n - T_m| \\
&< \epsilon.
\end{aligned}
$$

Q.E.D.

The Converse Is False

Convergence does not imply absolute convergence. The series

$$
1 - \frac{1}{2} + \frac{1}{3} - \frac{1}{4} + \cdots
$$

converges, but if we take the absolute values we get the harmonic series which does not converge. This is an example of a series that converges conditionally.

Definition: conditional convergence

We say that a series **converges conditionally** if it converges but does not converge absolutely.

Cauchy realized that while having the summands approach zero is not enough to guarantee convergence in all cases, it is sufficient when the summands decrease in size and alternate between positive and negative values.

Corollary 4.5 (Alternating Series Test). *If* $a_1, a_2, a_3, \ldots$ *are positive and decreasing* ($a_1 \geq a_2 \geq a_3 \geq \cdots \geq 0$), *then the alternating series*

$$
a_1 - a_2 + a_3 - a_4 + a_5 - a_6 + \cdots
$$

converges if and only if the summands approach zero. That is to say, we have convergence if and only if given any positive error ϵ, *we can find a subscript N such that for all* $n \geq N$, $a_n < \epsilon$.

Proof: Each time we add a summand with an odd subscript, we add back something less than or equal to what we just subtracted. Each time we subtract a summand with an even subscript, we subtract something less than or equal to what we just added. That means that all of the partial sums from the nth on lie between S_n and S_{n+1}. The absolute value of the difference between these two partial sums is precisely a_{n+1}, which we can make as small as we wish by taking $n+1$ sufficiently large. This series is Cauchy.

Q.E.D.

This corollary is a rich source of series that converge conditionally. For example,

$$\frac{1}{\ln 2} - \frac{1}{\ln 3} + \frac{1}{\ln 4} - \frac{1}{\ln 5} + \cdots$$

converges. If we turned the minus signs to plus signs, it would diverge. It does *not* help us determine the convergence of

$$\frac{\sin(2/100)}{\ln 2} + \frac{\sin(3/100)}{\ln 3} + \frac{\sin(4/100)}{\ln 4} + \frac{\sin(5/100)}{\ln 5} + \cdots$$

because the summands of this series do not alternate between positive and negative values.

Warning: The hypotheses of the alternating series test are *all* important. In particular, it is not enough that the signs alternate and the summands approach zero. Consider the series

$$1 - \frac{1}{2} + \frac{1}{2} - \frac{1}{4} + \frac{1}{3} - \frac{1}{8} + \cdots + \frac{1}{n} - \frac{1}{2^n} + \cdots .$$

The summands alternate, and the summands approach zero, but this series does not converge. If we take the first $2n$ summands, we know that this series is bounded below by $\ln n + \gamma - 1$, and so it diverges to infinity.

Exercises

The symbol $\boxed{\textbf{M\&M}}$ indicates that *Maple* and *Mathematica* codes for this problem are available in the **Web Resources** at **www.macalester.edu/aratra**.

4.1.1. How many terms of the series

$$1 + \frac{1}{2} + \frac{1}{4} + \frac{1}{8} + \cdots$$

do we need to take if we are to guarantee that we are within $\epsilon = 0.0001$ of the target value 2?

4.1.2. How many terms of the series in exercise 4.1.1 do we need to take if we are to guarantee that we are within $\epsilon = 10^{-1000000}$ of the target value 2?

4.1.3. How many terms of the series

$$1 - \frac{1}{2} + \frac{1}{3} - \frac{1}{4} + \cdots$$

do we need to take if we are to guarantee that we have an approximation to the target value, $\ln 2$, with 10-digit accuracy? with 20-digit accuracy? with 100-digit accuracy?

4.1.4. $\boxed{\textbf{M\&M}}$ Evaluate the partial sums

$$1 + \sum_{k=1}^{n} \frac{k!}{100^k},$$

for the multiples of 10 up to $n = 400$. Describe and discuss what you see happening.

4.1.5. (**M&M**) What is the smallest summand in the series in exercise 4.1.4?

4.1.6. d'Alembert would have described the series in exercise 4.1.4 as *converging* until we reach the smallest summand and then *diverging* after that point. What did he mean by the word "converging," and how does that differ from our modern understanding of the word?

4.1.7. Prove that

$$\frac{x}{e^x - 1} + \frac{x}{2}$$

is an even function, and therefore its power series only involves even powers of x.

4.1.8. (**M&M**) Let B_{2k} be the $2k$th Bernoulli number. Use the fact that

$$B_{2k} \approx (-1)^{k-1} \frac{2(2k)!}{(2\pi)^{2k}} \approx (-1)^{k-1} \frac{2(2k)^{2k} \sqrt{4\pi k} \, e^{-2k}}{(2\pi)^{2k}}$$

to find the summand with the smallest absolute value in the series

$$\frac{B_2}{1 \cdot 2 \cdot 10} + \frac{B_4}{3 \cdot 4 \cdot 10^3} + \frac{B_6}{5 \cdot 6 \cdot 10^5} + \cdots = \sum_{k=1}^{\infty} \frac{B_{2k}}{(2k-1)(2k) \, 10^{2k-1}}.$$

4.1.9. Prove that the series in exercise 4.1.8 does not converge.

4.1.10. Find the summand with smallest absolute value in the series

$$\sum_{k=1}^{\infty} \frac{B_{2k}}{(2k-1) \cdot (2k) \cdot 1000^{2k-1}}.$$

4.1.11. Prove that the series in exercise 4.1.10 does not converge.

4.1.12. Find a *divergent* series for which the first million partial sums, $S_1, S_2, \ldots, S_{1000000}$, all agree to ten significant digits.

4.1.13. (**M&M**) Calculate the partial sums

$$S_n = \sum_{k=2}^{n} \frac{\sin(k/100)}{\ln k}$$

up to at least $n = 2000$. Describe what you see happening. Make a guess of the approximate value to which this series is converging. Explain the rationale behind your guess.

4.1.14. (**M&M**) For each of the following series, explore the values of the partial sums to at least the first two thousand terms, then analyze the series to determine whether it converges absolutely, converges conditionally, or diverges. Justify your answer.

a. $\dfrac{1}{\ln 2} - \dfrac{1}{\ln 3} + \dfrac{1}{\ln 4} - \dfrac{1}{\ln 5} + \cdots = \displaystyle\sum_{k=2}^{\infty} \dfrac{(-1)^k}{\ln k}$

b. $\displaystyle\sum_{k=2}^{\infty} (-1)^k \dfrac{(\ln k)^2}{k}$

c. $\displaystyle\sum_{k=1}^{\infty} (-1)^k \sin(1/k)$

d. $\displaystyle\sum_{k=2}^{\infty} (-1)^k \dfrac{(\ln k)^{\ln k}}{k^2}$

4.1.15.   (**M&M**) For each of the following series, explore the values of the partial sums to at least the first two thousand terms, then analyze the series to determine whether it converges absolutely, converges conditionally, or diverges. Justify your answer.

a. $1 + \dfrac{1}{2} - \dfrac{1}{3} + \dfrac{1}{4} + \dfrac{1}{5} - \dfrac{1}{6} + \dfrac{1}{7} + \dfrac{1}{8} - \dfrac{1}{9} + \cdots = \displaystyle\sum_{n=1}^{\infty} \dfrac{(-1)^{n+1-3\lfloor (n+1)/3 \rfloor}}{n}$

b. $1 + \dfrac{1}{2} - \dfrac{1}{3} - \dfrac{1}{4} + \dfrac{1}{5} + \cdots = \displaystyle\sum_{n=1}^{\infty} \dfrac{(-1)^{\lfloor (n-1)/2 \rfloor}}{n}$

c. $-1 - \dfrac{1}{2} - \dfrac{1}{3} + \dfrac{1}{4} + \dfrac{1}{5} + \dfrac{1}{6} + \dfrac{1}{7} + \dfrac{1}{8} - \dfrac{1}{9} + \cdots = \displaystyle\sum_{n=1}^{\infty} \dfrac{(-1)^{\lfloor \sqrt{n} \rfloor}}{n}$

4.1.16.   Let

$$\epsilon_n = \begin{cases} +1 & \text{for } 2^{2k} \leq n < 2^{2k+1}, \\ -1 & \text{for } 2^{2k+1} \leq n < 2^{2k+2}, \end{cases}$$

where $k = 0, 1, 2, \ldots$ Determine whether the series

$$\sum_{n=1}^{\infty} \frac{\epsilon_n}{n}$$

converges absolutely, converges conditionally, or diverges.

4.1.17.   There are six inequalities in equation (4.7). Explain why each of them holds.

4.2   Comparison Tests

Underlying d'Alembert's treatment of the binomial series is the assumption that we can determine the convergence or divergence of a series by comparing it to another series whose convergence or divergence is known. We must be careful. This only works when the summands are positive, and it requires both skill and luck in choosing the right series with which to compare, but it is a powerful technique. Its justification rests on the Cauchy criterion.

Theorem 4.6 (The Comparison Test). *Let* $a_1 + a_2 + a_3 + \cdots$ *and* $b_1 + b_2 + b_3 + \cdots$ *be two series with summands that are greater than or equal to zero. We assume that each b_i is greater than or equal to the corresponding a_i:*

$$b_1 \geq a_1 \geq 0, \quad b_2 \geq a_2 \geq 0, \quad b_3 \geq a_3 \geq 0, \ldots$$

If $b_1 + b_2 + b_3 + \cdots$ converges, then so does $a_1 + a_2 + a_3 + \cdots$. If $a_1 + a_2 + a_3 + \cdots$ diverges, then so does $b_1 + b_2 + b_3 + \cdots$.

Proof: Let $S_n = a_1 + a_2 + \cdots + a_n$ and $T_n = b_1 + b_2 + \cdots + b_n$. If $m < n$, then

$$0 \leq S_n - S_m = a_{m+1} + a_{m+2} + \cdots + a_n \leq b_{m+1} + b_{m+2} + \cdots + b_n = T_n - T_m,$$

and so

$$|S_n - S_m| \leq |T_n - T_m|. \tag{4.8}$$

We assume the series $b_1 + b_2 + b_3 + \cdots$ converges. Given a positive bound ϵ, we have a response N. Equation (4.8) shows us that the same response will work for the series $a_1 + a_2 + a_3 + \cdots$.

The contrapositive of what we have just proven says that if $a_1 + a_2 + a_3 + \cdots$ diverges, then $b_1 + b_2 + b_3 + \cdots$ diverges.

Q.E.D.

The Ratio Test

The ratio and root test rely on comparing our series to a geometric series. They are very simple and powerful techniques that quickly yield one of three conclusions:

1. the series in question converges absolutely,
2. the series in question diverges, or
3. the results of this test are inconclusive.

It is the third possibility that is the principal drawback of these tests. The most interesting series mathematicians and scientists were encountering in the early 1800s all fell into category 3. Nevertheless, these tests are important because they are simple. Start with one of these tests, and move on to a more complicated test only if the results are inconclusive.

Theorem 4.7 (The Ratio Test). *Given a series with nonzero summands, $a_1 + a_2 + a_3 + \cdots$, we consider the ratio $r(n) = |a_{n+1}/a_n|$. If we can find a number $\alpha < 1$ and a subscript N such that for all $n \geq N$, $r(n)$ is less than or equal to α, then the series converges absolutely.*

If we can find a subscript N such that for all $n \geq N$, $r(n)$ is greater than or equal to 1, then the series diverges.

Proof: If $r(n)$ is less than or equal to $\alpha < 1$ when $n \geq N$, then the series of absolute values, $|a_1| + |a_2| + |a_3| + \cdots$, is dominated by the convergent series

$$|a_1| + \cdots + |a_N| + |a_N|\alpha + |a_N|\alpha^2 + |a_N|\alpha^3 + \cdots = |a_1| + \cdots + |a_{N-1}| + \frac{|a_N|}{1-\alpha}.$$

If $r(n)$ is greater than or equal to 1 when $n \geq N$, then $|a_n|$ is greater than or equal to $|a_N|$ and so does not approach zero as n gets larger.

Q.E.D.

In many cases, $r(n)$ approaches a limit as n gets very large. If this happens, there is a simpler form of the ratio test.

Corollary 4.8 (The Limit Ratio Test). *Given a series with nonzero summands and* $r(n) = |a_{n+1}/a_n|$, *if*

$$\lim_{n\to\infty} r(n) = L < 1,$$

then the series converges absolutely. If

$$\lim_{n\to\infty} r(n) = L > 1,$$

then the series diverges. If

$$\lim_{n\to\infty} r(n) = L = 1,$$

then this test is inconclusive.

Proof: Recall that

$$\lim_{n\to\infty} r(n) = L$$

means that given any positive error bound ϵ, we can find an N with which to reply such that if $n \geq N$, then $|r(n) - L| < \epsilon$. If $L < 1$ then we can use an ϵ that is small enough so that $L + \epsilon < 1$. We can then choose $L + \epsilon$ to be our α. If $n \geq N$, then $r(n) < L + \epsilon = \alpha$.

If $L > 1$, we can use an ϵ that is small enough so that $L - \epsilon \geq 1$. If $n \geq N$, then $r(n) > L - \epsilon \geq 1$.

If $L = 1$, then it might be the case that $r(n) \geq 1$ for all n sufficiently large, which implies that the series diverges. But if all we know is the value of this limit, then $r(n)$ could be less than 1 for all values of n. The ratio test is inconclusive.

Q.E.D.

The Root Test

Cauchy found an even better test that rests on a comparison with geometric series. We can view a geometric series as one for which the nth root of the nth summand is constant,

$$\sqrt[n]{|x^n|} = |x|.$$

This suggests taking the nth root of the absolute value of the nth summand in an arbitrary series. This test is often more complicated to apply than the ratio test, but it will give an answer in some cases where the ratio test is inconclusive.

Theorem 4.9 (The Root Test). *Given a series* $a_1 + a_2 + a_3 + \cdots$, *we consider*

$$\rho(n) = \sqrt[n]{|a_n|}.$$

If we can find a number $\alpha < 1$ *and a subscript* N *such that for all* $n \geq N$, $\rho(n)$ *is less than or equal to* α, *then the series converges absolutely.*
If for any subscript N *we can always find a larger* n *for which* $\rho(n)$ *is greater than or equal to 1, then the series diverges.*

Notice that while the convergence condition looks very much the same, the divergence condition has been liberalized a great deal. We do not have to go above 1 and stay there. It is enough if we can always find another $\rho(n)$ that climbs to or above 1. The ratio $r(n)$ and the root $\rho(n)$ are related. Exercises 4.2.8–4.2.11 show that if $\lim_{n \to \infty} r(n)$ exists, then so does $\lim_{n \to \infty} \rho(n)$ and the two will be equal. Whenever $\lim_{n \to \infty} r(n)$ exists, the root and ratio tests will always give the same response.

Proof: If $n \geq N$ implies that $\rho(n)$ is less than or equal to $\alpha < 1$, then the series of absolute values, $|a_1| + |a_2| + |a_3| + \cdots$, is dominated by the convergent series

$$|a_1| + |a_2| + \cdots + |a_{N-1}| + \alpha^N + \alpha^{N+1} + \alpha^{N+2} + \cdots$$

$$= |a_1| + |a_2| + \cdots + |a_{N-1}| + \frac{\alpha^N}{1 - \alpha}.$$

If $\rho(n) \geq 1$, then $|a_n| \geq 1$. If this happens for arbitrarily large values of n, then the summands do not approach zero, and so the series cannot converge.

Q.E.D.

Corollary 4.10 (The Limit Root Test). *Given a series with positive summands and* $\rho(n) = \sqrt[n]{|a_n|}$, *if*

$$\lim_{n \to \infty} \rho(n) = L < 1,$$

then the series converges absolutely. If

$$\lim_{n \to \infty} \rho(n) = L > 1,$$

then the series diverges. If

$$\lim_{n \to \infty} \rho(n) = L = 1,$$

then this test is inconclusive.

The proof of this corollary parallels that of Corollary 4.8 and is left as an exercise.

Examples

With all of this machinery in place, we can now answer the question of convergence for many series. We recall the series expansion for $(1 + x)^{1/2}$ at $x = 2/3$:

$$\left(1 + \frac{2}{3}\right)^{1/2} = 1 + (1/2)\frac{2}{3} + \frac{(1/2)(1/2 - 1)}{2!}\left(\frac{2}{3}\right)^2$$
$$+ \frac{(1/2)(1/2 - 1)(1/2 - 2)}{3!}\left(\frac{2}{3}\right)^3 + \cdots .$$

The absolute value of the ratios of successive terms is

$$r(n) = \left| \frac{1/2(1/2 - 1)\cdots(1/2 - n + 1)(2/3)^n/n!}{1/2(1/2 - 1)\cdots(1/2 - n + 2)(2/3)^{n-1}/(n - 1)!} \right|$$
$$= \left| \frac{(1/2 - n + 1)(2/3)}{n} \right| = \frac{2n - 3}{3n}.$$

This has a limit:

$$\lim_{n \to \infty} \frac{2n - 3}{3n} = \frac{2}{3} < 1.$$

This series converges absolutely.

A more interesting example is

$$1 + \frac{1!\,2!}{3!} + \frac{2!\,4!}{6!} + \frac{3!\,6!}{9!} + \cdots = 1 + \sum_{n=1}^{\infty} \frac{n!\,(2n)!}{(3n)!}.$$

The ratio is

$$r(n) = \frac{n!\,(2n)!/(3n)!}{(n - 1)!\,(2n - 2)!/(3n - 3)!}$$
$$= \frac{n(2n)(2n - 1)}{(3n)(3n - 1)(3n - 2)}$$
$$= \frac{4n^3 - 2n^2}{27n^3 - 27n^2 + 6n}.$$

The limit is

$$\lim_{n \to \infty} \frac{4n^3 - 2n^2}{27n^3 - 27n^2 + 6n} = \frac{4}{27} < 1,$$

and so this series converges absolutely.

We note that we obtain exactly the same limit if we use the nth root of the nth term. To make life a little simpler, let us ignore the first summand so that the nth summand is $n!\,(2n)!/(3n)!$. It is necessary to use Stirling's formula, $n! = n^n \sqrt{2\pi n}\, e^{-n + E(n)}$ where $E(n)$

approaches 0 as n gets large:

$$\rho(n) = \sqrt[n]{\frac{n!\,(2n)!}{(3n)!}}$$

$$= \left(\frac{n^n\sqrt{2\pi n}\,e^{-n+E(n)} \cdot (2n)^{2n}\sqrt{4\pi n}\,e^{-2n+E(2n)}}{(3n)^{3n}\sqrt{6\pi n}\,e^{-3n+E(3n)}}\right)^{1/n}$$

$$= \frac{n(2\pi n)^{1/2n}e^{-1+E(n)/n} \cdot 4n^2(4\pi n)^{1/2n}e^{-2+E(2n)/n}}{27n^3(6\pi n)^{1/2n}e^{-3+E(3n)/n}}$$

$$= \frac{4}{27}\left(\frac{4\pi n}{3}\right)^{1/2n} e^{[E(n)+E(2n)-E(3n)]/n}.$$

This also approaches $4/27 < 1$ as n gets arbitrarily large.

Still more interesting is

$$1 + \frac{2!}{2^2} + \frac{3!}{3^3} + \frac{4!}{4^4} + \cdots = \sum_{n=1}^{\infty} \frac{n!}{n^n}.$$

The ratio test gives us

$$r(n) = \frac{(n+1)!/(n+1)^{n+1}}{n!/n^n} = \frac{(n+1)n^n}{(n+1)^{n+1}} = \left(1 + \frac{1}{n}\right)^{-n}$$

which approaches $e^{-1} < 1$ as n gets arbitrarily large.

Web Resource: For a proof and exploration of the limit formula $\lim\limits_{n\to\infty}\left(1 + \dfrac{x}{n}\right)^n = e^x$, go to **Exponential function**.

The root test gives us

$$\rho(n) = \left(\frac{n^n\sqrt{2\pi n}\,e^{-n+E(n)}}{n^n}\right)^{1/n} = (2\pi n)^{1/2n}e^{-1+E(n)/n}$$

which also approaches $e^{-1} < 1$ as n gets arbitrarily large.

Sometimes it is easier to take the nth root rather than the nth ratio. Consider the series

$$(1+1)^{-1} + \left(1 + \frac{1}{2}\right)^{-4} + \left(1 + \frac{1}{3}\right)^{-9} + \left(1 + \frac{1}{4}\right)^{-16} + \cdots = \sum_{n=1}^{\infty}\left(1 + \frac{1}{n}\right)^{-n^2}.$$

The nth root of the nth summand is

$$\rho(n) = \left(1 + \frac{1}{n}\right)^{-n}$$

which approaches $e^{-1} < 1$ as n gets arbitrarily large. This series converges absolutely.

Limitations of the Root and Ratio Tests

While the root and ratio tests are usually the ones we want to use first, there are many important series for which they return an inconclusive result. Neither of these tests will confirm that the harmonic series diverges. For the limit ratio test we have

$$\lim_{n\to\infty} \frac{1/(n+1)}{1/n} = \lim_{n\to\infty} \frac{n}{n+1} = 1.$$

Similarly, the limit root test returns

$$\lim_{n\to\infty} n^{-1/n} = e^{\lim_{n\to\infty} -(\ln n)/n} = e^0 = 1.$$

Of course, we know that the harmonic series diverges. We can use this information with the comparison test. If $p \leq 1$, then $1/n^p \geq 1/n$ and so $\sum_{n=1}^{\infty} \frac{1}{n^p}$ diverges. What if p is greater than 1? Does

$$\sum_{n=1}^{\infty} \frac{1}{n^{1.01}}$$

converge or diverge? Can we find a divergent series with $a_n < 1/n$? What about $\sum_{n=2}^{\infty} \frac{1}{n \ln n}$?

Our last two tests enable us to answer these questions. They are both based on the observation that if the summands are positive, then the partial sums are increasing. If the partial sums are bounded, then they form a Cauchy sequence and so the series converges. If the partial sums are not bounded, then the series diverges to infinity.

Cauchy's Condensation Test

The first convergence test in Cauchy's *Cours d'analyse* is the root test. The second is the ratio test. The third is the condensation test.

Theorem 4.11 (Cauchy's Condensation Test). *Let $a_1 + a_2 + a_3 + \cdots$ be a series whose summands are eventually positive and decreasing. That is to say, there is a subscript N such that*

$$n \geq N \qquad \text{implies that} \qquad a_n \geq a_{n+1} \geq 0.$$

This series converges if and only if the series

$$a_1 + 2a_2 + 4a_4 + 8a_8 + \cdots + 2^k a_{2^k} + \cdots$$

converges.

This test is good enough to settle the convergence questions that the root and ratio tests could not handle. We shall state and prove the p-test after we have proven Cauchy's test. But first, we show that there is a series with smaller summands than the harmonic series but which still diverges. We consider

$$\frac{1}{2 \ln 2} + \frac{1}{3 \ln 3} + \frac{1}{4 \ln 4} + \cdots.$$

These summands are positive and decreasing. We can apply the condensation test, letting the first summand be 0 and treating $1/2 \ln 2$ as the second summand. We compare our series with

$$\frac{2}{2\ln 2} + \frac{4}{4\ln 4} + \frac{8}{8\ln 8} + \cdots = \sum_{k=1}^{\infty} \frac{2^k}{2^k \ln 2^k}$$

$$= \sum_{k=1}^{\infty} \frac{1}{k\ln 2}$$

$$= \frac{1}{\ln 2}\left(1 + \frac{1}{2} + \frac{1}{3} + \frac{1}{4} + \cdots\right).$$

We are comparing our original series with the harmonic series which we know diverges. It follows that $1/(2\ln 2) + 1/(3\ln 3) + \cdots$ also diverges.

Proof: We can assume that the summands are positive and decreasing beginning with the first summand. Otherwise, we chop off the initial portion containing the recalcitrant summands. This will change the value of the series (if it converges), but it will not change whether or not it converges.

If $a_1 + 2a_2 + 4a_4 + \cdots$ converges, then it has a value V. Given a partial sum of our original series,

$$S_n = a_1 + a_2 + a_3 + \cdots + a_n,$$

we choose the smallest integer m such that $n < 2^m$. We can compare S_n with the partial sum of the first m terms in the second series:

$$S_n = a_1 + (a_2 + a_3) + (a_4 + a_5 + a_6 + a_7)$$
$$+ (a_8 + a_9 + \cdots + a_{15}) + \cdots + (a_{2^{m-1}} + a_{2^{m-1}+1} + \cdots + a_n)$$
$$\leq a_1 + 2a_2 + 4a_4 + 8a_8 + \cdots + 2^{m-1}a_{2^{m-1}}$$
$$\leq V.$$

The partial sums are bounded and so they converge.

If $a_1 + a_2 + a_3 + \cdots$ converges, then it has a value W. Given a partial sum of the second series,

$$T_n = a_1 + 2a_2 + 4a_4 + \cdots + 2^n a_{2^n},$$

we can compare T_n with twice the partial sum of the first 2^n terms in the first series:

$$T_n \leq 2a_1 + 2a_2 + 2(a_3 + a_4) + 2(a_5 + a_6 + a_7 + a_8)$$
$$+ 2(a_9 + a_{10} + \cdots + a_{16}) + \cdots$$
$$+ 2(a_{2^{n-1}+1} + a_{2^{n-1}+2} + \cdots + a_{2^n})$$
$$\leq 2(a_1 + a_2 + \cdots + a_{2^n})$$
$$\leq 2W.$$

The partial sums are bounded and so they converge.

Q.E.D.

Corollary 4.12 (The p-Test). *The series*

$$\sum_{n=1}^{\infty} \frac{1}{n^p}$$

diverges for $p \leq 1$ and converges for $p > 1$.

Proof: We compare our series $\sum_{n=1}^{\infty} 1/n^p$ to

$$\sum_{n=1}^{\infty} \frac{2^n}{(2^n)^p} = \sum_{n=1}^{\infty} \frac{1}{\left(2^{p-1}\right)^n}.$$

This is a geometric series. It converges if and only if $2^{p-1} > 1$, which happens if and only if $p > 1$.

Q.E.D.

The Integral Test

When we first studied the harmonic series in section 2.4, we proved that $\sum_{n=1}^{\infty} 1/n$ diverges by comparing it to the improper integral $\int_{1}^{\infty}(1/x)\,dx$. This is an approach that works whenever a_n is the value of a function of n that is positive, decreasing, and asymptotic to 0 as n approaches infinity. The following test for convergence was published by Cauchy in 1827.

Theorem 4.13 (The Integral Test). *Let f be a positive, decreasing, integrable function for $x \geq 1$. The series*

$$\sum_{k=1}^{\infty} f(k)$$

converges if and only if we have convergence of the improper integral

$$\int_{1}^{\infty} f(x)\,dx.$$

Any time we see the symbol ∞, warning lights should go off. The improper integral actually means the limit

$$\lim_{n \to \infty} \int_{1}^{n} f(x)\,dx.$$

Proof: Since f is positive for $x \geq 1$, it is enough to show that when one of them converges, it provides an upper bound for the other.

Since f is decreasing, we know that (see Figure 4.2)

$$f(k+1) \leq \int_{k}^{k+1} f(x)\,dx \leq f(k).$$

Definition: improper integral (unbounded domain)

The **improper integral**

$$\int_1^\infty f(x)\,dx$$

is said to **converge** if there is a number V such that for any error bound ϵ, we can always find a response N for which

$$n \ge N \quad \textit{implies that} \quad \left| \int_1^n f(x)\,dx - V \right| < \epsilon.$$

The number V is called the **value** of the integral.

It follows that

$$\sum_{k=1}^N f(k+1) \le \sum_{k=1}^N \int_k^{k+1} f(x)\,dx = \int_1^{N+1} f(x)\,dx \le \sum_{k=1}^N f(k).$$

If the series converges, then the partial integrals are bounded:

$$\int_1^{N+1} f(x)\,dx \le \sum_{k=1}^N f(k) \le \sum_{k=1}^\infty f(k).$$

If the integral converges, then the partial sums are bounded:

$$\sum_{k=1}^{N+1} f(k) \le f(1) + \int_1^{N+1} f(x)\,dx \le f(1) + \int_1^\infty f(x)\,dx.$$

Q.E.D.

In section 2.4, we not only proved that the harmonic series diverges, we found an explicit formula for the difference between the partial sum of the first n terms and $\int_1^n dx/x = \ln n$. The same thing can be done whenever the summand is of the form $f(k)$ where f is an analytic function for $x > 0$. In the 1730's, Leonhard Euler and Colin Maclaurin

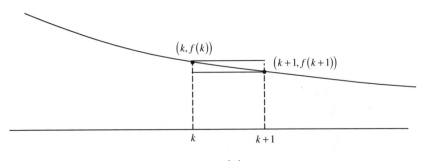

FIGURE 4.2. $f(k+1) \le \int_k^{k+1} f(x)\,dx \le f(k)$.

independently discovered this explicit connection, the **Euler–Maclaurin formula**:

$$\sum_{k=1}^{n} f(k) = \int_{1}^{n} f(x)\,dx + \frac{1}{2}[f(n) + f(1)] + \frac{B_2}{2!}[f'(n) - f'(1)]$$

$$+ \frac{B_4}{4!}[f'''(n) - f'''(1)] + \frac{B_6}{6!}[f^{(5)}(n) - f^{(5)}(1)] + \cdots, \qquad (4.9)$$

where the B_n are the Bernoulli numbers defined on page 119.

> To see a proof of the Euler–Maclaurin formula and to explore its consequences, Go to **Appendix A.4, The size of $n!$.**

Examples

The series

$$1 + \frac{1}{2\ln 2} + \frac{1}{3\ln 3} + \cdots = 1 + \sum_{k=2}^{\infty} \frac{1}{k\ln k}$$

is handled very efficiently by the integral test. We can ignore the first summand and consider the improper integral

$$\int_{2}^{\infty} \frac{dx}{x\ln x} = \lim_{n\to\infty} \int_{2}^{n} \frac{dx}{x\ln x}$$

$$= \lim_{n\to\infty} \ln(\ln x)\Big]_{2}^{n}$$

$$= \lim_{n\to\infty} (\ln\ln n - \ln\ln 2),$$

which is an infinite limit. The improper integral does not converge. It follows that the series also does not converge.

On the other hand, the series

$$1 + \frac{1}{2(\ln 2)^2} + \frac{1}{3(\ln 3)^2} + \cdots = 1 + \sum_{k=2}^{\infty} \frac{1}{k(\ln k)^2}$$

is compared with the improper integral

$$\int_{2}^{\infty} \frac{dx}{x(\ln x)^2} = \lim_{n\to\infty} \int_{2}^{n} \frac{dx}{x(\ln x)^2}$$

$$= \lim_{n\to\infty} \frac{-1}{\ln x}\Big]_{2}^{n}$$

$$= \lim_{n\to\infty} \left(\frac{1}{\ln 2} - \frac{1}{\ln n}\right)$$

$$= \frac{1}{\ln 2}.$$

Since the improper integral converges, this series must also converge.

Exercises

The symbol (**M&M**) indicates that *Maple* and *Mathematica* codes for this problem are available in the **Web Resources** at **www.macalester.edu/aratra**.

4.2.1. Prove that if $a_n > 0$ and $\sum_{n=1}^{\infty} a_n$ converges, then $\sum_{n=1}^{\infty} a_n^3$ must converge.

4.2.2. Show that the following series converges:

$$1 + \frac{1}{2\sqrt[3]{2}} + \frac{1}{2\sqrt[3]{2}} - \frac{1}{\sqrt[3]{2}} + \frac{1}{3\sqrt[3]{3}} + \frac{1}{3\sqrt[3]{3}} + \frac{1}{3\sqrt[3]{3}} - \frac{1}{\sqrt[3]{3}}$$

$$+ \cdots + \underbrace{\frac{1}{n\sqrt[3]{n}} + \frac{1}{n\sqrt[3]{n}} + \cdots + \frac{1}{n\sqrt[3]{n}}}_{n \text{ terms}} - \frac{1}{\sqrt[3]{n}} + \cdots . \tag{4.10}$$

4.2.3. Show that if b_k is the kth summand in the series given in (4.10), then $\sum_{k=1}^{\infty} b_k^3$ diverges. This gives us an example of a series for which $\sum b_k$ converges but $\sum b_k^3$ diverges. Why does this not contradict the result of exercise 4.2.1?

4.2.4. For each of the following series, determine whether it converges absolutely, converges conditionally, or diverges. Justify your answer.

a. $\dfrac{\arctan 1}{2} + \dfrac{\arctan 2}{2^2} + \cdots + \dfrac{\arctan n}{2^n} + \cdots$

b. $1 + \dfrac{1}{4} + \dfrac{2^2}{4^2} + \cdots + \dfrac{n^2}{4^n} + \cdots$

c. $1 + \dfrac{1}{2} + \dfrac{1}{3} + \cdots + \dfrac{1}{n} + \cdots$

d. $\dfrac{1}{1 \cdot 2} - \dfrac{1}{2 \cdot 3} + \cdots + (-1)^{n-1} \dfrac{1}{n(n+1)} + \cdots$

e. $\alpha_1 q^1 + \alpha_2 q^2 + \cdots + \alpha_n q^n + \cdots$, where $|q| < 1$ and $|\alpha_k| \le M$ for $k = 1, 2, \ldots$

f. $\dfrac{1}{2^2} + \dfrac{2}{3^2} + \cdots + \dfrac{n}{(n+1)^2} + \cdots$

4.2.5. For each of the following series, determine whether it converges absolutely, converges conditionally, or diverges. Justify your answer.

a. $\displaystyle\sum_{n=1}^{\infty} \left(\sqrt{n^2 + 1} - \sqrt[3]{n^3 + 1} \right)$

b. $\displaystyle\sum_{n=1}^{\infty} \left(\frac{n}{n+1} \right)^{n(n+1)}$

c. $\sum_{n=1}^{\infty} (1 - \cos(1/n))$

d. $\sum_{n=1}^{\infty} \left(\sqrt[n]{n} - 1\right)^n$

4.2.6. For each of the following series, find those values of a for which it converges absolutely, the values of a for which it converges conditionally, and the values of a for which it diverges. Justify your answers.

a. $\sum_{n=1}^{\infty} \left(\dfrac{an}{n+1}\right)^n$

b. $\sum_{n=1}^{\infty} \dfrac{1}{n+1} \left(\dfrac{a^2 - 4a - 8}{a^2 + 6a - 16}\right)^n, \quad a \neq -8, 2$

c. $\sum_{n=1}^{\infty} \dfrac{n^n}{a^{(n^2)}}$

4.2.7. Describe the region in the x, y-half-plane, $y > 0$, in which the series

$$\sum_{n=1}^{\infty} (-1)^n \frac{(\ln n)^x}{n^y}$$

converges absolutely, the region in which it converges conditionally, and the region in which it diverges.

4.2.8. Given a series $a_1 + a_2 + a_3 + \cdots$, assume that we can find a bound α and a subscript N such that $n \geq N$ implies that

$$\left|\frac{a_{n+1}}{a_n}\right| \leq \alpha.$$

Prove that given any positive error ϵ, there is a subscript M such that $n \geq M$ implies that

$$\sqrt[n]{|a_n|} < \alpha + \epsilon.$$

Show that this does *not* necessarily imply that $\sqrt[n]{|a_n|} \leq \alpha$.

4.2.9. Use the result of exercise 4.2.8 to prove that if the ratio test tells us that our series converges absolutely, then the root test will also tell us that our series converges absolutely.

4.2.10. Modify the argument in exercise 4.2.8 to prove that if we can find a bound β and a subscript N such that $n \geq N$ implies that

$$\left|\frac{a_{n+1}}{a_n}\right| \geq \beta,$$

then given any positive error ϵ, there is a subscript M such that $n \geq M$ implies

$$\sqrt[n]{|a_n|} > \beta - \epsilon.$$

4.2.11. Use the results from exercises 4.2.8 and 4.2.10 to prove that if $\lim_{n \to \infty} |a_{n+1}/a_n|$ exists, then

$$\lim_{n \to \infty} \sqrt[n]{|a_n|} = \lim_{n \to \infty} \left| \frac{a_{n+1}}{a_n} \right|. \tag{4.11}$$

4.2.12. Find an infinite series of positive summands for which the root test shows divergence but the ratio test is inconclusive. Explain why this example does not contradict the result of exercise 4.2.11.

4.2.13. Verify that the root test can be used in situations where the ratio test is inconclusive by applying both tests to the series

$$\frac{1}{3} + \frac{1}{2^2} + \frac{1}{3^3} + \frac{1}{2^4} + \frac{1}{3^5} + \frac{1}{2^6} + \frac{1}{3^7} + \frac{1}{2^8} + \cdots + \left(\frac{5 - (-1)^n}{2} \right)^{-n} + \cdots,$$

and to the series

$$\frac{1}{2} + 2^2 + \frac{1}{2^3} + 2^4 + \frac{1}{2^5} + 2^6 + \frac{1}{2^7} + 2^8 + \cdots + 2^{(-1)^n n} + \cdots.$$

4.2.14. Prove Corollary 4.10 on page 132.

4.2.15. (**M&M**) Find the partial sums

$$S_n = \sum_{k=1}^{n} \left(\frac{k}{2k - 1} \right)^k$$

for $n = 20, 40, \dots, 200$. Prove that this series converges.

4.2.16. (**M&M**) Find the partial sums

$$S_n(2) = \sum_{k=1}^{n} \left(\frac{k}{2k - 1} \right)^k 2^k \quad \text{and} \quad S_n(-2) = \sum_{k=1}^{n} \left(\frac{k}{2k - 1} \right)^k (-2)^k$$

for $n = 20, 40, \dots, 200$. Describe what you see happening. Do you expect that either or both of these converge? Prove your guesses about the convergence of the series in exercise 4.2.16.

4.2.17. (**M&M**) Find the partial sums

$$S_n = \sum_{k=1}^{n} \frac{k^k}{k!}$$

for $n = 20, 40, \dots, 200$. Prove that this series diverges.

4.2.18. (**M&M**) Find the partial sums

$$S_n(e^{-1}) = \sum_{k=1}^{n} \frac{k^k}{k!} e^{-k} \quad \text{and} \quad S_n(-e^{-1}) = \sum_{k=1}^{n} \frac{k^k}{k!} (-e)^{-k}$$

for $n = 20, 40, \ldots, 200$. Describe what you see happening. Do you expect that either or both of these converge? Prove your guesses about the convergence of the series in exercise 4.2.18.

4.2.19. (M&M) Find the partial sums

$$S_n = \sum_{k=1}^{n} \frac{2^k}{\sqrt{k}}$$

for $n = 20, 40, \ldots, 200$. Prove that this series diverges.

4.2.20. (M&M) Calculate the partial sum

$$\sum_{k=2}^{n} \frac{1}{k \ln k}$$

up to $n = 10,000$. Does it appear that this series is converging? Prove your assertion.

4.2.21. (M&M) Calculate the partial sum

$$\sum_{k=2}^{n} \frac{1}{k(\ln k)^{3/2}}$$

up to $n = 10,000$. Does it appear that this series is converging? Use both the integral test and the Cauchy condensation test to determine whether or not this series converges.

4.2.22. (M&M) Calculate the partial sum

$$\sum_{k=10}^{n} \frac{1}{k(\ln k)(\ln \ln k)}$$

up to $n = 10,000$. Does it appear that this series is converging? Use both the integral test and the Cauchy condensation test to determine whether or not this series converges.

4.2.23. For what values of α does

$$\sum_{n=10}^{\infty} \frac{1}{n(\ln n)(\ln \ln n)^{\alpha}}$$

converge?

4.2.24. Determine whether or not

$$\sum_{n=10}^{\infty} \frac{1}{n^{1+f(n)}}$$

converges when

$$f(n) = \frac{\ln \ln n + \ln \ln \ln n}{\ln n}.$$

Do we have convergence when

$$f(n) = \frac{\ln \ln n \; + \; 2 \ln \ln \ln n}{\ln n}?$$

4.2.25. For what values of α does

$$\int_1^\infty x^\alpha \, dx$$

converge?

4.2.26. Define

$$\ln_2 n = \ln \ln n, \quad \ln_3 n = \ln \ln \ln n, \quad \ldots, \quad \ln_k n = \ln(\ln_{k-1} n),$$

and let N_k be the smallest positive integer for which $\ln_k N_k > 0$. Prove that

$$\sum_{n=N_k}^\infty \frac{1}{n(\ln n)(\ln_2 n)(\ln_3 n)\cdots(\ln_k n)}$$

diverges.

4.2.27. Prove that

$$\sum_{n=N_k}^\infty \frac{1}{n(\ln n)(\ln_2 n)(\ln_3 n)\cdots(\ln_k n)^\alpha}$$

diverges for $\alpha \leq 1$ and converges for $\alpha > 1$.

4.2.28. Prove that

$$\sum_{n=1}^\infty \frac{(n!)^n}{(n^2)!}$$

converges. Find a function $f(n)$ that grows as fast as possible and such that

$$\sum_{n=1}^\infty \frac{(n!)^n}{(n^2)!} f(n)$$

still converges.

4.2.29. Use Cauchy condensation to determine whether the following series converge or diverge.

a. $\displaystyle\sum_{n=1}^\infty \frac{1}{2^{\sqrt{n}}}$

b. $\displaystyle\sum_{n=1}^\infty \frac{1}{2^{\ln n}}$

4.2.30. Prove that if a_n is a positive, decreasing sequence, then $\sum_{n=1}^{\infty} a_n$ converges if and only if $\sum_{n=0}^{\infty} 3^n a_{3^n}$ converges. Use this to determine whether the series

$$\sum_{n=1}^{\infty} \frac{1}{3^{\ln n}}$$

converges or diverges.

4.3 The Convergence of Power Series

We are concerned not just with infinite series but with infinite series of functions,

$$F(x) = f_1(x) + f_2(x) + f_3(x) + \cdots.$$

For our purposes, convergence is always **pointwise convergence**.

> **Definition: pointwise convergence**
>
> A series of functions $f_1 + f_2 + f_3 + \cdots$ converges **pointwise** to F if at each value of x, the value of F is the limit of the sum of the f_k evaluated at that value of x, $F(x) = f_1(x) + f_2(x) + f_3(x) + \cdots$.

> **Web Resource:** To see another type of convergence for infinite series of functions, go to **Convergence in norm**.

For Fourier's cosine series,

$$F(x) = \cos \frac{\pi x}{2} - \frac{1}{3} \cos \frac{3\pi x}{2} + \frac{1}{5} \cos \frac{5\pi x}{2} - \frac{1}{7} \cos \frac{7\pi x}{2} + \cdots,$$

the series at $x = 1$ is $0 - 0 + 0 - 0 + \cdots$ which converges to 0. If we evaluate the series at any x strictly between -1 and 1, we obtain a series that converges to $\pi/4$.

With a series of functions, the question is not whether or not it converges, but for which values of x it converges. In this section, we shall consider **power series** in which the summands are constant multiples of powers of x:

$$a_0 + \sum_{n=1}^{\infty} a_n x^n.$$

A power series might be shifted x_0 units to the right by replacing the variable x with $x - x_0$,

$$a_0 + \sum_{n=1}^{\infty} a_n (x - x_0)^n.$$

In the next section, we shall treat **trigonometric series** in which the summands are constant multiples of the sine and cosine of nx:

$$a_0 + \sum_{n=1}^{\infty} (a_n \cos nx + b_n \sin nx).$$

As we shall see, power series are well behaved. The set of x for which they converge is always an interval that is symmetric (except possibly for the endpoints) about the origin or about the value x_0 if it has been shifted. Trigonometric series are not always well behaved.

Some Examples

We begin with the most important of the power series, the binomial series:

$$(1 + x)^{\alpha} = 1 + \alpha x + \frac{\alpha(\alpha - 1)}{2!} x^2 + \frac{\alpha(\alpha - 1)(\alpha - 2)}{3!} x^3 + \cdots . \qquad (4.12)$$

As we saw in equation (2.54) on page 42, the absolute value of the ratios of successive terms is

$$r(n) = \left(1 - \frac{1 + \alpha}{n}\right) |x|.$$

This has a limit:

$$\lim_{n \to \infty} \left(1 - \frac{1 + \alpha}{n}\right) |x| = |x|.$$

By Corollary 4.8, the binomial series converges absolutely when $|x| < 1$, it diverges when $|x| > 1$, and we do not yet know what happens when $|x| = 1$.

The exponential series

$$e^x = 1 + x + \frac{x^2}{2!} + \frac{x^3}{3!} + \cdots$$

is another easy case. We have that

$$r(n) = \left| \frac{x^n / n!}{x^{n-1}/(n-1)!} \right| = \frac{|x|}{n}.$$

Regardless of the value of x, this approaches 0 as n gets arbitrarily large,

$$\lim_{n \to \infty} \frac{|x|}{n} = 0 < 1.$$

This series converges absolutely for all values of x.

We can also use the root test on this series, replacing $n!$ by Stirling's formula, $n! = n^n \sqrt{2\pi n}\, e^{-n+E(n)}$ where $E(n)$ approaches 0 as n gets large:

$$\rho(n) = \left| \frac{x^n}{n^n \sqrt{2\pi n}\, e^{-n+E(n)}} \right|^{1/n} = \frac{|x|}{n(2\pi n)^{1/2n} e^{-1+E(n)/n}}.$$

Again, this quantity approaches 0 as n gets large, and so the exponential series converges absolutely for all values of x.

Radius of Convergence

A power series will often have the property that the absolute value of the ratio of consecutive terms has a well-defined limit. The limit ratio test produces a bound on the absolute value of

x (or a bound on $|x - x_0|$ if the series has been shifted) within which the series converges. This bound is called the **radius of convergence**.

We apply the limit ratio test to

$$1 + 2x + 6x^2 + 15x^3 + \cdots + \binom{2n}{n} x^n + \cdots = 1 + \sum_{n=1}^{\infty} \binom{2n}{n} x^n,$$

where $\binom{2n}{n}$ is the binomial coefficient,

$$\binom{2n}{n} = \frac{(2n)!}{n!\, n!}.$$

By the limit ratio test, this converges absolutely for

$$\lim_{n \to \infty} \left| \frac{(2n+2)!\, x^{n+1}}{(n+1)!\,(n+1)!} \cdot \frac{n!\, n!}{(2n)!\, x^n} \right| = \lim_{n \to \infty} \left| \frac{(2n+1)(2n+2)\, x}{(n+1)(n+1)} \right| = 4|x| < 1.$$

The radius of convergence is $1/4$.

We can also apply the limit ratio test to

$$1 + 2x^2 + 6x^4 + 15x^6 + \cdots + \binom{2n}{n} x^{2n} + \cdots = 1 + \sum_{n=1}^{\infty} \binom{2n}{n} x^{2n}.$$

For this series, we have absolute convergence when

$$\lim_{n \to \infty} \left| \frac{(2n+2)!\, x^{2n+2}}{(n+1)!\,(n+1)!} \cdot \frac{n!\, n!}{(2n)!\, x^{2n}} \right| = \lim_{n \to \infty} \left| \frac{(2n+1)(2n+2)\, x^2}{(n+1)(n+1)} \right| = 4|x|^2 < 1.$$

The radius of convergence in this case is $1/2$.

Definition: radius of convergence

The **radius of convergence** of a power series $\sum_{n=1}^{\infty} a_n x^n$ is the bound B with the property that the series converges absolutely for $|x| < B$, and the series diverges for $|x| > B$.

As we shall see in the next few pages, part of the beauty and convenience of power series is that there will always be a radius of convergence. If

$$\lim_{n \to \infty} \left| \frac{a_{n+1} x^{n+1}}{a_n x^n} \right|$$

does not exist, perhaps because many of the a_n are zero, we can still use the root test. We have absolute convergence when the **upper limit** of $\sqrt[n]{|a_n x^n|}$, denoted by

$$\overline{\lim_{n \to \infty}} \sqrt[n]{|a_n x^n|} = \overline{\lim_{n \to \infty}} \sqrt[n]{|a_n|}\, |x|,$$

is strictly less than 1, divergence when it is strictly greater than 1. The **radius of convergence** is then

$$R = \frac{1}{\overline{\lim}_{n \to \infty} \sqrt[n]{|a_n|}}.$$

When the limit of a sequence exists, then the upper limit is simply the limit. The advantage of using the upper limit is that for any bounded sequence it *always* exists, even when the limit does not.

lim inf and lim sup (Limb Soup)

In the proof of the Cauchy criterion, starting on page 123, we took our bounded sequence of partial sums and considered the set of greatest lower bounds where L_k is the greatest lower bound of the set $\{S_k, S_{k+1}, S_{k+2}, \ldots\}$. We then took the least upper bound M of the set of L_k. In this case, because we had assumed that the sequence was Cauchy, M was the limit of the sequence $(S_1, S_2, S_3, \ldots)$. But all we needed in order to have a least upper bound of the sequence of greatest lower bounds was that our original sequence was bounded. Given a bounded sequence, we call this least upper bound of the sequence of greatest lower bounds the **lim inf** or **lower limit** of the original sequence. Similarly, the greatest lower bound of the sequence of least upper bounds is called the **lim sup** or **upper limit**.

Thus for the sequence

$$(0.9, 3.1, 0.99, 3.01, 0.999, 3.001, 0.9999, 3.0001, \ldots),$$

the lower limit is the least upper bound of $\{0.9, 0.99, 0.999, \ldots\}$ which is 1. The upper limit is the greatest lower bound of $\{3.1, 3.01, 3.001, \ldots\}$ which is 3.

Definition: upper and lower limits

The **upper limit** of a bounded sequence $(x_1, x_2, x_3, \ldots)$ is the greatest lower bound of the set $\{M_1, M_2, M_3, \ldots\}$ where M_k is the least upper bound of the set $\{x_k, x_{k+1}, x_{k+2}, \ldots\}$. The **lower limit** of the bounded sequence is the negative of the upper limit of $(-x_1, -x_2, -x_3, \ldots)$,

$$\underline{\lim_{n \to \infty}} \, x_n = -\overline{\lim_{n \to \infty}} \, (-x_n).$$

We owe the concept of the upper limit to Cauchy. He introduced it in his *Cours d'analyse* for exactly the reason we have used it here: to find the radius of convergence of an arbitrary power series. His definition was less precise than we would tolerate today. He spoke of it as "the limit towards which the greatest values converge."

Existence of Radius of Convergence

We consider the sequence $S = (|a_1|, \sqrt{|a_2|}, \sqrt[3]{|a_3|}, \sqrt[4]{|a_4|}, \ldots)$. If this sequence is unbounded, then for every $x \neq 0$, the sequence $(|a_1| \, |x|, \sqrt{|a_2|} \, |x|, \sqrt[3]{|a_3|} \, |x|, \sqrt[4]{|a_4|} \, |x|, \ldots)$ is also unbounded. By the root test (Theorem 4.9), the power series diverges at every value of x other than $x = 0$. In this case, the radius of convergence is zero. If S is bounded, then $\overline{\lim}_{n \to \infty} \sqrt[n]{|a_n|}$ will always be well defined and greater than or equal to zero. It still remains for us to prove that $R = 1/\overline{\lim}_{n \to \infty} \sqrt[n]{|a_n|}$ is a radius of convergence.

> **Theorem 4.14 (Existence of Radius of Convergence).** *Let* $a_0 + \sum_{n=1}^{\infty} a_n x^n$ *be an arbitrary power series and define*
>
> $$R = \frac{1}{\overline{\lim}_{n \to \infty} \sqrt[n]{|a_n|}}.$$
>
> *This series converges absolutely for* $|x| < R$ *and it diverges for* $|x| > R$*. The power series converges at all values of* x *when* $\overline{\lim}_{n \to \infty} \sqrt[n]{|a_n|} = 0$*, and it converges only at* $x = 0$ *when the upper limit is infinite.*

Proof: Let $\lambda = \overline{\lim}_{n \to \infty} \sqrt[n]{|a_n|}$. If $|x| < 1/\lambda$, then we can find an α just a little less than 1 and an ϵ just a little larger than zero so that we still have

$$|x| < \frac{\alpha}{\lambda + \epsilon}.$$

It follows that

$$\sqrt[n]{|a_n x^n|} = \sqrt[n]{|a_n|}\,|x| < \frac{\sqrt[n]{|a_n|}}{\lambda + \epsilon}\,\alpha.$$

By the definition of λ as the upper limit of $\sqrt[n]{|a_n|}$, this last term is strictly less than α for all sufficiently large values of n. The root test, Theorem 4.9, tells us that the series converges absolutely.

If $|x| > 1/\lambda$, then we can find an ϵ just a little larger than zero so that we still have

$$|x| > \frac{1}{\lambda - \epsilon}.$$

It follows that

$$\sqrt[n]{|a_n x^n|} = \sqrt[n]{|a_n|}\,|x| > \frac{\sqrt[n]{|a_n|}}{\lambda - \epsilon}.$$

From the definition of λ, there must be infinitely many elements of $\sqrt[n]{|a_n|}$ that equal or exceed $\lambda - \epsilon$. This means that there are infinitely many values of n for which $\sqrt[n]{|a_n x^n|} \geq 1$. The root test tells us that this series diverges.

Q.E.D.

Hypergeometric Series

What happens when $|x|$ *equals* the radius of convergence? The series might converge at both endpoints, diverge at both, or converge at only one of these values. If it converges at both, the convergence might be absolute or conditional. There is no single test that will return a conclusive answer for all power series, but in 1812 Carl Friedrich Gauss did publish a test that determines the convergence at the endpoints for every power series you are likely to encounter outside of a course in real analysis. It is a definitive test that works when the power series is **hypergeometric**.

The easiest infinite series with which to work is the geometric series,

$$1 + x + x^2 + x^3 + \cdots.$$

It converges to $1/(1-x)$ when $|x| < 1$, and it diverges when $|x| \geq 1$. In the seventeenth and eighteenth centuries, mathematicians began to appreciate a larger class of series that was almost as nice, the **hypergeometric series**. A geometric series is characterized by the fact that the ratio of two successive summands is constant. In a hypergeometric series, the ratio of two succesive nonzero summands is a rational function of the subscript.

Definition: hypergeometric series

A series $a_1 + a_2 + a_3 + \cdots$ is **hypergeometric** if

$$\frac{a_{n+1}}{a_n} = \frac{P(n)}{Q(n)}$$

where $P(n)$ and $Q(n)$ are polynomials in n.

For example, the exponential series is hypergeometric:

$$\frac{a_{n+1}}{a_n} = \frac{x^n/n!}{x^{n-1}/(n-1)!} = \frac{x}{n}.$$

The numerator is the constant x (constant with respect to n), and the denominator is the linear function n. The series for $\sin x$ is also hypergeometric:

$$\frac{a_{n+1}}{a_n} = \frac{(-1)^n x^{2n+1}/(2n+1)!}{(-1)^{n-1} x^{2n-1}/(2n-1)!} = \frac{-x^2}{(2n)(2n+1)}.$$

Again the numerator is constant, $-x^2$; the denominator is a quadratic function, $4n^2 + 2n$. The binomial series is also hypergeometric. Given that

$$a_n = \frac{\alpha(\alpha-1)\cdots(\alpha-n+2)}{(n-1)!} x^{n-1},$$

the ratio of consecutive terms is

$$\frac{a_{n+1}}{a_n} = \frac{(\alpha-n+1)x}{n}.$$

In this case, both numerator and denominator are linear functions of n. Even a series such as

$$1 + \frac{1}{2^2} + \frac{1}{3^2} + \frac{1}{4^2} + \cdots$$

is hypergeometric:

$$\frac{a_{n+1}}{a_n} = \frac{n^2}{(n+1)^2}.$$

It was quickly realized that most of the series people were finding were hypergeometric or could be expressed in terms of hypergeometric series. On page 39 we encountered Euler's differential equation that models a vibrating drumhead. Euler showed that the solution to this equation is given by the series

$$u(r) = r^\beta \left[1 - \frac{1}{(\beta + 1)} \left(\frac{\alpha r}{2} \right)^2 + \frac{1}{2! \, (\beta + 1)(\beta + 2)} \left(\frac{\alpha r}{2} \right)^4 \right.$$
$$\left. - \frac{1}{3! \, (\beta + 1)(\beta + 2)(\beta + 3)} \left(\frac{\alpha r}{2} \right)^6 + \cdots \right]. \tag{4.13}$$

The nth summand is

$$a_n = (-1)^{n-1} \frac{1}{(n-1)!(\beta + 1)(\beta + 2) \cdots (\beta + n - 1)} \left(\frac{\alpha r}{2} \right)^{2n-2}.$$

The ratio of successive summands is

$$\frac{a_{n+1}}{a_n} = \frac{-\alpha^2 r^2}{4n(\beta + n)}. \tag{4.14}$$

We again have a hypergeometric series. The numerator is constant (as a function of n), and the denominator is a quadratic polynomial.

Gauss's attention was turned to hypergeometric series by problems in astronomy. Like Euler, he found that the solutions he was obtaining were power series that satisfied the hypergeometric condition. In 1812, he presented a thorough study of these series entitled "Disquisitionis generales circa seriem infinitam $1 + \frac{\alpha\beta}{1 \cdot \gamma} x + \frac{\alpha(\alpha+1)\beta(\beta+1)}{1 \cdot 2 \cdot \gamma(\gamma+1)} xx + \frac{\alpha(\alpha+1)(\alpha+2)\beta(\beta+1)(\beta+2)}{1 \cdot 2 \cdot 3 \cdot \gamma(\gamma+1)(\gamma+2)} x^3 + $ etc."

The Question of Convergence

A hypergeometric series is custom-made for the ratio test—or rather, the ratio test is custom-made for hypergeometric series. We can always make sense of the limit

$$\lim_{n \to \infty} \left| \frac{a_{n+1}}{a_n} \right| = \lim_{n \to \infty} \left| \frac{P(n)}{Q(n)} \right|.$$

We observe that if the degree of $P(n)$ is larger than the degree of $Q(n)$, then $|P(n)/Q(n)|$ gets arbitrarily large as n increases, and so the series diverges. If the degree of $P(n)$ is less than the degree of $Q(n)$, then our ratio approaches zero as n increases and so the series is absolutely convergent. In both of these cases, our conclusion is independent of the choice of x.

The exponential function, the sine, and the cosine all fall into this second category. These are functions for which the radius of convergence is infinite, and so there are no endpoints of the interval of convergence. If x is an endpoint of the interval of convergence, then we know that the series evaluated at this point satisfies

$$\lim_{n \to \infty} \left| \frac{a_{n+1}}{a_n} \right| = \lim_{n \to \infty} \left| \frac{P(n)}{Q(n)} \right| = 1.$$

This happens if and only if P and Q are polynomials of the same degree with leading coefficients that have the same absolute value:

$$\frac{P(n)}{Q(n)} = \frac{C_t n^t + C_{t-1} n^{t-1} + \cdots + C_0}{c_t n^t + c_{t-1} n^{t-1} + \cdots + c_0},$$

where $C_t = \pm c_t$.

On the Radius of Convergence

Gauss found a test that is absolutely sharp for all hypergeometric series for which $\lim_{n \to \infty} |P(n)/Q(n)| = 1$. It never returns an inconclusive answer. Nine years before Cauchy published his *Cours d'analyse*, Gauss demonstrated an understanding of the question of convergence that was decades ahead of its time. Twenty years later, in 1832, J. L. Raabe was to publish a test for convergence that could be applied to hypergeometric series but which was less effective than Gauss's test, leaving some situations indeterminate. Gauss was so far ahead of his contemporaries that few realized what he had accomplished. It was not until other mathematicians began to rediscover his test that it was recognized that Gauss had already been there. Not only was Gauss the first to arrive, his proof is a model of clarity and precision. One sees in it the hand of the master.

Web Resource: To see Gauss's proof of his test as well as additional information on it, go to **Gauss's test**.

Theorem 4.15 (Gauss's Test). *Let* $a_0 + a_1 + a_2 + \cdots$ *be a hypergeometric series for which*

$$\frac{a_{n+1}}{a_n} = \frac{C_t n^t + C_{t-1} n^{t-1} + \cdots + C_0}{c_t n^t + c_{t-1} n^{t-1} + \cdots + c_0},$$

where $C_t = \pm c_t$. *Set* $B_j = C_j / C_t$ *and* $b_j = c_j / c_t$ *so that the resulting polynomials are* **monic** *(the coefficient of the highest term is 1). The test is as follows:*

1. *If* $B_{t-1} > b_{t-1}$, *then the absolute values of the summands grow without limit and the series cannot converge.*

2. *If* $B_{t-1} = b_{t-1}$, *then the absolute values of the summands approach a finite nonzero limit and the series cannot converge.*

3. *If* $B_{t-1} < b_{t-1}$, *then the absolute values of the summands approach zero. If the series is alternating, then it converges.*

4. *If* $B_{t-1} \geq b_{t-1} - 1$, *then the series is not absolutely convergent.*

5. *If* $B_{t-1} < b_{t-1} - 1$, *then the series is absolutely convergent.*

We note that if the question is simply one of convergence, then there are three cases:

1. If $B_{t-1} \geq b_{t-1}$, then the series does not converge.

2. If $b_{t-1} > B_{t-1} \geq b_{t-1} - 1$, then the series converges if and only if it is an alternating series.

3. If $b_{t-1} - 1 > B_{t-1}$, then the series is absolutely convergent.

We can use Gauss's test to determine the convergence of $1 + \sum_{n=1}^{\infty} \binom{2n}{n} x^n$ at the endpoints of the interval of convergence, $x = \pm 1/4$. At $x = 1/4$, we have

$$
\frac{a_{n+1}}{a_n} = \frac{(2n+2)! \, (1/4)^{n+1}}{(n+1)! \, (n+1)!} \cdot \frac{n! \, n!}{(2n)! \, (1/4)^n}
$$

$$
= \frac{(2n+2)(2n+1)}{(n+1)(n+1)4} = \frac{4n^2 + 6n + 2}{4n^2 + 8n + 4} = \frac{n^2 + (3/2)n + (1/2)}{n^2 + 2n + 1}.
$$

We see that $B_1 = 3/2$, $b_1 = 2$ and we are in the situation where

$$
b_1 = 2 > B_1 = \frac{3}{2} \geq b_1 - 1 = 1.
$$

The series converges if and only if it is alternating. At $x = 1/4$, the series does not alternate, and so it diverges. At $x = -1/4$, the summands of the series do alternate in sign, and so the series converges conditionally. The interval of convergence is $[-1/4, 1/4)$.

For a more general example, we consider the binomial series

$$
(1+x)^{\alpha} = 1 + \alpha x + \frac{\alpha(\alpha-1)}{2!} x^2 + \frac{\alpha(\alpha-1)(\alpha-2)}{3!} x^3 + \cdots,
$$

$$
\frac{a_{n+1}}{a_n} = \frac{(-n+\alpha)x}{n+1}.
$$

The radius of convergence is $|-1/1| = 1$. If $x = \pm 1$, then the rational function that determines convergence is

$$
\frac{n-\alpha}{n+1}.
$$

We see that $t = 1$, $B_0 = -\alpha$, $b_0 = 1$.

1. If $-\alpha \geq 1$ ($\alpha \leq -1$), then the summands either grow without limit ($\alpha < -1$) or all have absolute value 1 ($\alpha = -1$). In either case, the series does not converge.

2. If $1 > -\alpha > 0$ ($-1 < \alpha < 0$), then the summands approach zero. The series converges when the summands alternate in sign which happens when $x = 1$. It diverges when $x = -1$. (Note that the case $\alpha = 0$ is degenerate: $(1+x)^0 = 1$. This is true for all values of x.)

3. If $0 > -\alpha$ ($\alpha > 0$), then the series converges absolutely. It converges for both $x = 1$ and $x = -1$.

Exercises

The symbol (**M&M**) indicates that *Maple* and *Mathematica* codes for this problem are available in the **Web Resources** at **www.macalester.edu/aratra**.

4.3.1. Determine the domain of convergence of the power series given below.

a. $\displaystyle\sum_{n=1}^{\infty} n^3 x^n$

b. $\displaystyle\sum_{n=1}^{\infty} \frac{2^n}{n!} x^n$

c. $\displaystyle\sum_{n=1}^{\infty} \frac{2^n}{n^2} x^n$

d. $\displaystyle\sum_{n=1}^{\infty} (2 + (-1)^n)^n x^n$

e. $\displaystyle\sum_{n=1}^{\infty} \left(\frac{2 + (-1)^n}{5 + (-1)^{n+1}} \right)^n x^n$

f. $\displaystyle\sum_{n=1}^{\infty} 2^n x^{n^2}$

g. $\displaystyle\sum_{n=1}^{\infty} 2^{n^2} x^{n!}$

h. $\displaystyle\sum_{n=1}^{\infty} \left(1 + \frac{1}{n} \right)^{(-1)^n n^2} x^n$

4.3.2. Find the domain of convergence of the following series.

a. $\displaystyle\sum_{n=1}^{\infty} \frac{(x-1)^{2n}}{2^n n^3}$

b. $\displaystyle\sum_{n=1}^{\infty} \frac{n}{n+1} \left(\frac{2x+1}{x} \right)^n$

c. $\displaystyle\sum_{n=1}^{\infty} \frac{n \, 4^n}{3^n} x^n (1-x)^n$

d. $\displaystyle\sum_{n=1}^{\infty} \frac{(n!)^2}{(2n)!} (x-1)^n$

e. $\displaystyle\sum_{n=1}^{\infty} \sqrt{n} \, (\tan x)^n$

f. $\displaystyle\sum_{n=1}^{\infty} (\arctan(1/x))^{n^2}$

4.3.3. Find the radius of convergence R of $\sum_{n=0}^{\infty} a_n x^n$ if

a. there are α and $L > 0$ such that $\lim_{n\to\infty} |a_n n^\alpha| = L$,

b. there exist positive α and L such that $\lim_{n\to\infty} |a_n \alpha^n| = L$,

c. there is a positive L such that $\lim_{n\to\infty} |a_n n!| = L$.

4.3.4. Suppose that the radius of convergence of $\sum_{n=0}^{\infty} a_n x^n$ is R, $0 < R < \infty$. Evaluate the radius of convergence of the following series.

a. $\displaystyle\sum_{n=0}^{\infty} 2^n a_n x^n$

b. $\displaystyle\sum_{n=0}^{\infty} n^n a_n x^n$

c. $\displaystyle\sum_{n=0}^{\infty} \frac{n^n}{n!} a_n x^n$

d. $\displaystyle\sum_{n=0}^{\infty} a_n^2 x^n$

4.3.5. Find the radius of convergence for

$$\sum_{k=1}^{\infty} \left(\frac{k}{2k-1} \right)^k x^k.$$

4.3.6. (**M&M**) Graph the partial sums

$$S_n(x) = \sum_{k=1}^{n} \left(\frac{k}{2k-1} \right)^k x^k$$

for $n = 3, 6, 9$, and 12. Describe what you see. Do you expect convergence at either or both of the endpoints of the interval of convergence of the infinite series. Prove your assertions.

4.3.7. Find the radius of convergence for

$$\sum_{k=1}^{\infty} \frac{k^k}{k!} x^k.$$

4.3.8. (**M&M**) Graph the partial sums

$$S_n(x) = \sum_{k=1}^{n} \frac{k^k}{k!} x^k$$

for $n = 3, 6, 9$, and 12. Describe what you see. Do you expect convergence at either or both of the endpoints of the interval of convergence of the infinite series. Prove your assertions.

4.3.9. Find the radius of convergence for

$$\sum_{k=1}^{\infty} \frac{2^k}{\sqrt{k}} x^k.$$

4.3.10. (**M&M**) Graph the partial sums

$$S_n(x) = \sum_{k=1}^{n} \frac{2^k}{\sqrt{k}} x^k$$

for $n = 3, 6, 9,$ and 12. Describe what you see. Do you expect convergence at either or both of the endpoints of the interval of convergence of the infinite series. Prove your assertions.

4.3.11. (**M&M**) For the series

$$\frac{2}{3} x + \frac{2 \cdot 4}{3 \cdot 5} x^2 + \frac{2 \cdot 4 \cdot 6}{3 \cdot 5 \cdot 7} x^3 + \cdots,$$

graph the partial sums of the first 3 terms, the first 6 terms, the first 9 terms, and the first 12 terms over the domain $-2 \le x \le 2$. Find the radius of convergence R for this series.

4.3.12. Using the series in exercise 4.3.11, decide whether or not this series converges when $x = R$ and when $x = -R$. Explain your answers.

4.3.13. Use Stirling's formula to prove that

$$1 \cdot 3 \cdot 5 \cdots (2n - 1) = 2^{n+1/2} n^n e^{-n + F(n)} \tag{4.15}$$

where $F(n)$ is an error term that approaches zero as n gets large.

4.3.14. (**M&M**) Graph the partial sums of the first 3 terms, the first 6 terms, the first 9 terms, and the first 12 terms of the series

$$\sum_{k=1}^{\infty} \frac{k^k}{1 \cdot 3 \cdot 5 \cdots (2k - 1)} x^k,$$

over the domain $-2 \le x \le 2$. Find the radius of convergence for this series.

4.3.15. (**M&M**) For each of the following series:
  (i) Verify that the series is hypergeometric.
 (ii) Graph the polynomial approximations that are obtained from the first three, six, and nine terms of the series. Describe what you see and where it appears that each of these gives a reasonable approximation to the function represented by this series.
(iii) Find the radius of convergence.
(iv) Use Gauss's test to determine whether or not the series converges at each endpoint.

a. $x + \dfrac{x^2}{4} + \dfrac{x^3}{9} + \cdots = \displaystyle\sum_{k=1}^{\infty} \dfrac{x^k}{k^2}$

b. $1 + \dfrac{2!}{1 \cdot 1} x + \dfrac{4!}{2! \cdot 2!} x^2 + \dfrac{6!}{3! \cdot 3!} x^3 + \cdots = 1 + \displaystyle\sum_{k=1}^{\infty} \dfrac{(2k)!}{k! \cdot k!} x^k$

c. $1 + \dfrac{1}{3!}x + \dfrac{(2!)^3}{6!}x^2 + \dfrac{(3!)^3}{9!}x^3 + \cdots = 1 + \displaystyle\sum_{k=1}^{\infty} \dfrac{(k!)^3}{(3k)!}x^k$

d. $1 + \dfrac{3}{1}x + \dfrac{3 \cdot 5}{2!}x^2 + \dfrac{3 \cdot 5 \cdot 7}{3!}x^3 + \cdots = 1 + \displaystyle\sum_{k=1}^{\infty} \dfrac{3 \cdot 5 \cdots (2k+1)}{k!}x^k$

e. $\dfrac{3}{4}x^2 + \dfrac{3 \cdot 8}{4 \cdot 9}x^3 + \dfrac{3 \cdot 8 \cdot 15}{4 \cdot 9 \cdot 16}x^4 + \cdots = \displaystyle\sum_{k=2}^{\infty} \dfrac{3 \cdot 8 \cdots (k^2-1)}{4 \cdot 9 \cdots k^2}x^k$

f. $1 + x + \dfrac{1^2 \cdot 3^2}{(2!)^2}x^2 + \dfrac{1^2 \cdot 3^2 \cdot 5^2}{(3!)^2}x^3 + \cdots = 1 + \displaystyle\sum_{k=1}^{\infty} \dfrac{1^2 \cdot 3^2 \cdots (2k-1)^2}{(k!)^2}x^k$

g. $1 + \dfrac{3!}{1 \cdot 2!}x + \dfrac{6!}{2! \cdot 4!}x^2 + \dfrac{9!}{3! \cdot 6!}x^3 + \cdots = 1 + \displaystyle\sum_{k=1}^{\infty} \dfrac{(3k)!}{k! \cdot (2k)!}x^k$

4.3.16. Explain why the following series is *not* a hypergeometric series:

$$x + \dfrac{x^2}{2} - \dfrac{x^3}{3} + \dfrac{x^4}{4} + \dfrac{x^5}{5} - \dfrac{x^6}{6} + \cdots .$$

4.3.17. The power series in exercise 4.3.16 can be expressed as a difference of two hypergeometric series. What are they?

4.3.18. Find the upper and lower limits of the following sequences.

a. $a_n = n\alpha - \lfloor n\alpha \rfloor, \quad \alpha \in \mathbb{Q}$

b. $a_n = n\alpha - \lfloor n\alpha \rfloor, \quad \alpha \notin \mathbb{Q}$

c. $a_n = \sin(n\pi\alpha), \quad \alpha \in \mathbb{Q}$

d. $a_n = \sin(n\pi\alpha), \quad \alpha \notin \mathbb{Q}$

4.3.19. Prove that $\overline{\lim}_{n\to\infty} a_n = A$ if and only if given any $\epsilon > 0$, there exists a response N so that for any $n \geq N$, $a_n < A + \epsilon$ and there is an $m \geq n$ with $a_m > A - \epsilon$.

4.3.20. Prove that if a_n and b_n are bounded sequences, then

$$\underline{\lim_{n\to\infty}} a_n + \underline{\lim_{n\to\infty}} b_n \leq \underline{\lim_{n\to\infty}} (a_n + b_n) \leq \underline{\lim_{n\to\infty}} a_n + \overline{\lim_{n\to\infty}} b_n$$
$$\leq \overline{\lim_{n\to\infty}} (a_n + b_n) \leq \overline{\lim_{n\to\infty}} a_n + \overline{\lim_{n\to\infty}} b_n.$$

For each of these inequalities, give an example of sequences $\{a_n\}$ and $\{b_n\}$ for which weak inequality ($\leq$) becomes strict inequality ($<$).

4.3.21. Prove that if $\lim_{n\to\infty} a_n = a$, then

$$\overline{\lim_{n\to\infty}} (a_n + b_n) = a + \overline{\lim_{n\to\infty}} b_n.$$

4.3.22. Prove that if $\lim_{n\to\infty} a_n = a > 0$ and $b_n \geq 0$ for all n sufficiently large, then

$$\overline{\lim_{n\to\infty}} (a_n \cdot b_n) = a \cdot \overline{\lim_{n\to\infty}} b_n.$$

4.3.23. Prove that if $a_n > 0$ then

$$\varliminf_{n\to\infty} \frac{a_{n+1}}{a_n} \le \varliminf_{n\to\infty} \sqrt[n]{a_n} \le \varlimsup_{n\to\infty} \sqrt[n]{a_n} \le \varlimsup_{n\to\infty} \frac{a_{n+1}}{a_n}. \tag{4.16}$$

4.3.24. Let f be a continuous function for all x and let $\{x_n\}$ be a bounded sequence. Prove or disprove:

$$\varliminf_{n\to\infty} f(x_n) = f\left(\varliminf_{n\to\infty} x_n\right) \quad \text{and} \quad \varlimsup_{n\to\infty} f(x_n) = f\left(\varlimsup_{n\to\infty} x_n\right).$$

4.3.25. Let f be a continuous and increasing function for all x and let $\{x_n\}$ be a bounded sequence. Prove that

$$\varliminf_{n\to\infty} f(x_n) = f\left(\varliminf_{n\to\infty} x_n\right) \quad \text{and} \quad \varlimsup_{n\to\infty} f(x_n) = f\left(\varlimsup_{n\to\infty} x_n\right).$$

4.3.26. Let f be a continuous and decreasing function for all x and let $\{x_n\}$ be a bounded sequence. Prove or disprove:

$$\varliminf_{n\to\infty} f(x_n) = f\left(\varlimsup_{n\to\infty} x_n\right) \quad \text{and} \quad \varlimsup_{n\to\infty} f(x_n) = f\left(\varliminf_{n\to\infty} x_n\right).$$

4.3.27. (**M&M**) Evaluate the sum of the first thousand terms of

$$\sum_{k=1}^{\infty} \left(\frac{2 \cdot 5 \cdots (3k-1)}{3 \cdot 6 \cdots (3k)}\right)^m$$

when $m = 1, 2, 3$, and 4. Use Gauss's test to determine those values of k for which this series converges.

4.3.28. (**M&M**) Evaluate the sum of the first thousand terms of

$$\sum_{j=1}^{\infty} \left(\frac{1 \cdot 4 \cdots (3j-2)}{3 \cdot 6 \cdots (3j)}\right)^m$$

when $m = 1, 2, 3$, and 4. Use Gauss's test to determine those values of k for which this series converges.

4.4 The Convergence of Fourier Series

None of the convergence tests that we have examined so far can help us with the question of convergence of the Fourier series that we met in the first chapter:

$$\cos(\pi x/2) - \frac{1}{3}\cos(3\pi x/2) + \frac{1}{5}\cos(5\pi x/2) - \frac{1}{7}\cos(7\pi x/2) + \cdots. \tag{4.17}$$

Table 4.3. Comparison of summands and partial sums—summations are over k odd, $1 \leq k \leq n$.

n	$n^{-1}\cos(.15\,n\pi)$	$\sum \pm k^{-1}\cos(.15\,k\pi)$	$\sum \pm \cos(.15\,k\pi)$
1	0.8910070	0.891007	0.891007
3	−0.0521448	0.838862	0.734572
5	−0.1414210	0.697440	0.027465
7	0.1410980	0.838539	1.015150
9	−0.0504434	0.788095	0.561163
11	−0.0412719	0.746823	0.107173
13	0.0759760	0.822799	1.094860
15	−0.0471405	0.775659	0.387754
17	−0.0092020	0.766457	0.231320
19	0.0468951	0.813352	1.122330
21	−0.0424289	0.770923	0.231320
23	0.0068015	0.777725	0.387754
25	0.0282843	0.806009	1.094860
27	−0.0365810	0.769428	0.107173
29	0.0156548	0.785083	0.561163
31	0.0146449	0.799728	1.015150
33	−0.0299299	0.769798	0.027465
35	0.0202031	0.790001	0.734572
37	0.0042280	0.794229	0.891007
39	−0.0228463	0.771382	0.000000
41	0.0217319	0.793114	0.891007
43	−0.0036380	0.789476	0.734572
45	−0.0157135	0.773763	0.027465
47	0.0210146	0.794777	1.015150
49	−0.0092651	0.785512	0.561163
51	−0.0089018	0.776611	0.107173
53	0.0186356	0.795246	1.094860
55	−0.0128565	0.782390	0.387754
57	−0.0027445	0.779645	0.231320
59	0.0151018	0.794747	1.122330

It does *not* converge absolutely; at $x = 0$ it becomes the alternating series

$$1 - \frac{1}{3} + \frac{1}{5} - \frac{1}{7} + \cdots .$$

On the other hand, for most values of x it does not alternate. Table 4.3. shows summands and partial sums when $x = 0.3$ (the significance of the last column will be explained after Abel's lemma). The sign of the summands displays an interesting pattern:

$$+ - - + - - + - - +$$
$$- + + - + + - + + -$$
$$+ - - + - - + - - +$$
$$- + + \cdots ,$$

but this is not an alternating series.

Joseph Fourier had shown that this particular series converges for all values of x, but it was Niels Henrik Abel (1802–1829) who, in 1826, published results on the analysis of such series, enabling the construction of a simple and useful test for the convergence of Fourier series. Abel was a Norwegian, born in Findö. In 1825, the Norwegian government paid for him to travel through Europe to meet and study with the great mathematicians of the time. He arrived in Paris in the summer of 1826. He had already done great mathematics: his primary accomplishment was the proof that the roots of a general fifth degree (quintic) polynomial cannot be expressed in terms of algebraic operations on the coefficients (for arbitrary quadratic, cubic, and biquadratic polynomials, the roots *can* be expressed in terms of the coefficients).

Abel stayed only six months in Paris. Almost all of the great mathematicians of the time were there, but it was difficult to get to know them and he felt very isolated. He described this world in a letter written to his former teacher and mentor, Bernt Holmboe, back in Norway:

> I am so anxious to hear your news! You have no idea. Don't let me down, send me a few consoling lines in this isolation where I find myself because, to tell you the truth, this, the busiest capital on the continent, now feels to me like a desert. I know almost no one; that's because in the summer months everyone is in the country, and no one can be found. Up until now, I have only met Mr. Legendre, Cauchy and Hachette, and several less famous but very capable mathematicians: Mr. Saigey, the editor of the Bulletin of the Sciences, and Mr. Lejeune-Dirichlet, a Prussian who came to see me the other day believing I was a compatriot. He's a mathematician of penetrating ability. With Mr. Legendre he has proven the impossibility of a solution in integers to the equation $x^5 + y^5 = z^5$, and some other very beautiful things. Legendre is extremely kind, but unfortunately very old. Cauchy is crazy, and there is no way of getting along with him, even though right now he is the only one who knows how mathematics should be done. What he is doing is excellent, but very confusing. At first I understood almost nothing; now I see a little more clearly. He is publishing a series of memoirs under the title *Exercises in mathematics*. I'm buying them and reading them assiduously. Nine issues have appeared since the beginning of this year. Cauchy is, at the moment,

the only one concerned with pure mathematics. Poisson, Fourier, Ampère, etc. are working exclusively on magnetism and other physical subjects. Mr. Laplace is writing nothing, I think. His last work was a supplement to his *Theory of probabilities*. I've often seen him at the Institute. He's a very jolly little man. Poisson is a short gentleman; he knows how to carry himself with a great deal of dignity; the same for Mr. Fourier. Lacroix is really old. Mr. Hachette is going to introduce me to some of these men.

Dirichlet was actually from Düren in Germany, near Bonn and Cologne where he had attended school. In 1822, at the age of seventeen, he had come to Paris to study. Dirichlet also left Paris at the end of 1826, going to a professorship at the university in Breslau. Both of these men play a role in the mathematics that we shall see in this section. Abel's traveling allowance was not generous, and most of it was sent home to support his widowed mother and younger siblings. His living conditions were mean. While in Paris, he was diagnosed with tuberculosis. In January of 1829, it killed him.

Abel's Lemma

Theorem 4.16 (Abel's Lemma). *We consider a series of the form*

$$a_1 b_1 + a_2 b_2 + a_3 b_3 + \cdots$$

where the b's are positive and decreasing: $b_1 \geq b_2 \geq b_3 \geq \cdots \geq 0$. Let S_n be the nth partial sum of the a's:

$$S_n = \sum_{k=1}^{n} a_k.$$

If these partial sums stay bounded—that is to say, if there is some number M for which $|S_n| \leq M$ for all values of n—then

$$\left| \sum_{k=1}^{n} a_k b_k \right| \leq M b_1. \tag{4.18}$$

We note that this theorem is applicable to Fourier series such as the series given in (4.17). We take

$$b_1 = 1, \quad b_2 = \frac{1}{3}, \quad b_3 = \frac{1}{5}, \quad b_4 = \frac{1}{7}, \quad \ldots,$$

$$a_1 = \cos(\pi x/2), \quad a_2 = -\cos(3\pi x/2), \quad a_3 = \cos(5\pi x/2), \quad a_4 = -\cos(7\pi x/2), \ldots.$$

While it will still take some work to prove that the partial sums of these a's stay bounded, a little experimentation shows that whenever x is rational the partial sums are periodic (see the last column of Table 4.3. and Figure 4.3).

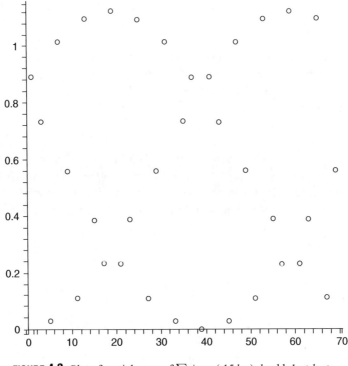

FIGURE 4.3. Plot of partial sums of $\sum \pm \cos(.15\,k\pi)$, k odd, $1 \le k \le n$.

We notice that the sum of the a's does not have to converge. When $x = 0$, we have $a_1 = 1, a_2 = -1, a_3 = 1, a_4 = -1, \dots$,

$$S_n = \sum_{k=1}^{n} (-1)^{k-1} = \begin{cases} 1, & \text{if } n \text{ is odd,} \\ 0, & \text{if } n \text{ is even.} \end{cases}$$

This series does not converge, but the partial sums are bounded by $M = 1$.

Proof: We use the fact that $a_k = S_k - S_{k-1}$ and do a little rearranging of the partial sum of the $a_k b_k$:

$$\sum_{k=1}^{n} a_k b_k = S_1 b_1 + (S_2 - S_1)b_2 + \cdots + (S_n - S_{n-1})b_n$$

$$= (S_1 b_1 + S_2 b_2 + \cdots + S_n b_n) - (S_1 b_2 + S_2 b_3 + \cdots + S_{n-1} b_n)$$

$$= S_1(b_1 - b_2) + S_2(b_2 - b_3) + \cdots + S_{n-1}(b_{n-1} - b_n) + S_n b_n$$

$$= \sum_{k=1}^{n-1} S_k(b_k - b_{k+1}) + S_n b_n. \tag{4.19}$$

We take absolute values and use the fact that the absolute value of a sum is less than or equal to the sum of the absolute values. We then use our assumptions that $|S_k| \le M$ and

$b_k - b_{k+1} \geq 0$:

$$\left| \sum_{k=1}^{n} a_k b_k \right| \leq \sum_{k=1}^{n-1} \left| S_k(b_k - b_{k+1}) \right| + |S_n b_n|$$

$$= \sum_{k=1}^{n-1} |S_k| (b_k - b_{k+1}) + |S_n| b_n$$

$$\leq \sum_{k=1}^{n-1} M(b_k - b_{k+1}) + M b_n$$

$$= M(b_1 - b_2 + b_2 - b_3 + \cdots + b_{n-1} - b_n + b_n)$$

$$= M b_1. \tag{4.20}$$

Q.E.D.

The Convergence Test

As it stands, Abel's lemma does not seem to be much help. The partial sums of the a's in our example never exceed $1.112233\ldots$, but since $b_1 = 1$, Abel's lemma only tells us that for every odd integer n,

$$\left| \cos(0.15\,\pi) - \frac{1}{3} \cos(0.45\pi) + \frac{1}{5} \cos(0.75\pi) - \frac{1}{7} \cos(1.05\pi) \right.$$

$$\left. + \cdots \pm \frac{1}{n} \cos(0.15\,n\pi) \right| \leq 1.112234. \tag{4.21}$$

That may be nice to know, but it does not prove convergence.

Abel proved his lemma in order to answer questions about the convergence of power series. His paper of 1826 was the first fully rigorous treatment of the binomial series for all values (real and complex) of x. We are *almost* to a convergence test for Fourier series, but it was Dirichlet who was the first to publicly point out how to pass from Abel's lemma to the convergence test that we shall apply to Fourier series.

The key is to use the Cauchy criterion. A series converges if and only if the partial sums can be brought arbitrarily close together by taking sufficient terms. The difference between two partial sums is simply a partial sum that starts much farther out. We observe that

$$\left| \sum_{k=m+1}^{n} a_k \right| = \left| \sum_{k=1}^{n} a_k - \sum_{k=1}^{m} a_k \right|$$

$$\leq |S_n| + |S_m|$$

$$\leq 2M. \tag{4.22}$$

If $T_n = \sum_{k=1}^{n} a_k b_k$ and the a's and b's satisfy the conditions of Abel's lemma, then

$$T_n - T_m = \sum_{k=m+1}^{n} a_k b_k \leq 2M b_{m+1}. \tag{4.23}$$

If the b's actually are approaching 0—notice that we did not need to assume this for Abel's lemma—then the difference between the partial sums can be made arbitrarily small and the series must converge.

Corollary 4.17 (Dirichlet's Test). *We consider a series of the form*

$$a_1b_1 + a_2b_2 + a_3b_3 + \cdots$$

where the b's are positive, decreasing, **and approaching 0,**

$$b_1 \geq b_2 \geq b_3 \geq \ldots \geq 0.$$

Let S_n be the nth partial sum of the a's:

$$S_n = \sum_{k=1}^{n} a_k.$$

If these partial sums stay bounded—that is to say, if there is some number M for which $|S_n| \leq M$ for all values of n—then the series converges.

Proof: We must demonstrate that we can win an ϵ–N game: given any error bound ϵ, we must always have a response N for which $N \leq m < n$ implies that

$$\left| \sum_{k=m+1}^{n} a_k b_k \right| < \epsilon.$$

As we have seen, Abel's lemma implies that

$$\left| \sum_{k=m+1}^{n} a_k b_k \right| \leq 2M b_{m+1}.$$

We use the fact that the b's are approaching 0. We can find an N for which $n > N$ implies that $b_n < \epsilon/2M$. This is our response. If $m + 1$ is larger than N, then

$$\left| \sum_{k=m+1}^{n} a_k b_k \right| \leq 2M b_{m+1} < 2M \frac{\epsilon}{2M} = \epsilon.$$

Q.E.D.

A Trigonometric Identity

We have proven that

$$\sum_{k=1}^{\infty} \frac{(-1)^{k-1}}{2k - 1} \cos\left(\frac{(2k - 1)\pi x}{2} \right)$$

converges when $x = 0.3$. What about other values of x? In order to apply Dirichlet's test, we must prove that once we have chosen x, the absolute value of the partial sum of the a's,

$$\left| \sum_{k=1}^{n} (-1)^{k-1} \cos\left(\frac{(2k - 1)\pi x}{2} \right) \right|,$$

stays bounded for all n.

We let y stand for $\pi x/2$. We would like to find a trigonometric identity that enables us to simplify

$$\cos y - \cos 3y + \cos 5y - \cdots - (-1)^n \cos(2n-1)y.$$

Such an identity can be found by using the fact that

$$\cos A + i \sin A = e^{iA}. \qquad (4.24)$$

If we add

$$i \sin y - i \sin 3y + i \sin 5y - \cdots - (-1)^n i \sin(2n-1)y$$

to our summation, we can rewrite it as a finite geometric series which we know how to sum:

$$[\cos y + i \sin y] - [\cos 3y + i \sin 3y] + [\cos 5y + i \sin 5y] - \cdots$$

$$-(-1)^n[\cos(2n-1)y + i \sin(2n-1)y]$$

$$= e^{iy} - e^{3iy} + e^{5iy} - \cdots - (-1)^n e^{(2n-1)iy}$$

$$= e^{iy} \left(1 + z + z^2 + \cdots + z^{n-1}\right)$$

$$= e^{iy} \frac{1 - z^n}{1 - z}, \qquad (4.25)$$

where $z = -e^{2iy}$.

We want to separate the real and imaginary parts of our formula. To do this, we need to make the denominator real. If we multiply it by $1 + z$, we get

$$1 - z^2 = z(z^{-1} - z)$$

$$= z(-\cos 2y + i \sin 2y + \cos 2y + i \sin 2y)$$

$$= z(2i \sin 2y).$$

Multiplying numerator and denominator by $1 + z$ yields

$$e^{iy} \frac{1 - z^n}{1 - z} = e^{iy} \frac{(1 - z^n)(1 + z)}{1 - z^2}$$

$$= -iz^{-1} e^{iy} \frac{(1 - z^n)(1 + z)}{2 \sin 2y}$$

$$= ie^{-iy} \frac{(1 - z^n)(1 - e^{2iy})}{2 \sin 2y}$$

$$= i \frac{(1 - z^n)(e^{-iy} - e^{iy})}{2 \sin 2y}$$

$$= i \frac{(1 - z^n)(-2i \sin y)}{2 \sin 2y}$$

$$= \frac{1 - z^n}{2\cos y}$$

$$= \frac{1 - (-1)^n[\cos(2ny) + i\sin(2ny)]}{2\cos y}$$

$$= \frac{1 - (-1)^n \cos 2ny}{2\cos y} + i\,\frac{(-1)^{n+1}\sin 2ny}{2\cos y}. \tag{4.26}$$

We get two identities. The real part will equal the sum of the cosines. The imaginary part will be i times the sum of the sines:

$$\sum_{k=1}^{n}(-1)^{k-1}\cos\left(\frac{(2k-1)\pi x}{2}\right) = \frac{1 - (-1)^n \cos(\pi nx)}{2\cos(\pi x/2)}, \tag{4.27}$$

$$\sum_{k=1}^{n}(-1)^{k-1}\sin\left(\frac{(2k-1)\pi x}{2}\right) = \frac{(-1)^{n+1}\sin(\pi nx)}{2\cos(\pi x/2)}. \tag{4.28}$$

Since $|\cos \pi nx| \le 1$, we have a bound on the partial sums of the a's:

$$\left|\sum_{k=1}^{n}(-1)^{k-1}\cos\left(\frac{(2k-1)\pi x}{2}\right)\right| \le |\sec(\pi x/2)|. \tag{4.29}$$

This gives us a bound provided x is not an integer. When x *is* an integer, we have that $\cos\left((2k-1)\pi x/2\right) = 0$, and so the partial sums are 0. Dirichlet's test can be invoked to imply the convergence of Fourier's series,

$$\cos(\pi x/2) - \frac{1}{3}\cos(3\pi x/2) + \frac{1}{5}\cos(5\pi x/2) - \frac{1}{7}\cos(7\pi x/2) + \cdots,$$

for *all* values of x. Lagrange was wrong. It does converge.

An Observation

We recall that Dirichlet's test requires a bound on the partial sums that does not depend on n. The bound that we found in inequality (4.29) satisfies this requirement, but it *does* depend on x. Choosing a specific value for x, we get a specific bound and so can apply Dirichlet's test. If we graph this bound as a function of x (Figure 4.4), we see that the graph is not bounded as x approaches an odd integer. Something very curious is happening; our bound is not bounded.

    This is significant. We do not need a bounded bound in order to get convergence, but we do need a bounded bound if we want *the sum of these continuous functions to again be continuous*. This strange behavior of the bound is directly related to the strange behavior of the Fourier series and the fact that it is discontinuous at odd integers. We shall explore this topic further in Chapter 5.

Exercises

The symbol (**M&M**) indicates that *Maple* and *Mathematica* codes for this problem are available in the **Web Resources** at **www.macalester.edu/aratra**.

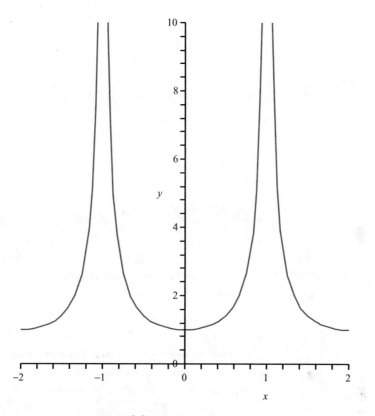

FIGURE 4.4. The graph of $|\sec \pi x/2|$.

4.4.1. (**M&M**) Construct a table of the values of the partial sums of

$$\sum_{k=1}^{\infty} (-1)^{k-1} \cos\left[(2k-1)\pi x/2\right]$$

when $x = 1/2, 2/3, 3/5$, and $5/18$. How large do you have to take n in order to see whatever patterns are present? Describe what you see. What happens when x is irrational?

4.4.2. Prove that if x is rational, then the values taken on by the partial sums in exercise 4.4.1 are periodic. What is the period? How many values do you go through before they repeat?

4.4.3. (**M&M**) Plot the values of the partial sums

$$T_n(x) = \sum_{k=1}^{n} \frac{(-1)^{k-1}}{2k-1} \cos\left[(2k-1)\pi x/2\right]$$

for $1 \leq n \leq 200$ when $x = 1/2, 2/3, 9/10$, and $99/100$. Describe what you see.

4.4.4. Let $T_n(x)$ be the partial sum defined in exercise 4.4.3. We are given an error bound of $\epsilon = 0.1$. For $x = 1/2, 2/3, 9/10$, and $99/100$, use Dirichlet's test to determine the size of a response N such that if $N \leq m < n$, then

$$|T_n - T_m| < 0.1.$$

4.4.5. Let $T_n(x)$ be the partial sum defined in exercise 4.4.3. We are given an error bound of $\epsilon = 0.001$. For $x = 1/2, 2/3, 9/10$, and $99/100$, use Dirichlet's test to determine the size of a response N such that if $N \leq m < n$, then

$$|T_n - T_m| < 0.001.$$

4.4.6. (**M&M**) Graph the partial sums

$$U_n(x) = \sum_{k=1}^{n} \frac{(-1)^{k-1}}{2k-1} \sin\left[(2k-1)\pi x/2\right]$$

over $-2 \leq x \leq 2$ for $n = 3, 6, 9$, and 12. Describe what you see. For what values of x does this series converge? Use Dirichlet's test to prove your assertion.

4.4.7. (**M&M**) Show that

$$z + z^2 + z^3 + \cdots + z^n = \frac{(1+z)(1-z^n)}{z^{-1} - z}.$$

Use this to prove that if x is not a multiple of π, then

$$\sin x + \sin 2x + \sin 3x + \cdots + \sin nx = \frac{\sin x}{2}\left(\frac{1 - \cos nx}{1 - \cos x}\right) + \frac{\sin nx}{2}. \qquad (4.30)$$

Graph this function of x over the domain $-\pi \leq x \leq \pi$ for $n = 10, 20, 100$, and 1000.

4.4.8. If x is held constant, does $\sin x + \sin 2x + \sin 3x + \cdots + \sin nx$ stay bounded for all values of n?

4.4.9. Prove that the series that we met in section 4.1,

$$\sum_{k=2}^{\infty} \frac{\sin(k/100)}{\ln k},$$

does converge.

4.4.10. Use Dirichlet's test to estimate the number of terms of the series in exercise 4.4.9 that we must take if we are to insure that the partial sums are within $\epsilon = 0.01$ of the value of this series.

4.4.11. (**M&M**) Graph the partial sums

$$V_n(x) = \sum_{k=1}^{n} \frac{(-1)^{k-1}}{k} \sin(k\pi x/2)$$

over $-2 \le x \le 2$ for $n = 3, 6, 9,$ and 12. Describe what you see. For what values of x does this series converge? Use Dirichlet's test to prove your assertion.

4.4.12. The term **radius of convergence** was coined because of its applicability to power series in which x is allowed to take on complex values. In this case, the series converges absolutely for every x inside the circle with center at the origin and radius R, and it diverges at every x outside this circle. The situation on the circle of radius R was investigated by Abel in his paper on the convergence of binomial series. For the following problems, $c_0 + \sum_{k=1}^{\infty} c_k x^k$ is a power series with radius of convergence R. For the sake of simplicity, we assume that $c_k R^k \ge c_{k+1} R^{k+1} \ge 0$ for all k.

a. Show that if $\lim_{k \to \infty} c_k R^k \ne 0$, then this series does not converge for any x on the circle of radius R.

b. Show that if $c_0 + \sum_{k=1}^{\infty} c_k R^k$ converges, then the power series converges for every x on the circle of radius R.

c. Show that if $\lim_{k \to \infty} c_k R^k = 0$ but $c_0 + \sum_{k=1}^{\infty} c_k R^k$ diverges, then the power series converges for every x on the circle of radius R except $x = R$.

5

Understanding Infinite Series

As we have seen, infinite series are not summations with lots of terms. Many of the nice things that hold for sums of functions fall apart when we look at series of functions. But they do not *always* fall apart. Sometimes, we can regroup or rearrange a series without affecting its value. Sometimes, an infinite summation of continuous functions will be continuous and can be differentiated or integrated following the rules that hold for finite summations. It is precisely because infinite series can often be treated as if they were finite sums that so much progress was made in the eighteenth century.

Trigonometric series such as those introduced by Joseph Fourier can be troublesome. Once Fourier's series were accepted, the question that came to the fore was why some series behaved well and others did not. By understanding why, it became possible to predict when a series could be rearranged without changing its value, when it was safe to differentiate each summand and claim that the resulting series was the derivative of the original series.

Most of the problems that we shall investigate reduce to a basic question: *when are we allowed to interchange limits?* Continuity, differentiation, and integration are each defined in terms of limiting processes. So is infinite summation. In exercise 1.2.6 of section 1.2, we saw that these limiting processes are not always interchangeable. If we use $\lim_{x \to 1^-}$ to designate the limit as x approaches 1 from the left, then

$$\lim_{x \to 1^-} \left(\lim_{n \to \infty} \frac{4}{\pi} \sum_{k=1}^{n} \frac{(-1)^{k-1}}{2k-1} \cos \frac{(2k-1)\pi x}{2} \right) = \lim_{x \to 1^-} 1$$
$$= 1; \qquad (5.1)$$

$$\lim_{n \to \infty} \left(\lim_{x \to 1^-} \frac{4}{\pi} \sum_{k=1}^{n} \frac{(-1)^{k-1}}{2k-1} \cos \frac{(2k-1)\pi x}{2} \right) = \lim_{n \to \infty} \frac{4}{\pi} \sum_{k=1}^{n} 0$$
$$= 0. \qquad (5.2)$$

171

FIGURE 5.1. $z = (x^2 - y^2)/(x^2 + y^2)$.

On a more basic level, we can see why interchanging limits is potentially dangerous. Consider the function defined by

$$f(x, y) = \frac{x^2 - y^2}{x^2 + y^2}, \quad (x, y) \neq (0, 0); \qquad f(0, 0) = 0. \tag{5.3}$$

If we want to find the value f approaches as (x, y) approaches $(0, 0)$, it makes a great deal of difference how we approach $(0, 0)$:

$$\lim_{x \to 0} \left(\lim_{y \to 0} \frac{x^2 - y^2}{x^2 + y^2} \right) = \lim_{x \to 0} \frac{x^2}{x^2} = 1, \tag{5.4}$$

$$\lim_{y \to 0} \left(\lim_{x \to 0} \frac{x^2 - y^2}{x^2 + y^2} \right) = \lim_{x \to 0} \frac{-y^2}{y^2} = -1. \tag{5.5}$$

The reason for the difference is transparent when we look at the graph of $z = f(x, y)$ (Figure 5.1). In the first case, we moved to the ridgeline when we took the limit $y \to 0$. We then stayed on this ridge as we approached the origin. In the second case, the limit $x \to 0$ took us to the bottom of the valley which we followed toward the origin.

   We shall see lots of examples where geometric representations will help us understand what can go wrong when limits are interchanged, but such pictures are not always available. When in doubt, the safest route will be to rely on the ϵ–δ and ϵ–N definitions.

5.1   Groupings and Rearrangements

In section 2.1 we saw that while the operation of addition is associative and commutative, these properties often disappear from infinite series. The standard example of the lack of associativity is the divergent series whose value Leibniz and Euler fixed at 1/2. If we were free to regroup at will, then we would have

$$1 - 1 + 1 - 1 + \cdots = (1 - 1) + (1 - 1) + \cdots \ = \ 0 \tag{5.6}$$
$$= 1 - (1 - 1) - (1 - 1) - \cdots \ = \ 1. \tag{5.7}$$

Regouping does seem to be allowed, however, when the series converges. For example, regrouping yields another series that represents $\ln 2$:

$$\begin{aligned}
\ln 2 &= 1 - \frac{1}{2} + \frac{1}{3} - \frac{1}{4} + \cdots \\[2mm]
&= \left(1 - \frac{1}{2}\right) + \left(\frac{1}{3} - \frac{1}{4}\right) + \cdots \\[2mm]
&= \frac{1}{1 \cdot 2} + \frac{1}{3 \cdot 4} + \frac{1}{5 \cdot 6} + \cdots \\[2mm]
&= 1 - \left(\frac{1}{2} - \frac{1}{3}\right) - \left(\frac{1}{4} - \frac{1}{5}\right) - \cdots \\[2mm]
&= 1 - \frac{1}{2 \cdot 3} - \frac{1}{4 \cdot 5} - \frac{1}{6 \cdot 7} - \cdots .
\end{aligned}$$

$$\tag{5.8}$$
$$\tag{5.9}$$

This is easily justified using our definition of convergence.

Theorem 5.1 (Regrouping Convergent Series). *Given a convergent series $a_1 + a_2 + a_3 + \cdots = A$, we can regroup consecutive summands without changing the value of the series. In other words, the associative law holds for convergent series.*

Proof: We consider an arbitrary series, $b_1 + b_2 + b_3 + \cdots$, formed from our original series by regrouping the summands so that b_1 is the sum of one or more of the initial terms in our series, b_2 is obtained by adding one or more of the next terms to come along, and so on. We are not allowed to change the order of the summands, only to regroup. For convenience, we choose k_1 to denote the subscript of the last a in b_1, k_2 the subscript of the last a in b_2, and so on:

$$\begin{aligned}
b_1 &= a_1 + a_2 + \cdots + a_{k_1}, \\
b_2 &= a_{k_1+1} + a_{k_1+2} + \cdots + a_{k_2}, \\
b_3 &= a_{k_2+1} + a_{k_2+2} + \cdots + a_{k_3}, \\
&\ \ \vdots
\end{aligned}$$

We note that $k_m \geq m$. If $S_n = a_1 + a_2 + \cdots + a_n$ is the partial sum of the a's and $T_m = b_1 + b_2 + \cdots + b_m$ is the partial sum of the b's, then

$$T_m \ = \ b_1 + b_2 + \cdots + b_m \ = \ a_1 + a_2 + \cdots + a_{k_m} \ = \ S_{k_m}. \tag{5.10}$$

We need to show that we can win the ϵ–N game for the b's. Given a positive error bound ϵ, we must find a response N so that $m \geq N$ implies that $|T_m - A| < \epsilon$. We know that

$$|T_m - A| = |S_{k_m} - A|,$$

and there is a response for the a's; call it N. We use this response. If m is greater than or equal to N, then so is k_m:

$$|T_m - A| = |S_{k_m} - A| < \epsilon.$$

Q.E.D.

Rearrangements

Determining when it is safe to rearrange an infinite series is going to be harder. We have seen convergent series that change their value when rearranged. For example,

$$1 - \frac{1}{3} + \frac{1}{5} - \frac{1}{7} + \frac{1}{9} - \frac{1}{11} + \cdots \tag{5.11}$$

converges to $\pi/4$. If we rearrange it to

$$1 + \frac{1}{5} - \frac{1}{3} + \frac{1}{9} + \frac{1}{13} - \frac{1}{7} + \frac{1}{17} + \frac{1}{21} - \frac{1}{11} + \cdots, \tag{5.12}$$

then we have exactly the same summands, but this series converges to a number near 1.0, well above $\pi/4$.

What is happening? Different rearrangements can lead to different answers, but not always. Taking the series

$$1 + \frac{1}{2} + \frac{1}{4} + \frac{1}{8} + \cdots$$

and experimenting with different rearrangements, we see that it *always* approaches 2. The fact that all of these summands are positive is significant. The previous series grew larger after rearrangement because we kept postponing those negative summands that would bring the value back down. But it is not just the presence of negative summands that spoils rearrangements. The series

$$1 - \frac{1}{2} + \frac{1}{4} - \frac{1}{8} + \frac{1}{16} - \frac{1}{32} + \cdots = \frac{2}{3} \tag{5.13}$$

will also be the same no matter how we rearrange it. What is the difference between the series in (5.11) and the one in (5.13)?

Bernhard Riemann

The first complete answer to this question did not appear until 1867 in Bernhard Riemann's posthumous work *Über die Darstellbarkeit einer Function durch eine trigonometrische Reihe* (*On the representability of a function by a trigonometric series*). Georg Friedrich Bernhard Riemann was born in 1826, the same fall that Abel and Dirichlet had met in Paris. Riemann entered Göttingen in 1846 at the age of nineteen, stayed one year, and then transferred to Berlin where he studied with Dirichlet (who had gone to Berlin in 1828), Eisenstein, Jacobi, and Steiner. He remained there two years, and then transferred back to Göttingen to finish his studies with Gauss, now an old man whose sparse praise for others became effusive when he saw Riemann's work.

In the fall of 1852, Dirichlet visited Gauss and spent much of the time talking about series with young Riemann. Riemann was later to credit many of his insights into trigonometric series to these discussions. Gauss died in 1855. Dirichlet succeeded to his chair at Göttingen. He had only four years to live. In 1859, Riemann became heir to what was now the world's most prestigious position in mathematics. Riemann died in 1866. Like Abel, he was killed by tuberculosis.

He had not published his work on trigonometric series. It was his friend and colleague Richard Dedekind who, after Riemann's death, recognized that it had to be published. It revolutionized our understanding of these series.

The Difference

As Riemann realized, the difference between the series in (5.11) and in (5.13) lies in the summation formed from just the positive (or negative) terms. In (5.13), the positive summands give a convergent series:

$$1 + \frac{1}{4} + \frac{1}{16} + \frac{1}{64} + \cdots = \frac{4}{3}.$$

No matter how we rearrange our positive terms, they will never take us above 4/3. The negative terms in this series will always subtract 2/3. Any rearrangement will leave us with $4/3 - 2/3 = 2/3$.

On the other hand, the series in (5.11) has positive terms whose sum diverges:

$$1 + \frac{1}{5} + \frac{1}{9} + \frac{1}{13} + \cdots .$$

The only thing that keeps the whole series from diverging is the presence of the negative terms that constantly compensate. The sum of the negative terms, taken on their own, must also diverge, otherwise they would not be sufficient to compensate for the diverging sum of positive terms. The difference between our series is that (5.13) is absolutely convergent and (5.11) is not.

> **Theorem 5.2 (Rearranging Convergent Series).** *Given an absolutely convergent series $a_1 + a_2 + a_3 + \cdots = A$, any rearrangement of this series yields another convergent series, converging to the same value. In other words, the commutative law holds for absolutely convergent series.*

Proof: We begin with the simplest case. We assume that all of the summands are positive. Let $b_1 + b_2 + b_3 + \cdots$ be a rearrangement of our series, still all positive but put into a different order. As we saw in section 4.1, a series of positive summands converges if and only if there is an upper bound on the set of partial sums. The least upper bound is then the value of this series.

We look at any partial sum of the b's:

$$T_n = b_1 + b_2 + \cdots + b_n.$$

The value A is larger than any partial sum of the a's, and we can always find some partial sum of the a's that includes everything in T_n, so A must be larger than T_n. This tells us that the partial sums of the b's are bounded. They must converge to something that is less than or equal to A:

$$b_1 + b_2 + b_3 + \cdots = B \leq A.$$

We turn this argument around. Any partial sum of the a's must be less than B, and so B is an upper bound for these partial sums. It follows that A is less than or equal to B,

$$A \leq B \leq A.$$

This tells us that $A = B$. The values are the same.

   We now consider the case where the a's are not all positive, but we do have absolute convergence:

$$|a_1| + |a_2| + |a_3| + \cdots = A^\star.$$

As before, we let $b_1 + b_2 + b_3 + \cdots$ be some rearrangement of the a's. By what we proved in the first part, we at least know that

$$|b_1| + |b_2| + |b_3| + \cdots = A^\star$$

is also convergent, and so the series of b's is absolutely convergent. Let $T_m = b_1 + b_2 + \cdots + b_m$ be a partial sum of the b's. We must show that we can win the ϵ–N game, that given any positive ϵ, we can always find a response N such that $m \geq N$ implies that $|A - T_m| < \epsilon$.

   We let $S_n = a_1 + a_2 + \cdots + a_n$ be a partial sum of the a's. We know that we can force S_n close to A:

$$\begin{aligned}
|A - T_m| &= |A - S_n + S_n - T_m| \\
&\leq |A - S_n| + |S_n - T_m|.
\end{aligned} \tag{5.14}$$

We choose an N_1 so that $n \geq N_1$ implies that $|A - S_n| < \epsilon/2$. It remains to choose an $n \geq N_1$ and a lower bound on m so that $|S_n - T_m|$ is less than $\epsilon/2$.

   Let $\mathcal{S}_n = \{a_1, a_2, \ldots, a_n\}$ and $\mathcal{T}_m = \{b_1, b_2, \ldots, b_m\}$. Using the Cauchy criterion and the fact that the sum of the b's converges absolutely, we can find an N_2 for which

$$m > \ell \geq N_2 \quad \text{implies that} \quad |b_{\ell+1}| + |b_{\ell+2}| + \cdots + |b_m| < \epsilon/2.$$

We find an n large enough so that

$$\mathcal{T}_{N_2} \subseteq \mathcal{S}_n.$$

If necessary, we make n slightly larger so that it is at least as big as N_1. We now find an N large enough that

$$\mathcal{S}_n \subseteq \mathcal{T}_N.$$

If $m \geq N$, then

$$\mathcal{T}_{N_2} \subseteq \mathcal{S}_n \subseteq \mathcal{T}_N \subseteq \mathcal{T}_m,$$

and so the summands that appear in $T_m - S_n$ lie in $T_m - T_{N_2}$. That is to say, they are taken from the b's with subscript less than or equal to m and strictly greater than N_2. This implies that

$$|T_m - S_n| \leq |b_{N_2+1}| + |b_{N_2+2}| + \cdots + |b_m| < \epsilon/2. \tag{5.15}$$

Our response is N.

Q.E.D.

Rearrangement with Conditional Convergence

If a series converges conditionally, then a rearrangement can change its value. How many possible values are there? Riemann realized that *every real number can be obtained by rearranging* such a series. You want to rearrange the series in (5.13) so that it converges to 1? We can do it. To 10.35? No problem. Sum it up to $-68 + \sqrt{3} - e^\pi$? A piece of cake.

> **Theorem 5.3 (Riemann Rearrangement Theorem).** *If the series $a_1 + a_2 + a_3 + \cdots$ converges conditionally, then for any real number r, we can find a rearrangement of this series that converges to r.*

Rather than a formal proof, we shall see how this is done with an example. We shall take the series

$$1 - \frac{1}{3} + \frac{1}{5} - \frac{1}{7} + \cdots$$

and rearrange it so that it converges to $\ln 2$ instead of $\pi/4$. We separate the positive summands from the negative ones, keeping their relative order, and note that if we added up just the positive summands, the series would diverge. The same must be true for the negative summands. This is important.

We have a target value T; in this case $T = \ln 2 = .693147\ldots$. We add positive summands until we are at or over this target:

$$1 = 1.$$

We know that sooner or later we shall reach or exceed the target because the positive summands diverge. We now add the negative summands until we are below the target:

$$1 - \frac{1}{3} = .6666\ldots.$$

Again, the negative summands diverge so that eventually we shall move below the target. We now put in more positive summands until we are above the target:

$$1 - \frac{1}{3} + \frac{1}{5} = .8666\ldots,$$

and then add negative summands until we are below again:

$$1 - \frac{1}{3} + \frac{1}{5} - \frac{1}{7} - \frac{1}{11} = .63290\ldots,$$

and so on. No matter how far along the series we may be, there are always enough positive or negative terms remaining to move us back to the other side of the target. Every summand will eventually be inserted.

$$\ln 2 = 1 - \frac{1}{3} + \frac{1}{5} - \frac{1}{7} - \frac{1}{11} + \frac{1}{9} - \frac{1}{15} + \frac{1}{13} - \frac{1}{19} - \frac{1}{23}$$
$$+ \frac{1}{17} - \frac{1}{27} + \frac{1}{21} - \frac{1}{31} - \frac{1}{35} + \cdots . \tag{5.16}$$

Let $b_1 + b_2 + b_3 + \cdots$ be the reordered series and $S_m = b_1 + b_2 + \cdots + b_m$ be the mth partial sum.

How do we know that we are converging to the target value and not just bouncing around it? We have to show that when we are given a positive error ϵ, we always have a response N such that

$$m \geq N \quad \text{implies that} \quad |T - S_m| < \epsilon.$$

We know that our original series converges, and so the summands are approaching zero. This means that there is a finite list of summands with absolute value greater than or equal to ϵ. We move down our reordered series, $b_1 + b_2 + b_3 + \cdots$, until we have included all summands with absolute value greater than or equal to ϵ. We continue moving down the series until we come to the next pair of consecutive partial sums that lie on opposite sides of the target T:

$$S_N < T \leq S_{N+1} \quad \text{or} \quad S_N \geq T > S_{N+1}.$$

The subscript N is our response. We know that all of the summands from here on have absolute value less than ϵ. Since T lies between S_N and S_{N+1}, it must differ from each by less than ϵ. Each time we add a new summand we are either moving closer to T (the difference is getting smaller) or we are jumping by an amount less than ϵ to the other side of T (we are still within ϵ of T).

Other Results

There is another result on rearranging series that we shall need later in this chapter. When can we add two series by adding the corresponding summands? When are we allowed to say that

$$(a_1 + a_2 + a_3 + \cdots) + (b_1 + b_2 + b_3 + \cdots) = (a_1 + b_1) + (a_2 + b_2) + (a_3 + b_3) + \cdots ? \tag{5.17}$$

> **Theorem 5.4 (Addition of Series).** *If* $a_1 + a_2 + a_3 + \cdots = A$ *and* $b_1 + b_2 + b_3 + \cdots = B$ *both converge, then* $(a_1 + b_1) + (a_2 + b_2) + (a_3 + b_3) + \cdots$ *converges to* $A + B$.

Proof: Let

$$S_n = a_1 + a_2 + \cdots + a_n, \quad T_m = b_1 + b_2 + \cdots + b_m.$$

Given an error bound ϵ, we must find an N such that

$$n \geq N \quad \text{implies that} \quad |(A + B) - (S_n + T_n)| < \epsilon.$$

We use the fact that

$$|(A + B) - (S_n + T_n)| \leq |A - S_n| + |B - T_n|,$$

and split our allowable error between these differences. We find an N_1 such that

$$n \geq N_1 \quad \text{implies that} \quad |A - S_n| < \epsilon/2$$

and an N_2 such that

$$n \geq N_2 \quad \text{implies that} \quad |B - T_n| < \epsilon/2.$$

Our response is the larger of N_1 and N_2.

Q.E.D.

We need one more basic result.

> **Theorem 5.5 (Distributive Law for Series).** *If $a_1 + a_2 + a_3 + \cdots$ converges to A and if c is any constant, then $ca_1 + ca_2 + ca_3 + \cdots$ converges to cA.*

The proof of this theorem is similar and is left as an exercise.

Exercises

The symbol (**M&M**) indicates that *Maple* and *Mathematica* codes for this problem are available in the **Web Resources** at **www.macalester.edu/aratra**.

5.1.1. Evaluate the series

$$1 - \frac{1}{3} + \frac{1}{9} - \frac{1}{27} + \frac{1}{81} - \frac{1}{243} + \cdots$$

in two different ways: first as a geometric series with initial term 1 and ratio $-1/3$, then by combining each positive term with the succeeding negative term.

5.1.2. (**M&M**) Use regrouping to evaluate the series

$$1 + \frac{1}{2} - \frac{1}{4} + \frac{1}{8} + \frac{1}{16} - \frac{1}{32} + \frac{1}{64} + \frac{1}{128} - \frac{1}{256} + \cdots.$$

Use numerical calculation to check your answer.

5.1.3. Prove that

$$1 + \frac{1}{5} + \frac{1}{9} + \frac{1}{13} + \cdots$$

diverges.

5.1.4. (**M&M**) The series in (5.12) can be regrouped so that it forms a series of positive summands:

$$\left(1 + \frac{1}{5} - \frac{1}{3}\right) + \left(\frac{1}{9} + \frac{1}{13} - \frac{1}{7}\right) + \left(\frac{1}{17} + \frac{1}{21} - \frac{1}{11}\right) + \cdots$$
$$+ \left(\frac{1}{8n-7} + \frac{1}{8n-3} - \frac{1}{4n-1}\right) + \cdots$$
$$= \frac{13}{15} + \frac{37}{819} + \frac{61}{3927} + \cdots + \frac{24n-11}{(8n-7)(8n-3)(4n-1)} + \cdots .$$

Calculate the partial sum of the first thousand terms of this series and so find a lower bound for the value of the rearrangement in (5.12).

5.1.5. (**M&M**) We can also regroup the series in (5.12) so that it is 6/5 plus a series of negative summands:

$$1 + \frac{1}{5} - \left(\frac{1}{3} - \frac{1}{9} - \frac{1}{13}\right) - \left(\frac{1}{7} - \frac{1}{17} - \frac{1}{21}\right) - \cdots$$
$$- \left(\frac{1}{4n-1} - \frac{1}{8n+1} - \frac{1}{8n+5}\right) - \cdots .$$

Find the general summand of this regrouping and calculate the partial sum of the first thousand terms of this new series, thereby finding an upper bound for the value of the rearrangement in (5.12).

5.1.6. Consider the following two evaluations of the series $1/2 \cdot 3 + 1/3 \cdot 4 + \cdots + 1/(k+1)(k+2) + \cdots$. Which of these is correct? Where is the flaw in the one that is wrong? Justify the reasoning for the one that is correct.

$$\frac{1}{2 \cdot 3} + \frac{1}{3 \cdot 4} + \cdots + \frac{1}{(k+1)(k+2)}$$
$$= \left(\frac{1}{2} - \frac{1}{3}\right) + \left(\frac{1}{3} - \frac{1}{4}\right) + \cdots + \left(\frac{1}{k+1} - \frac{1}{k+2}\right) + \cdots = \frac{1}{2},$$
$$\frac{1}{2 \cdot 3} + \frac{1}{3 \cdot 4} + \cdots + \frac{1}{(k+1)(k+2)}$$
$$= \left(1 - \frac{5}{6}\right) + \left(\frac{5}{6} - \frac{3}{4}\right) + \cdots + \left(\frac{k+3}{2k+2} - \frac{k+4}{2k+4}\right) + \cdots = 1.$$

5.1.7. (**M&M**) Find the first 200 summands in the rearrangements of

$$1 - \frac{1}{2} + \frac{1}{3} - \frac{1}{4} + \frac{1}{5} - \frac{1}{6} + \cdots$$

that approach 1.5 and 0.5, respectively. Is it possible to pick up any patterns that will continue forever?

5.1.8. (**M&M**) Find the first 200 summands in the rearrangements of

$$1 - \frac{1}{3} + \frac{1}{5} - \frac{1}{7} + \frac{1}{9} - \frac{1}{11} + \cdots$$

that approach 1.5 and 0.5, respectively. Is it possible to pick up any patterns that will continue forever?

5.1.9. Find a *different* rearrangement of the series in exercise 5.1.7 that approaches 1.5.

5.1.10. If a series converges conditionally, how many distinct rearrangements of that series are there that yield the same value? Can you describe all possible rearrangements that yield the same value?

5.1.11. Find an example of two divergent series $a_1 + a_2 + a_3 + \cdots$ and $b_1 + b_2 + b_3 + \cdots$ for which the sum $(a_1 + b_1) + (a_2 + b_2) + (a_3 + b_3) + \cdots$ converges.

5.1.12. Is it possible for $a_1 + a_2 + a_3 + \cdots$ to converge, $b_1 + b_2 + b_3 + \cdots$ to diverge, and the sum $(a_1 + b_1) + (a_2 + b_2) + (a_3 + b_3) + \cdots$ to converge? Either give an example of such series or prove it is impossible.

5.1.13. Prove Theorem 5.5.

5.1.14. Prove that if a series converges conditionally, then we can find a rearrangement that diverges.

5.2   Cauchy and Continuity

On page 120 of his *Cours d'analyse*, Cauchy proves his first theorem about infinite series. Let S be an infinite series of continuous functions,

$$S(x) = f_1(x) + f_2(x) + f_3(x) + \cdots,$$

let S_n be the partial sum of the first n terms,

$$S_n(x) = f_1(x) + f_2(x) + \cdots + f_n(x),$$

and let R_n be the remainder,

$$R_n(x) = S(x) - S_n(x) = f_{n+1}(x) + f_{n+2}(x) + \cdots.$$

Just as questions of convergence are investigated by considering the sequence of partial sums, so also in this chapter we shall look at questions of continuity, differentiability, and integrability in terms of the sequence of partial sums. Cauchy remarks that S_n, a finite sum of continuous functions, must be continuous, and then goes on to state:

> Let us consider the changes in these three functions when we increase x by an infinitely small value α. For all possible values of n, the change in $S_n(x)$ will be infinitely small; the change in $R_n(x)$ will be as insignificant[1] as the size of $R_n(x)$ when n is made very large. It follows that the change in the function $S(x)$ can

---
[1] Literally: insensible.

only be an infinitely small quantity. From this remark, we immediately deduce the following proposition:

THEOREM I—*When the terms of a series are functions of a single variable x and are continuous with respect to this variable in the neighborhood of a particular value where the series converges, the sum S(x) of the series is also, in the neighborhood of this particular value, a continuous function of x.*

Cauchy has proven that any infinite series of continuous functions is continuous.

There is only one problem with this theorem. It is wrong. The Fourier series

$$\cos(\pi x/2) - \frac{1}{3}\cos(3\pi x/2) + \frac{1}{5}\cos(5\pi x/2) - \frac{1}{7}\cos(7\pi x/2) + \cdots$$

is an infinite series of continuous functions. As we have seen, it is not continuous at $x = 1$. No one seems to have noticed this contradiction until 1826 when Niels Abel pointed it out in a footnote to his paper on infinite series.

Even though Dirichlet definitively established the validity of Fourier series in 1829, it was 1847 before anyone was able to make progress on resolving the contradiction between Cauchy's theorem and the properties of Fourier series. The first light was shed by George Stokes (1819–1903). A year later, Dirichlet's student Phillip Seidel (1821–1896) went a long way toward clarifying Cauchy's error. Cauchy corrected his error in 1853, but the conditions required for the continuity of an infinite series were not generally recognized until the 1860s when Weierstrass began to emphasize their importance.

Cauchy's Proof

Before we search for the flaw in Cauchy's argument, we need to restate it more carefully using our definitions of continuity and convergence. The simple act of putting it into precise language may reveal the problem.

To prove the continuity of $S(x)$ at $x = a$, we must show that for any given $\epsilon > 0$, there is a δ such that as long as x stays within δ of a, $S(x)$ will be within ϵ of $S(a)$:

$$|x - a| < \delta \quad \text{implies that} \quad |S(x) - S(a)| < \epsilon.$$

Cauchy's analysis begins with the observation that

$$|S(x) - S(a)| = |S_n(x) + R_n(x) - S_n(a) - R_n(a)|$$
$$\leq |S_n(x) - S_n(a)| + |R_n(x)| + |R_n(a)|. \tag{5.18}$$

We can divide the allowable error three ways, giving $\epsilon/3$ to each of the terms in the last line. The continuity of $S_n(x)$ guarantees that we can make

$$|S_n(x) - S_n(a)| < \epsilon/3.$$

The convergence of $S(x)$ at $x = a$ and at all points close to a tells us that the remainders can each be made arbitrarily small:

$$|R_n(x)| < \epsilon/3 \quad \text{and} \quad |R_n(a)| < \epsilon/3.$$

If you still do not see what is wrong with this proof, you should not be discouraged. It took mathematicians over a quarter of a century to find the error.

An Example

It is easiest to see where Cauchy went wrong by analyzing an example of an infinite series of continuous functions that is itself discontinuous. Fourier series are rather complicated. We shall use a simpler example:

$$S(x) = \sum_{k=1}^{\infty} \frac{x^2}{(1 + kx^2)(1 + (k-1)x^2)}. \tag{5.19}$$

Each of the summands is a continuous function of x. The partial sums are particularly easy to work with. We observe that

$$\frac{x^2}{(1 + kx^2)(1 + (k-1)x^2)} = \frac{1}{1 + (k-1)x^2} - \frac{1}{1 + kx^2},$$

and therefore

$$S_n(x) = \left(1 - \frac{1}{1 + x^2}\right) + \left(\frac{1}{1 + x^2} - \frac{1}{1 + 2x^2}\right)$$
$$+ \left(\frac{1}{1 + 2x^2} - \frac{1}{1 + 3x^2}\right) + \cdots + \left(\frac{1}{1 + (n-1)x^2} - \frac{1}{1 + nx^2}\right)$$
$$= 1 - \frac{1}{1 + nx^2}$$
$$= \frac{nx^2}{1 + nx^2}. \tag{5.20}$$

We see that $S_n(0) = 0$ for all values of n, and so $S(0) = 0$. If x is not zero, then

$$S_n(x) = \frac{x^2}{n^{-1} + x^2}$$

which approaches 1 as n gets large,

$$S(x) = 1, \quad x \neq 0.$$

The series is definitely discontinuous at $x = 0$.

We can see what is happening if we look at the graphs of the partial sums (Figure 5.2). As n increases, the graphs become steeper near $x = 0$. In the limit, we get a vertical jump.

Where is the Mistake?

Cauchy must be making some unwarranted assumption in his proof. To see what it might be, we return to his proof and use our specific example:

$$S(x) = \begin{cases} 0, & \text{if } x = 0, \\ 1, & \text{if } x \neq 0, \end{cases} \tag{5.21}$$

$$S_n(x) = \frac{nx^2}{1 + nx^2}, \tag{5.22}$$

$$R_n(x) = \begin{cases} 0, & \text{if } x = 0, \\ 1/(1 + nx^2), & \text{if } x \neq 0. \end{cases} \tag{5.23}$$

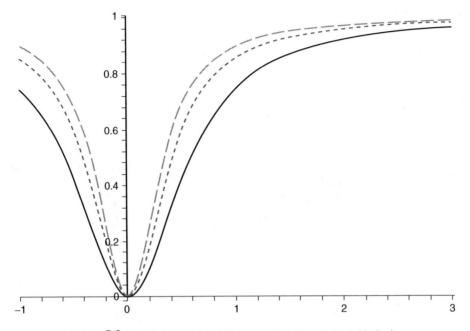

FIGURE 5.2. Graphs of $S_3(x)$ (solid), $S_6(x)$ (dotted), and $S_9(x)$ (dashed).

The critical point at which we want to investigate continuity is $a = 0$. If x is close to but not equal to 0, then inequality (5.18) becomes

$$|S(x) - S(0)| \leq |S_n(x) - S_n(0)| + |R_n(x)| + |R_n(0)|$$

$$= \left| \frac{nx^2}{1 + nx^2} - 0 \right| + \left| \frac{1}{1 + nx^2} \right| + |0| \tag{5.24}$$

$$= \frac{nx^2}{1 + nx^2} + \frac{1}{1 + nx^2}$$

$$= 1. \tag{5.25}$$

Something is wrong with the assertion that we can make each of the pieces in line (5.24) arbitrarily small.

We make the first piece small by taking x close to 0. How close does it have to be? We want

$$\frac{nx^2}{1 + nx^2} < \frac{\epsilon}{3}. \tag{5.26}$$

Multiplying through by $1 + nx^2$ and then solving for x^2, we see that

$$nx^2 < (\epsilon/3)(1 + nx^2),$$
$$x^2(n - n\epsilon/3) < \epsilon/3,$$
$$x^2 < \frac{\epsilon/3}{n - n\epsilon/3} = \frac{\epsilon}{n(3 - \epsilon)},$$
$$|x| < \sqrt{\epsilon/(3n - \epsilon n)}. \tag{5.27}$$

The size of our response δ depends on n. As n gets larger, δ must get smaller. This makes sense if we think of the graph in Figure 5.2. If $\epsilon = 0.1$ so that we want $S_n(x) < 0.1$, we need to take a much tighter interval when $n = 9$ than we do when $n = 3$.

To make the second piece small,

$$\frac{1}{1 + nx^2} < \frac{\epsilon}{3}, \tag{5.28}$$

we have to take a large value of n. If we solve this inequality for n, we see that we need

$$1 < (\epsilon/3)(1 + nx^2),$$
$$\frac{3}{\epsilon} - 1 < nx^2,$$
$$\frac{3 - \epsilon}{\epsilon x^2} < n. \tag{5.29}$$

The size of n depends on x. As $|x|$ gets smaller, n must be taken larger. This also makes sense when we look at the graph. If we take an x that is very close to 0, then we need a very large value of n before we are near $S(x) = 1$.

Here is our difficulty. The size of x depends on n, and the size of n depends on x. We can make the first piece small by making x small, but that increases the size of the second piece. If we increase n to make the second piece small, the first piece increases. We are in a vicious cycle. *We cannot make both pieces small simultaneously.*

Fixing it up with Uniform Convergence

Part of the reason that Cauchy made his mistake is that many infinite series of continuous functions *are* continuous. Having found what is wrong with Cauchy's proof, we can attempt to find criteria that will identify infinite series that are continuous. If we are going to be able to break our cycle, then either the size of the first piece does not depend on n or the size of the second piece does not depend on x.

The usual solution is the second: that the size of $|R_n(x)|$ does not depend on x. When this happens, we say that the series is **uniformly convergent**. Specifically, we have the following definition.

Definition: uniform convergence

Given a series of functions, $S = f_1 + f_2 + f_3 + \cdots$, which converges for all x in an interval I, we let $\{S_1, S_2, S_3, \ldots\}$ denote the sequence of partial sums: $S_n = f_1 + f_2 + \cdots + f_n$. We say that this series **converges uniformly over I** if given any positive error bound ϵ, we always have a response N such that

$$n \geq N \quad \text{implies that} \quad |S(x) - S_n(x)| < \epsilon.$$

The same N must work for all $x \in I$.

Graphically, this implies that if we put an **envelope** extending distance ϵ above and below S (Figure 5.3), then there is a response N such that $n \geq N$ implies that the graph of S_n lies entirely inside this envelope. Using the example from equation (5.19) (Figure 5.4), we see

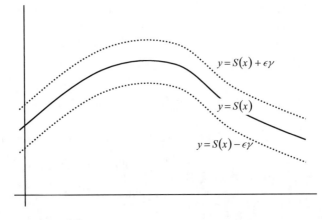

FIGURE 5.3. The ϵ envelope around the graph of $y = S(x)$.

that when ϵ is small (less than 1/2), none of the partial sums stay inside the ϵ envelope. This example was not uniformly convergent.

Proof: We repeat Cauchy's proof, being careful to choose n first. We choose any $a \in (\alpha, \beta)$ and use inequality (5.18):

$$|S(x) - S(a)| \leq |S_n(x) - S_n(a)| + |R_n(x)| + |R_n(a)|.$$

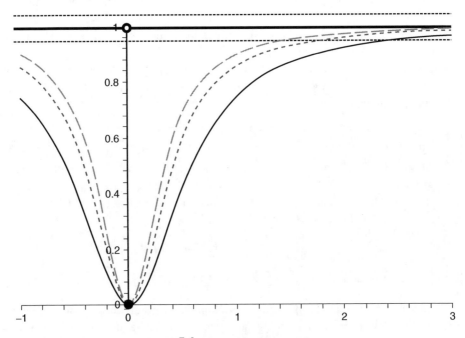

FIGURE 5.4. Figure 5.2 with ϵ envelope.

> **Theorem 5.6 (Continuity of Infinite Series).** *If* $S = f_1 + f_2 + f_3 + \cdots$ *converges uniformly over the interval* (α, β), *and if each of the summands is continuous at every point in* (α, β), *then the series* S *is continuous at every point in* (α, β).

As before, we assign a third of our error bound to each of these terms. Using the uniform convergence, we can find an n for which both $|R_n(x)|$ and $|R_n(a)|$ are less than $\epsilon/3$, regardless of our choice of x. Once n is chosen, we turn to the first piece and use the continuity of $S_n(x)$ to find a δ for which

$$|x - a| < \delta \quad \text{implies that} \quad |S_n(x) - S_n(a)| < \epsilon/3.$$

This is now the δ that we can use as our response,

$$|x - a| < \delta \quad \text{implies that} \quad |S(x) - S(a)| < \epsilon/3 + \epsilon/3 + \epsilon/3 = \epsilon.$$

Q.E.D.

A Nice Example

As an example of the use of uniform convergence, we consider the **dilogarithm** shown in Figure 5.5:

$$\text{Li}_2(x) = \sum_{k=1}^{\infty} \frac{x^k}{k^2}. \tag{5.30}$$

This series has radius of convergence 1. Using either Gauss's test or an appropriate comparison test, we see that it converges for all $x \in [-1, 1]$.

> **Web Resource:** To learn more about the dilogarithm, go to **The Dilogarithm**.

This series converges uniformly over $[-1, 1]$ as we can see by comparing the remainder $R_n(x)$ with a bounding integral:

$$
\begin{aligned}
|R_n(x)| &\leq \frac{|x|^{n+1}}{(n+1)^2} + \frac{|x|^{n+2}}{(n+2)^2} + \frac{|x|^{n+3}}{(n+3)^2} + \cdots \\
&\leq \frac{1}{(n+1)^2} + \frac{1}{(n+2)^2} + \frac{1}{(n+3)^2} + \cdots \\
&< \int_n^{\infty} \frac{dt}{t^2} = \frac{1}{n}.
\end{aligned}
\tag{5.31}
$$

Given an error bound ϵ, we can respond with any integer $N \geq 1/\epsilon$. If $n \geq N$, then $|R_n(x)| < 1/n \leq \epsilon$ regardless of which x we choose from $[-1, 1]$.

Theorem 5.6 assumes that we are working over an *open* interval. It only implies that $\text{Li}_2(x)$ is continuous at every $x \in (-1, 1)$. The behavior of this function at the endpoints is left as an exercise.

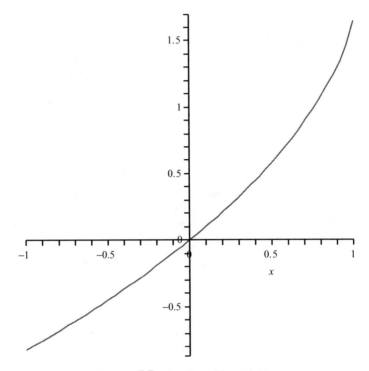

FIGURE 5.5. The dilogarithm, $Li_2(x)$.

Continuity without Uniform Convergence

Uniform convergence is sufficient to patch up Cauchy's theorem. It is not necessary. It is possible that the series is continuous even when we do not have uniform convergence. An example of this is the series

$$S(x) = \sum_{k=1}^{\infty} \frac{x + x^3(k - k^2)}{(1 + k^2 x^2)(1 + (k-1)^2 x^2)}. \tag{5.32}$$

Observing that

$$\frac{x + x^3(k - k^2)}{(1 + k^2 x^2)(1 + (k-1)^2 x^2)} = \frac{kx}{1 + k^2 x^2} - \frac{(k-1)x}{1 + (k-1)^2 x^2},$$

we can evaluate the partial sums,

$$S_n(x) = \frac{x}{1 + x^2} + \left(\frac{2x}{1 + 4x^2} - \frac{x}{1 + x^2}\right) + \left(\frac{3x}{1 + 9x^2} - \frac{2x}{1 + 4x^2}\right)$$

$$+ \cdots + \left(\frac{nx}{1 + n^2 x^2} - \frac{(n-1)x}{1 + (n-1)^2 x^2}\right)$$

$$= \frac{nx}{1 + n^2 x^2}. \tag{5.33}$$

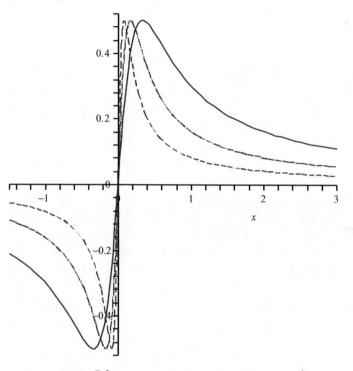

FIGURE 5.6. $S_n(x) = nx/(1 + n^2x^2)$, $n = 3, 6, 12$.

For *any* value of x, these partial sums approach 0 as n increases and so

$$S(x) = 0 \quad \text{for all } x. \tag{5.34}$$

Graphing our partial sums and a small ϵ envelope (Figure 5.6), we see that we do not have uniform convergence over any interval containing $x = 0$. The remainder is

$$R_n(x) = \frac{-nx}{1 + n^2x^2}.$$

If we have been given an error $\epsilon < 1/2$, then we can always find a large integer n for which $1/n$ or $-1/n$ is inside the interval, but

$$|R_n(\pm 1/n)| = \left| \frac{\pm 1}{1 + 1} \right| = \frac{1}{2} > \epsilon.$$

Nevertheless, the constant function $S(x) = 0$ *is* continuous.

Exercises

The symbol $\boxed{\textbf{M\&M}}$ indicates that *Maple* and *Mathematica* codes for this problem are available in the **Web Resources** at **www.macalester.edu/aratra**.

5.2.1. (**M&M**) Graph the partial sums to 3, 6, 9, and 12 terms of

$$S(x) = \sum_{n=1}^{\infty} x^2(1 - x^2)^{n-1}, \quad -1 \le x \le 1.$$

Either prove that this series converges uniformly on $[-1, 1]$, or explain why it cannot converge uniformly over this interval.

5.2.2. (**M&M**) Calculate the partial sums of the series in equation (5.19):

$$S_n(x) = \sum_{k=1}^{n} \frac{x^2}{(1 + kx^2)(1 + (k-1)x^2)}$$

when $x = 1/10$, $1/100$, and $1/1000$. How many terms are needed in order to get the value of $S_n(x)$ within 0.01 of $S(x) = 1$? Explain the reasoning that leads to your answer.

5.2.3. (**M&M**) Calculate the partial sums of the series in equation (5.32):

$$S_n(x) = \sum_{k=1}^{n} \frac{x + x^3(k - k^2)}{(1 + k^2x^2)(1 + (k-1)^2x^2)}$$

when $x = 1/10$, $1/100$, and $1/1000$. How many terms are needed in order to get the value of $S_n(x)$ within 0.01 of $S(x) = 0$? Explain the reasoning that leads to your answer.

5.2.4. Consider the power series expansion for the sine:

$$\sin x = x - \frac{x^3}{3!} + \frac{x^5}{5!} - \cdots = \sum_{k=1}^{\infty} (-1)^{k-1} \frac{x^{2k-1}}{(2k-1)!}.$$

Show that this series converges uniformly over the interval $[-\pi, \pi]$. How many terms must you take if the partial sum is to lie within the ϵ envelope when $\epsilon = 1/2$, $1/10$, $1/100$?

5.2.5. Prove that the power series expansion for the sine converges uniformly over the interval $[-2\pi, 2\pi]$. How many terms must you take if the partial sum is to lie within the ϵ envelope when $\epsilon = 1/2$, $1/10$, $1/100$?

5.2.6. Is the power series expansion for the sine uniformly convergent over the set of all real numbers? Explain your answer.

5.2.7. Euler proved (see Appendix A.3) that

$$\text{Li}_2(1) = 1 + \frac{1}{4} + \frac{1}{9} + \frac{1}{16} + \cdots = \frac{\pi^2}{6}.$$

Find the value of $\text{Li}_2(-1)$.

5.2.8. What is the relationship between the series

$$\sum_{k=1}^{\infty} \frac{x^k}{k}$$

and the natural logarithm? Why do you think that $\sum_{k=1}^{\infty} x^k/k^2$ is called the *di*logarithm?

5.2.9. Prove that as x approaches 1 or -1 from inside the interval $(-1, 1)$, the value of $\text{Li}_2(x)$ approaches $\text{Li}_2(1)$ or $\text{Li}_2(-1)$, respectively. For $a = 1$, you have to show that for any given error bound ϵ, there is always a response δ such that

$$1 - \delta < x < 1 \quad \text{implies that} \quad |\text{Li}_2(1) - \text{Li}_2(x)| < \epsilon.$$

How large is δ when $\epsilon = 1/4$?

5.2.10. The graph of $\text{Li}_2(x)$ and the fact that it is analogous to the natural logarithm both suggest that we should be able to define this function for values of x that are less than -1. Show that if term-by-term integration of power series is allowed over the domain of convergence, then

$$\text{Li}_2(x) = \int_{x}^{0} \frac{\ln(1 - t)}{t}\, dt$$

for $-1 \leq x \leq 1$, and this integral is defined for all $x < 1$.

5.2.11. We arrived at the notion of uniform convergence by breaking the second part of the cycle that we encountered on page 185. We found an N that was independent of x for which $n \geq N$ implies that $|R_n(x)| < \epsilon/3$. Discuss what it would mean to break the first part of the cycle, to find a δ independent of n for which $|x - a| < \delta$ implies that $|S_n(x) - S_n(a)| < \epsilon/3$. Find an example of such a series. Why is this not the route that is usually chosen?

5.3   Differentiation and Integration

As we saw in section 3.2, it is not always safe to differentiate a series by differentiating each term. For example, the Fourier series

$$F(x) = \frac{4}{\pi}\left[\cos(\pi x/2) - \frac{1}{3}\cos(3\pi x/2) + \frac{1}{5}\cos(5\pi x/2) - \frac{1}{7}\cos(7\pi x/2) + \cdots\right],$$

is equal to 1 for $-1 < x < 1$. Its derivative is zero at each x between -1 and 1. If we try to differentiate each term, we obtain the series

$$-2\left[\sin(\pi x/2) - \sin(3\pi x/2) + \sin(5\pi x/2) - \sin(7\pi x/2) + \cdots\right],$$

which does not converge unless x is an even integer.

Worse than this can happen. Trying to differentiate a series by differentiating each summand can give us a series that converges to the *wrong* answer. Consider the series

$$F(x) = \sum_{k=1}^{\infty} f_k(x) = \sum_{k=1}^{\infty} \frac{x^3}{(1+kx^2)(1+(k-1)x^2)}. \tag{5.35}$$

The derivative of the kth summand is

$$f_k'(x) = \frac{3x^2}{(1+kx^2)(1+(k-1)x^2)} - \frac{2kx^4}{(1+kx^2)^2(1+(k-1)x^2)}$$
$$- \frac{2(k-1)x^4}{(1+kx^2)(1+(k-1)x^2)^2}.$$

We see that

$$f_k'(0) = 0$$

for all values of k. If we try to find $F'(0)$ by differentiating each term and then setting $x = 0$, we get 0—the wrong answer.

To see what the derivative should be, we look at the partial sums and observe that

$$\frac{x^3}{(1+kx^2)(1+(k-1)x^2)} = \frac{kx^3}{1+kx^2} - \frac{(k-1)x^3}{1+(k-1)x^2}.$$

We let $F_n(x)$ denote the partial sum of the first n summands:

$$F_n(x) = f_1(x) + f_2(x) + \cdots + f_n(x)$$
$$= \frac{x^3}{1+x^2} + \left(\frac{2x^3}{1+2x^2} - \frac{x^3}{1+x^2}\right) + \left(\frac{3x^3}{1+3x^2} - \frac{2x^3}{1+2x^2}\right)$$
$$+ \cdots + \left(\frac{nx^3}{1+nx^2} - \frac{(n-1)x^3}{1+(n-1)x^2}\right)$$
$$= \frac{nx^3}{1+nx^2}. \tag{5.36}$$

As n gets large, $F_n(x)$ approaches x for all values of x. Our series is

$$F(x) = x, \qquad F'(x) = 1. \tag{5.37}$$

Figure 5.7 shows what is happening. We see that it is possible for the slope of the partial sums at a particular point to bear no relationship whatsoever to the slope of the infinite series. The series that we are using even converges uniformly. The graph suggests that it should. We can confirm this algebraically. Given an error bound ϵ, we want to find N (independent of x) for which $n \geq N$ implies that

$$\epsilon > |F(x) - F_n(x)| = \left|x - \frac{nx^3}{1+nx^2}\right| = \left|\frac{x}{1+nx^2}\right| = \frac{|x|}{1+nx^2}. \tag{5.38}$$

Solving this inequality for n, we see that we want

$$n > \frac{|x| - \epsilon}{\epsilon x^2}. \tag{5.39}$$

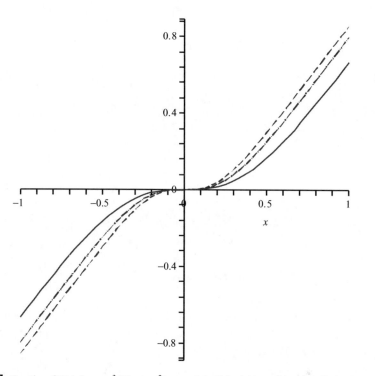

FIGURE 5.7. Graphs of $F_n(x) = nx^3/(1 + nx^2)$, $n = 2$ (solid), 4 (dotted), and 6 (dashed), with graph of $y = x$ included.

It appears that the right-hand side depends on x, but if we graph $(|x| - \epsilon)/\epsilon x^2$ (Figure 5.8), we see that it has an absolute maximum of $1/4\epsilon^2$ at $x = \pm 2\epsilon$. As long as $N > 1/4\epsilon^2$, the error will be within the allowed bounds.

When is Term-by-term Differentiation Legitimate?

An example like this should make the prospects of being able to differentiate a series by differentiating each summand seem very dim. In fact, in most of the series you are likely to encounter, it is safe to differentiate each summand. This can be a very powerful technique. For example, once you know that

$$\sin x = x - \frac{x^3}{3!} + \frac{x^5}{5!} - \frac{x^7}{7!} + \cdots,$$

then, *provided it is legal to differentiate this series by differentiating each summand*, we can conclude that

$$\cos x = 1 - \frac{x^2}{2!} + \frac{x^4}{4!} - \frac{x^6}{6!} + \cdots.$$

To find conditions under which it is safe to differentiate each term, we return to the definition of the derivative given in section 3.2. To say that $f_k(x)$ is differentiable at $x = a$ means that

$$E_k(x, a) = f_k'(a) - \frac{f_k(x) - f_k(a)}{x - a} \qquad (5.40)$$

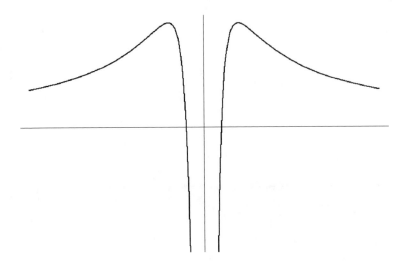

FIGURE 5.8. Graph of $(|x| - \epsilon)/\epsilon x^2$.

can be made arbitrarily small by taking x sufficiently close to a. We know that any finite sum of differentiable functions is differentiable, and so there is a comparable error term that corresponds to the partial sum $F_n(x)$. This error,

$$\mathcal{E}_n(x, a) = F_n'(a) - \frac{F_n(x) - F_n(a)}{x - a}, \tag{5.41}$$

can be made arbitrarily small by taking x sufficiently close to a. If $\sum f_k(x)$ converges for all x close to a and if $\sum f_k'(a)$ converges, then we have that

$$\begin{aligned}
\mathcal{E}(x, a) &= \left(\sum_{k=1}^{\infty} f_k'(a) \right) - \frac{F(x) - F(a)}{x - a} \\
&= \left(\sum_{k=1}^{\infty} f_k'(a) \right) - \frac{\sum_{k=1}^{\infty} f_k(x) - \sum_{k=1}^{\infty} f_k(a)}{x - a} \\
&= \sum_{k=1}^{\infty} \left(f_k'(a) - \frac{f_k(x) - f_k(a)}{x - a} \right) \\
&= \sum_{k=1}^{\infty} E_k(x, a). \tag{5.42}
\end{aligned}$$

Our series is differentiable at $x = a$ and the derivative is equal to $\sum_{k=1}^{\infty} f_k'(a)$ if and only if $\mathcal{E}(x, a) = \sum_{k=1}^{\infty} E_k(x, a)$ can be made arbitrarily small by taking x sufficiently close to a.

A Glance at Our Example

For the series given in equation (5.35) on page 192, we see that

$$E_k(x, 0) = 0 - \frac{f_k(x) - 0}{x - 0} = \frac{-x^2}{(1 + kx^2)(1 + (k - 1)x^2)}. \tag{5.43}$$

This should look familiar. It is precisely the summand that we saw in the last section where we showed that

$$\mathcal{E}(x, 0) = \sum_{k=1}^{\infty} E_k(x, 0) = \sum_{k=1}^{\infty} \frac{-x^2}{(1 + kx^2)(1 + (k-1)x^2)} = -1,$$

(since x is not 0). No matter how close we take x to 0, $\mathcal{E}(x, 0)$ will remain -1. It cannot be made arbitrarily small. This confirms what we already knew; we cannot differentiate this series at $x = 0$ by differentiating each summand.

The Solution

Uniform convergence of $\sum_{k=1}^{\infty} f_k(x)$ is not enough to guarantee that term-by-term differentiation can be used. Uniform convergence of the series of *derivatives*, $\sum_{k=1}^{\infty} f_k'(x)$, is sufficient.

Theorem 5.7 (Term-by-term Differentiation). *Let $f_1 + f_2 + f_3 + \cdots$ be a series of functions that converges at $x = a$ and for which the series of derivatives, $f_1' + f_2' + f_3' + \cdots$, converges uniformly over an open interval I that contains a. It follows that*

1. *$F = f_1 + f_2 + f_3 + \cdots$ converges uniformly over the interval I,*
2. *F is differentiable at $x = a$, and*
3. *for all $x \in I$, $F'(x) = \sum_{k=1}^{\infty} f_k'(x)$.*

Proof: The key to this proof is defining the function

$$g_k(x) = \frac{f_k(x) - f_k(a)}{x - a}, \quad x \neq a. \tag{5.44}$$

We can make this function continuous at $x = a$ by setting $g_k(a) = f_k'(a)$, though, in fact, we are only interested in it for $x \neq a$, $x \in I$. We will show that $\sum_{k=1}^{\infty} g_k(x)$ converges uniformly over I. As you think about how we might be able to do this (hint: think mean value theorem), notice what uniform convergence will do for us.

First, we can express $f_k(x)$ in terms of $g_k(x)$ and $f_k(a)$,

$$f_k(x) = (x - a)g_k(x) + f_k(a).$$

Using Theorems 5.4 and 5.5, it follows that $\sum f_k(x)$ converges. It is not hard to see (exercise 5.3.3) that if $\sum g_k$ converges uniformly over I, then so must $\sum f_k$.

Next, we let F_n denote the partial sum of the first n functions:

$$F_n = f_1 + f_2 + \cdots + f_n.$$

We denote by $\mathcal{E}_n(x, a)$ the size of the error when the average rate of change of F_n between x and a is replaced by the derivative of F_n at a:

$$\mathcal{E}_n(x, a) = F_n'(a) - \frac{F_n(x) - F_n(a)}{x - a}.$$

Given any positive error ϵ, our task is to find a response δ for which $0 < |x - a| < \delta$ implies that

$$|\mathcal{E}(x, a)| = \left| \sum_{k=1}^{\infty} f_k'(a) - \frac{F(x) - F(a)}{x - a} \right| < \epsilon.$$

We know that we can control the size of $\mathcal{E}_n(x, a)$, though we must keep in mind that the δ response could depend on n. We rewrite the quantity to be bounded as

$$
\begin{aligned}
|\mathcal{E}(x, a)| &= \left| \sum_{k=1}^{\infty} f_k'(a) - \frac{F(x) - F(a)}{x - a} \right| \\
&= \left| \sum_{k=1}^{\infty} f_k'(a) - F_n'(a) - \frac{F(x) - F(a)}{x - a} + \frac{F_n(x) - F_n(a)}{x - a} \right. \\
&\quad \left. + F_n'(a) - \frac{F_n(x) - F_n(a)}{x - a} \right| \\
&\leq \left| \sum_{k=1}^{\infty} f_k'(a) - F_n'(a) \right| + \left| \frac{F(x) - F(a)}{x - a} - \frac{F_n(x) - F_n(a)}{x - a} \right| \\
&\quad + \left| F_n'(a) - \frac{F_n(x) - F_n(a)}{x - a} \right| \\
&= \left| \sum_{k=n+1}^{\infty} f_k'(a) \right| + \left| \sum_{k=n+1}^{\infty} g_k(x) \right| + \left| \mathcal{E}_n(x, a) \right|. \tag{5.45}
\end{aligned}
$$

We split our error three ways and choose an n so that each of the first two pieces is less than $\epsilon/3$. We then choose our δ so that the third piece is also less than $\epsilon/3$. We can do this because of the convergence of $\sum_{k=1}^{\infty} f_k'(a)$, the uniform convergence of $\sum_{k=1}^{\infty} g_k(x)$ (so that the choice of n does not depend on x), and the differentiability of F_n, a finite sum of differentiable functions.

So it all comes down to the uniform convergence of $\sum g_k$. Have you figured it out yet? We can use the Cauchy criterion to establish uniform convergence. A series such as

$$\sum_{k=1}^{\infty} g_k(x)$$

converges uniformly over the interval I if and only if given an error bound ϵ, there is a response N *independent of* x for which

$$N \leq m < n \quad \text{implies that} \quad \left| \sum_{k=m+1}^{n} g_k(x) \right| < \epsilon.$$

For the series under consideration, the difference between the partial sums is

$$
\begin{aligned}
\sum_{k=m+1}^{n} g_k(x) &= \frac{F_n(x) - F_n(a)}{x - a} - \frac{F_m(x) - F_m(a)}{x - a} \\
&= \frac{[F_n(x) - F_m(x)] - [F_n(a) - F_m(a)]}{x - a}. \tag{5.46}
\end{aligned}
$$

Applying the mean value theorem to the function $F_n(x) - F_m(x)$, we see that this is equal to

$$\sum_{k=m+1}^{n} g_k(x) = F_n'(t) - F_m'(t) = \sum_{k=m+1}^{n} f_k'(t) \tag{5.47}$$

for some t between x and a. This t must also lie in I. By the uniform convergence of $\sum_{k=1}^{\infty} f_k'(t)$, we can find a response N that forces

$$\left| \sum_{k=m+1}^{n} g_k(x) \right|$$

to be as small as we wish regardless of the choice of $x \in I$. It follows that $\sum_{k=1}^{\infty} g_k(x)$ is also uniformly convergent.

We have only proven that $F'(a) = \sum_{k=1}^{\infty} f_k'(a)$, but now that we know that $\sum_{k=1}^{\infty} f_k(x)$ converges for all x in I, we can replace a by any x in I.

Q.E.D.

Integration

In his derivation of the formula for the coefficients of a Fourier series, Joseph Fourier assumed that the integral of a series is the sum of the integrals. This is a questionable procedure that will sometimes fail. It *is* correct, however, when the series in question converges uniformly over the interval of integration.

Theorem 5.8 (Term-by-term Integration). *Let $f_1 + f_2 + f_3 + \cdots$ be uniformly convergent over the interval $[a, b]$, converging to F. If each f_k is integrable over $[a, b]$, then so is F and*

$$\int_a^b F(x)\,dx = \sum_{k=1}^{\infty} \int_a^b f_k(x)\,dx.$$

Before proceeding with the proof, we need to face one major obstacle: we have not yet defined integration. The reason for this is that defining integration is not easy. It requires a very profound understanding of the nature of the real number line. In fact, it will not be until the sequel to this book, *A Radical Approach to Lebesgue's Theory of Integration*, that we do justice to the question of integration. The modern definition was not determined until the 20th century.

In the meantime, you will have to rely on whatever definition of integration you prefer. Fortunately, to prove this theorem we only need two properties of the integral:

$$\int_a^b \left(f(x) + g(x) \right) dx = \int_a^b f(x)\,dx + \int_a^b g(x)\,dx, \tag{5.48}$$

$$\left| \int_a^b f(x)\,dx \right| \leq \int_a^b \left| f(x) \right| dx. \tag{5.49}$$

In the next chapter we shall discuss integration as defined by Cauchy and Riemann. It is not hard to show that their integrals satisfy these properties.

Proof: We have to show that given any $\epsilon > 0$, we can find an N for which

$$\left| \int_a^b F(x)\,dx - \sum_{k=1}^n \int_a^b f_k(x)\,dx \right| < \epsilon$$

when n is at least N. From equation (5.48), any finite sum of integrals over the same interval is the integral of the sum. We can rewrite our difference as

$$\left| \int_a^b F(x)\,dx - \sum_{k=1}^n \int_a^b f_k(x)\,dx \right| = \left| \int_a^b F(x)\,dx - \int_a^b F_n(x)\,dx \right|$$

$$= \left| \int_a^b \left(F(x) - F_n(x) \right)\,dx \right|$$

$$\leq \int_a^b \left| F(x) - F_n(x) \right|\,dx. \qquad (5.50)$$

Since our series converges uniformly, we can find an N for which $n \geq N$ implies that $|F(x) - F_n(x)| < \epsilon/(b-a)$. Substituting this bound, we see that

$$\left| \int_a^b F(x)\,dx - \sum_{k=1}^n \int_a^b f_k(x)\,dx \right| < \int_a^b \frac{\epsilon}{b-a}\,dx = \epsilon. \qquad (5.51)$$

Q.E.D.

An Example

An example of a series that cannot be integrated by integrating each summand is given by

$$\sum_{k=1}^\infty \left[k^2 x^k (1-x) - (k-1)^2 x^{k-1}(1-x) \right]$$

whose partial sums are (see Figure 5.9)

$$S_n(x) = n^2 x^n (1-x).$$

As n increases, the hump in the graph of S_n gets pushed further to the right. For any x in $[0, 1]$, $S_n(x)$ approaches 0 as n gets larger, and so

$$\int_0^1 \left(\sum_{k=1}^\infty \left[k^2 x^k (1-x) - (k-1)^2 x^{k-1}(1-x) \right] \right) dx = \int_0^1 0\,dx = 0. \qquad (5.52)$$

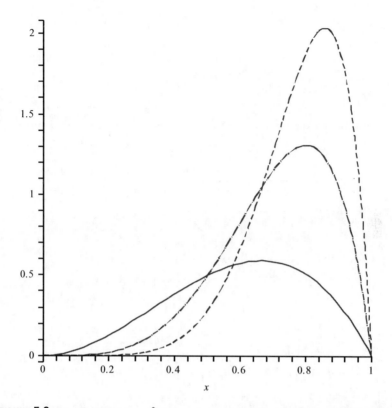

FIGURE 5.9. Graphs of $S_n(x) = n^2 x^n(1 - x)$, $n = 2$ (solid), 4 (dotted), and 6 (dashed).

But the area under $S_n(x)$ approaches 1 as n gets larger:

$$\lim_{n\to\infty} \sum_{k=1}^{n} \left(\int_0^1 \left[k^2 x^k (1 - x) - (k - 1)^2 x^{k-1}(1 - x) \right] dx \right)$$

$$= \lim_{n\to\infty} \sum_{k=1}^{n} \left(\frac{k^2}{(k + 1)(k + 2)} - \frac{(k - 1)^2}{k(k + 1)} \right)$$

$$= \lim_{n\to\infty} \frac{n^2}{(n + 1)(n + 2)}$$

$$= 1. \tag{5.53}$$

In this example, the integral of the sum is not the same as the sum of the integrals.

Not Enough!

Theorem 5.8 is often useful, but it is not what we need for Fourier series. Looking back to the technique introduced by Joseph Fourier for finding the coefficients in the cosine expansion of an even function, we see that he began by assuming that his function had a cosine expansion:

$$f(x) = a_1 \cos(\pi x/2) + a_2 \cos(3\pi x/2) + a_3 \cos(5\pi x/2) + \cdots$$
$$= \sum_{k=1}^{\infty} a_k \cos\left(\frac{(2k-1)\pi x}{2}\right).$$

This is a dangerous assumption, but we shall accept it for the moment. Fourier observed that

$$\int_{-1}^{1} \cos\left(\frac{(2k-1)\pi x}{2}\right) \cos\left(\frac{(2m-1)\pi x}{2}\right) dx = \begin{cases} 0, & \text{if } k \neq m, \\ 1, & \text{if } k = m. \end{cases}$$

He then argued as follows:

$$\int_{-1}^{1} f(x) \cos\left(\frac{(2m-1)\pi}{2} x\right) dx$$
$$= \int_{-1}^{1} \left[\sum_{k=1}^{\infty} a_k \cos\left(\frac{(2k-1)\pi}{2} x\right)\right] \cos\left(\frac{(2m-1)\pi}{2} x\right) dx$$
$$= \sum_{k=1}^{\infty} a_k \left[\int_{-1}^{1} \cos\left(\frac{(2k-1)\pi}{2} x\right) \cos\left(\frac{(2m-1)\pi}{2} x\right) dx\right] \quad (5.54)$$
$$= a_1 \cdot 0 + a_2 \cdot 0 + \cdots + a_{m-1} \cdot 0 + a_m \cdot 1 + a_{m+1} \cdot 0 + \cdots$$
$$= a_m. \quad (5.55)$$

As we see, his argument rests on integrating the series by integrating each summand. If the cosine series converges uniformly, then we are completely justified in doing this. But one of the most important series that we have to deal with is the cosine expansion of the constant 1 between -1 and 1. As we have seen, its expansion,

$$\frac{4}{\pi}\left(\cos(\pi x/2) - \frac{1}{3}\cos(3\pi x/2) + \frac{1}{5}\cos(5\pi x/2) - \frac{1}{7}\cos(7\pi x/2) + \cdots\right),$$

does *not* converge uniformly over $[-1, 1]$.

Fortunately, Theorem 5.8 gives a condition that is sufficient but not necessary. Even if the series does not converge uniformly, it may be permissible to integrate by integrating each summand. The search in the late 19th century for necessary as well as sufficient conditions will be an important part of the story in *A Radical Approach to Lebesgue's Theory of Integration*.

Exercises

The symbol (**M&M**) indicates that *Maple* and *Mathematica* codes for this problem are available in the **Web Resources** at **www.macalester.edu/aratra**.

5.3.1. Give an example of a series for which each summand f_k is differentiable at every x in an interval I and $\sum_{k=1}^{\infty} f_k'$ converges uniformly over I, but $\sum_{k=1}^{\infty} f_k(x)$ does not converge for any x in I.

5.3.2. Prove that

$$f(x) = \sum_{n=1}^{\infty} \frac{1}{n^2 + x^2}$$

is differentiable for all values of x.

5.3.3. Prove that if $\sum g_k$ converges uniformly over the interval I, then

$$\sum_{k=1}^{\infty} f_k(x) = (x - a) \sum_{k=1}^{\infty} g_k(x) + \sum_{k=1}^{\infty} f_k(a)$$

also converges uniformly over I.

5.3.4. ⬡ **M&M** Graph the partial sums of the first 5, the first 10, and the first 20 terms of

$$-2\left[\sin(\pi x/2) - \sin(3\pi x/2) + \sin(5\pi x/2) - \sin(7\pi x/2) + \cdots\right].$$

Prove that this series converges if and only if x is an even integer.

5.3.5. ⬡ **M&M** Consider the series

$$G(x) = \sum_{k=1}^{\infty} g_k(x) = \sum_{k=1}^{\infty} \frac{x^2 \sin x}{(1 + kx^2)(1 + (k - 1)x^2)}.$$

Evaluate the partial sum of this series to at least a thousand terms when $x = \pi/6, \pi/4$, and $\pi/2$.

5.3.6. ⬡ **M&M** Graph the partial sums

$$G_n(x) = \sum_{k=1}^{n} \frac{x^2 \sin x}{(1 + kx^2)(1 + (k - 1)x^2)}$$

for $-\pi \le x \le \pi$ and $n = 3, 6, 9$, and 12. Discuss what you see. Prove that

$$G_n(x) = \frac{nx^2 \sin x}{1 + nx^2}.$$

What is $G(x)$?

5.3.7. Prove that

$$G(x) = \sum_{k=1}^{\infty} \frac{x^2 \sin x}{(1 + kx^2)(1 + (k - 1)x^2)}$$

converges uniformly for all values of x.

5.3.8. Show that if

$$g_k(x) = \frac{x^2 \sin x}{(1 + kx^2)(1 + (k - 1)x^2)},$$

then

$$\sum_{k=1}^{\infty} g_k'(0) = 0.$$

Using the result from exercise 5.3.6, find $G'(0)$. This is a series that is differentiable but which we cannot differentiate term-by-term. This series *does* converge uniformly. Explain why this does not contradict Theorem 5.7.

5.3.9. Prove that the Cauchy criterion can be used for uniform convergence:

Let $f_1 + f_2 + f_3 + \cdots$ be a series of functions converging to F for all x in the interval I, and let $F_n = f_1 + f_2 + \cdots + f_n$ be the partial sum. This series converges uniformly over I if and only if given any error bound ϵ, there is a response N (valid for all $x \in I$) such that $N \le m < n$ implies that $|F_m(x) - F_n(x)| < \epsilon$.

5.3.10. (M&M) Show that

$$\sum_{k=1}^{N} \left(kxe^{-kx^2} - (k-1)xe^{-(k-1)x^2} \right) = Nxe^{-Nx^2},$$

and use this to prove that

$$\sum_{k=1}^{\infty} \left(kxe^{-kx^2} - (k-1)xe^{-(k-1)x^2} \right) = 0$$

for all values of x (including $x = 0$). Graph the partial sums for $N = 5$, 10, and 20.

5.3.11. Using the result of exercise 6.3.7, evaluate

$$\int_0^1 \left(\sum_{k=1}^{\infty} kxe^{-kx^2} - (k-1)xe^{-(k-1)x^2} \right) dx.$$

5.3.12. Show that

$$\int_0^1 \left(kxe^{-kx^2} - (k-1)xe^{-(k-1)x^2} \right) dx = -\frac{1}{2}e^{-k} + \frac{1}{2}e^{-(k-1)}.$$

Use this to evaluate

$$\sum_{k=1}^{\infty} \int_0^1 \left(kxe^{-kx^2} - (k-1)xe^{-(k-1)x^2} \right) dx.$$

5.3.13. The last two exercises should have yielded different results. This tells us that the convergence of $\sum_{k=1}^{\infty} kxe^{-kx^2} - (k-1)xe^{-(k-1)x^2}$ cannot be uniform over the interval $[0, 1]$. Where is it that this series does not converge uniformly?

5.4 Verifying Uniform Convergence

The importance of uniform convergence was not generally recognized until the 1860s. Once it was accepted as a critical property of "nice" series, the question that came to the fore was how to determine whether or not a series converged uniformly over a given interval. Three names stand out among those associated with the tests for uniform convergence: Gustav Lejeune Dirichlet and Niels Henrik Abel whose work of forty years earlier turned out to be applicable to this new question, and Karl Weierstrass (1815–1897).

Weierstrass had gone to the University of Bonn at the age of nineteen to study law. Instead, he became noted for his drinking and fencing. He left after four years without earning a degree. After convincing the authorities that he had reformed himself, he was allowed to enter the university at Münster to seek a teaching certificate. There he had the good fortune to be taught by Christof Gudermann (1798–1852). It was with Gudermann that Weierstrass began his life-long love of analysis.

In 1841, at the age of 26, Weierstrass received his certification and began to teach high school[2] mathematics. In his spare time, he worked on questions of analysis, concentrating on the writings of Abel and building upon them. His first papers appeared in 1854. They excited the entire mathematical community. Weierstrass was granted an honorary doctorate by the University of Königsberg. Two years later he was made a professor at the University of Berlin. To him we owe the first truly clear vision of the nature and significance of uniform convergence.

The Weierstrass *M*-test

One of the simplest and most useful tests for uniform convergence was published by Weierstrass in 1880, the *M*-test. It is based on the following analog of the comparison test.

Theorem 5.9 (Dominated Uniform Convergence). *If the series $g_1 + g_2 + g_3 + \cdots$ is uniformly convergent over the interval I and if $|f_k(x)| \leq g_k(x)$ for all $x \in I$ and for every positive integer k, then $f_1 + f_2 + f_3 + \cdots$ is also uniformly convergent over I.*

Proof: We use the Cauchy criterion for uniform convergence. Given an error bound ϵ, we must find a response N, independent of x, such that $N \leq m < n$ implies that

$$\left| \sum_{k=m+1}^{n} f_k(x) \right| < \epsilon.$$

We know that

$$\left| \sum_{k=m+1}^{n} f_k(x) \right| \leq \sum_{k=m+1}^{n} \left| f_k(x) \right| \leq \sum_{k=m+1}^{n} g_k(x).$$

---

[2] The German *gymnasium*.

The uniform convergence of $g_1(x) + g_2(x) + g_3(x) + \cdots$ guarantees an N, independent of x, for which this sum is less than ϵ when $N \le m < n$.

Q.E.D.

This theorem has several immediate corollaries, including the M-test.

Corollary 5.10 (Absolute Uniform Convergence). *If $|f_1| + |f_2| + |f_3| + \cdots$ converges uniformly over I, then so does $f_1 + f_2 + f_3 + \cdots$.*

Corollary 5.11 (Variation on Dominated Uniform Convergence). *If the series $g_1 + g_2 + g_3 + \cdots$ is uniformly convergent over the interval I and if $|f_k(x)| \le g_k(x)$* **for every k greater than or equal to some fixed integer N and for all $x \in I$,** *then $f_1(x) + f_2(x) + f_3(x) + \cdots$ is also uniformly convergent over I.*

Corollary 5.12 (Weierstrass M-test). *If we can find a sequence of constants M_1, M_2, M_3, . . . such that*

$$|f_k(x)| \le M_k$$

for every k greater than or equal to some fixed integer N and for all $x \in I$, and if $M_1 + M_2 + M_3 + \cdots$ converges, then $f_1 + f_2 + f_3 + \cdots$ converges uniformly over I.

The first and third corollaries follow from the theorem by taking $g_k(x) = |f_k(x)|$, $g_k(x) = M_k$, respectively. The second corollary is simply the observation that convergence, and thus uniform convergence, is not affected by changing a finite number of summands.

Why Power Series are so Nice

The M-test has an important consequence for power series:

Corollary 5.13 (Uniform Convergence of Power Series, I). *If*

$$a_0 + a_1 x + a_2 x^2 + a_3 x^3 + \cdots$$

is a power series with finite radius of convergence $R > 0$ and if $0 < \alpha < R$, then this series converges uniformly over $[-\alpha, \alpha]$. If the radius of convergence is infinite, then the power series converges uniformly over $[-\alpha, \alpha]$ for any finite positive value of α.

Proof: From the definition of the radius of convergence, R, we know that if R is finite, then

$$\limsup_{k \to \infty} \sqrt[k]{|a_k R^k|} = 1. \tag{5.56}$$

For any positive error ϵ we can find an N such that $k \geq N$ implies that

$$\sqrt[k]{|a_k R^k|} < 1 + \epsilon.$$

If $0 < |x| < \alpha < R$ and we choose $\epsilon = (\alpha - |x|)/|x|$, then

$$\sqrt[k]{|a_k x^k|} = \frac{|x|}{R} \sqrt[k]{|a_k R^k|} < \frac{|x|}{R}(1 + \epsilon) = \frac{\alpha}{R} < 1, \tag{5.57}$$

for $k \geq N$. We can apply the Weierstrass M-test to this series, using

$$M_k = \left(\frac{\alpha}{R}\right)^k.$$

We have proven that the convergence is uniform on the open interval $(-\alpha, \alpha)$. Exercise 5.4.5 asks you to finish this part of the proof by explaining why the convergence must be uniform on the closed interval $[-\alpha, \alpha]$.

If R is infinite, then

$$\limsup_{k \to \infty} \sqrt[k]{|a_k|} = 0. \tag{5.58}$$

We can find an N such that $k \geq N$ implies that $\sqrt[k]{|a_k|} < 1/2\alpha$. If $0 \leq |x| < \alpha$, then

$$\sqrt[k]{|a_k x^k|} < \frac{|x|}{2\alpha} < \frac{1}{2},$$

for $k \geq N$. We can apply the Weierstrass M-test to this series, using $M_k = 1/2^k$.

Q.E.D.

We note that this corollary does not permit us to take uniform convergence all the way out to the end of the radius of convergence. We have to stop at some $\alpha < R$. This should not be too surprising as we do not always have convergence when $x = \pm R$, much less uniform convergence on $[-R, R]$.

We also note that the radius of convergence of any power series is the same as the radius of convergence of the series of derivatives. This is because

$$\lim_{k \to \infty} \sqrt[k]{k} = 1,$$

and so

$$\limsup_{k \to \infty} \sqrt[k]{|a_k|} = \limsup_{k \to \infty} \sqrt[k]{|a_k| k}. \tag{5.59}$$

We can now see why power series are so very nice and never gave any indication that there might be problems with continuity or how to differentiate or integrate infinite series. At each point *inside* the interval of convergence, we are inside an interval in which the series and the series of derivatives both converge uniformly. Power series will always be continuous functions; differentiation and integration can always be accomplished by differentiating or integrating each term in the series. Power series are always "nice."

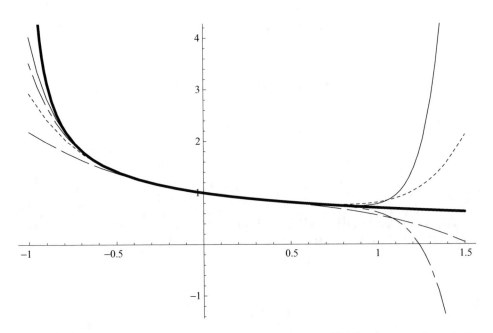

FIGURE 5.10. Partial sums of the graphs of the expansions of $1/\sqrt{1+x}$ (thick) to 3 (long dash), 6 (dots), 9 (short-long), and 12 (short-short-short-long) terms.

An Example

Let us return to Newton's binomial series and in particular look at the expansion of $1/\sqrt{1+x}$:

$$(1+x)^{-1/2} = 1 + (-1/2)x + \frac{(-1/2)(-3/2)}{2!}x^2 + \frac{(-1/2)(-3/2)(-5/2)}{3!}x^3 + \cdots .$$
$$(5.60)$$

As we have seen, this has radius of convergence $R = 1$. It converges at $x = 1$ but not at $x = -1$. Figure 5.10 shows the graphs of the partial sums of the first 3, 6, 9, and 12 terms of this series. We see that the graphs are spreading further apart near $x = -1$. Even though our series converges at every point in $(-1, 0)$, it appears that there is no hope for uniform convergence over this interval. On the other hand, we see that the graphs do seem to be coming closer near $x = 1$ (see Figures 5.11 and 5.12). It looks as if it should be possible, given an ϵ envelope around $1/\sqrt{1+x}$ over the interval $[0, 1]$, to find an N so that all of the partial sums with at least N terms have graphs that lie entirely inside the ϵ envelope.

In fact, the behavior that we see here is typical of any power series. If the series converges at the end of the radius of convergence, then it converges uniformly up to and including that point. If it does not converge at the end of the radius of convergence, then we cannot maintain uniform convergence over the entire open interval. In the next two subsections, we shall prove these assertions.

To simplify our arguments, we are only going to consider what happens at the right-hand endpoint, $x = R$, where R is the radius of convergence. If $a_0 + a_1 x + a_2 x^2 + a_3 x^3 + \cdots$ is a power series with radius of convergence R and we want to look at its behavior at

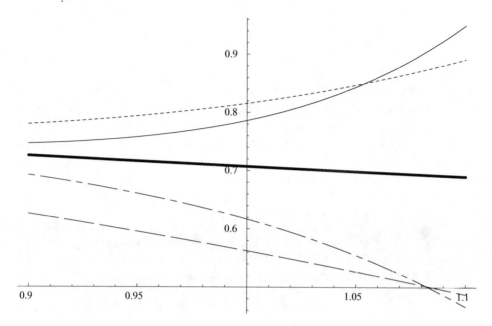

FIGURE 5.11. Close-up near $x = 1$ of graphs of partial sums from Figure 5.10.

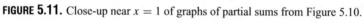

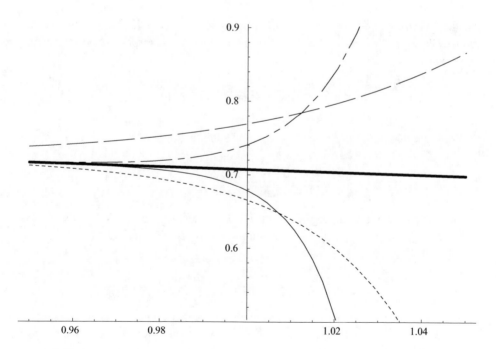

FIGURE 5.12. Close-up near $x = 1$ of graphs of $1/\sqrt{1 + x}$ (thick) and of partial sums with 20 (long dash), 45 (dots), 70 (short-long), and 95 (short-short-short-long) terms.

$x = -R$, then this is the same as the behavior at $x = R$ of

$$a_0 - a_1 x + a_2 x^2 - a_3 x^3 + \cdots.$$

Non-uniform Convergence

Given a power series $a_0 + a_1 x + a_2 x^2 + a_3 x^3 + \cdots$ with a finite radius of convergence $R > 0$, we assume that the series does not converge at $x = R$. We want to show that even though we have convergence at every point in $(0, R)$, we cannot have uniform convergence over this interval. It is easier to look at this problem from the other direction: if we *do* have uniform convergence over $(0, R)$, then the series must converge at $x = R$. This result is not unique to power series. It holds for any series of continuous functions.

> **Theorem 5.14 (Continuity & Uniform Conv. $\Longrightarrow$ Conv. at Endpoints).** *Let $F = f_1 + f_2 + f_3 + \cdots$ be a series that is uniformly convergent over (a, b). If each of the summands f_k is continuous at every point in $[a, b]$, then this series converges at a and at b and is uniformly convergent over $[a, b]$.*

Proof: The difficult part is proving that the series converges at $x = a$ and $x = b$. Once we have shown convergence, uniform convergence follows (see exercise 5.4.5).

We need to prove that $f_1(b) + f_2(b) + f_3(b) + \cdots$ converges. Since we do not have any handle on $F(b)$ (we do not yet know that $F(x)$ is continuous on $[a, b]$), we shall use the Cauchy criterion for convergence. We must show how to find a response N to any positive error bound ϵ so that

$$N \le m < n \quad \text{implies that} \quad \left| \sum_{k=m+1}^{n} f_k(b) \right| < \epsilon.$$

We know that if $a < x < b$, then we can find an N that forces

$$\left| \sum_{k=m+1}^{n} f_k(x) \right|$$

to be as small as we want, regardless of which x we have chosen. We also know that $f_{m+1}(x) + f_{m+2}(x) + \cdots + f_n(x)$ is continuous over $[a, b]$. We can make

$$\left| \sum_{k=m+1}^{n} f_k(b) - \sum_{k=m+1}^{n} f_k(x) \right|$$

as small as we want by choosing an appropriate x. Note that x *will* depend on our choice of m and n. We have to be careful. We cannot choose x until after we have specified m and n.

The key inequality is

$$\left| \sum_{k=m+1}^{n} f_k(b) \right| \le \left| \sum_{k=m+1}^{n} f_k(b) - \sum_{k=m+1}^{n} f_k(x) \right| + \left| \sum_{k=m+1}^{n} f_k(x) \right|. \tag{5.61}$$

We first choose an N so that

$$N \leq m < n \quad \text{implies that} \quad \left| \sum_{k=m+1}^{n} f_k(x) \right| < \epsilon/2,$$

regardless of the choice of $x \in (a, b)$. We now look at such a pair, m, n, and choose an x close enough to b that

$$\left| \sum_{k=m+1}^{n} f_k(b) - \sum_{k=m+1}^{n} f_k(x) \right| < \epsilon/2.$$

Combining these two bounds with our inequality (5.61) gives us the desired result. The same argument works at $x = a$.

Q.E.D.

Uniform Convergence

We want to prove that when a power series converges at $x = R$, then it converges uniformly over $[0, R]$. We need something stronger than the Weierstrass M-test. If the power series converges absolutely at $x = R$, then the M-test can be used. But there are many examples like our expansion of $1/\sqrt{1+x}$ for which the convergence at $x = 1$ is not absolute. We need to prove that even in this case, we still have uniform convergence over $[0, R]$.

The key to proving this is the work that Abel published in 1826 on the binomial expansion. In particular, we shall use Theorem 4.16 (Abel's lemma) stated in section 4.5: if

$$b_1 \geq b_2 \geq b_3 \geq \cdots \geq 0$$

and if

$$\left| \sum_{k=1}^{n} c_k \right| < M \quad \text{for all } n,$$

then

$$\left| \sum_{k=1}^{n} c_k b_k \right| \leq M b_1. \tag{5.62}$$

We shall see exactly how it is used after stating the theorem.

> **Theorem 5.15 (Uniform Convergence of Power Series, II).** *Let $a_0 + a_1 x + a_2 x^2 + a_3 x^3 + \cdots$ be a power series with a finite radius of convergence $R > 0$. If this series converges at $x = R$, then it converges uniformly over $[0, R]$.*

Proof: Again we use the Cauchy criterion. Given an error bound ϵ, we must find a response N so that

$$N \leq m < n \quad \text{implies that} \quad \left| \sum_{k=m+1}^{n} a_k x^k \right| < \epsilon,$$

regardless of the choice of x, $0 \leq x \leq R$. We use equation (5.62) with

$$b_k = \left(\frac{x}{R}\right)^{m+k}$$

and

$$c_k = a_{m+k} R^{m+k}.$$

Since $0 \leq x \leq R$, we have

$$\left(\frac{x}{R}\right)^{m+1} \geq \left(\frac{x}{R}\right)^{m+2} \geq \left(\frac{x}{R}\right)^{m+3} \geq \left(\frac{x}{R}\right)^{m+4} \geq \cdots \geq 0.$$

We fix an integer m and let M_m be the least upper bound of the absolute values of the partial sums that begin with the $m + 1$st term and are evaluated at R. In other words, for all $n > m$:

$$\left| a_{m+1} R^{m+1} + a_{m+2} R^{m+2} + \cdots + a_n R^n \right| \leq M_m.$$

By the Cauchy criterion, the convergence of $a_0 + a_1 x + a_2 x^2 + a_3 x^3 + \cdots$ at $x = R$ is equivalent to the statement that we can make M_m as small as we want by taking m sufficiently large.

Equation (5.62) implies that

$$\left| \sum_{k=m+1}^{n} a_k x^k \right| = \left| \sum_{k=m+1}^{n} a_k R^k \cdot \left(\frac{x}{R}\right)^k \right| \leq M_m \cdot \left(\frac{x}{R}\right)^{m+1} \leq M_m.$$

Our response is any N for which $m \geq N$ guarantees that M_m is less than ϵ.

Q.E.D.

Fourier Series

We know that we do not always have uniform convergence for Fourier series. We know of Fourier series that converge but are not continuous. Nevertheless, we can hope to find some Fourier series that do converge uniformly. To find conditions under which a Fourier series converges uniformly, we return to Dirichlet's test (Corollary 4.17) in section 4.4 and replace $a_1, a_2, a_3, \ldots$ by functions of x: $a_1(x), a_2(x), a_3(x), \ldots$. The condition that $S_n = \sum_{k=1}^{n} a_k$ is bounded is replaced by the requirement that $S_n(x) = \sum_{k=1}^{n} a_k(x)$ is **uniformly bounded**. That is to say, for all x in some interval I, there is a bound M *independent of x* for which

$$\left| S_n(x) \right| = \left| \sum_{k=1}^{n} a_k(x) \right| \leq M.$$

> **Theorem 5.16 (Dirichlet's Test for Uniform Convergence).** *We consider a series of the form*
>
> $$a_1(x)b_1(x) + a_2(x)b_2(x) + a_3(x)b_3(x) + \cdots$$
>
> *where the a_k and b_k are functions defined for all x in the interval I, where for each $x \in I$ the values of $b_k(x)$ are positive, decreasing, and approaching 0,*
>
> $$b_1(x) \geq b_2(x) \geq b_3(x) \geq \cdots \geq 0, \qquad \lim_{k \to \infty} b_k(x) = 0,$$
>
> *and for which there exists a sequence $(B_k)_{k=1}^{\infty}$ approaching 0 such that $B_k \geq b_k(x)$ for all $x \in I$. Let $S_n(x)$ be the nth partial sum of the $a_k(x)$'s:*
>
> $$S_n(x) = \sum_{k=1}^{n} a_k(x).$$
>
> *If these partial sums are uniformly bounded over I—that is to say, if there is some number M for which $\left| S_n(x) \right| \leq M$ for all values of n and all $x \in I$—then the series converges uniformly over I.*

The proof follows that of Dirichlet's test (Corollary 4.17) and is left as exercise 5.4.9.

Example

As an example, consider the Fourier sine series that we met in section 4.1,

$$F(x) = \sum_{k=2}^{\infty} \frac{\sin kx}{\ln k}. \tag{5.63}$$

We looked at this at $x = 0.01$. Now we want to consider its behavior at an arbitrary value of x.

Our b_k's are

$$b_k = \frac{1}{\ln k} \quad (k \geq 2).$$

These are positive, decreasing, and approaching zero. Our a_k's are

$$a_k(x) = \sin kx.$$

In exercise 4.4.7 on page 168, you were asked to prove that

$$\sin x + \sin 2x + \sin 3x + \cdots + \sin nx = \frac{\sin x}{2}\left(\frac{1 - \cos nx}{1 - \cos x}\right) + \frac{\sin nx}{2} \tag{5.64}$$

when x is not a multiple of π. It is zero when x is a multiple of π. Since $|1 - \cos nx| \leq 2$ and $|\sin nx| \leq 1$, we have a bound of

$$\left| \sum_{k=1}^{n} \sin kx \right| \leq \left| \frac{\sin x}{1 - \cos x} \right| + \frac{1}{2}$$

when x is not a multiple of π. This bound is independent of n. Together with Theorem 5.12, it implies that our series converges for any value of x. We cannot find a bound that is independent of x. As x approaches any even multiple of π, our bound is unbounded. We do have uniform convergence on any interval that stays a positive distance away from any even multiple of π. It appears—although we have not actually proven it—that this series is not uniformly convergent on any interval that contains or comes arbitrarily close to an even multiple of π. For example, we would not expect it to be uniformly convergent on $(0, \pi/2)$. Figure 5.13 shows the graphs of some of the partial sums of this series. It appears that this Fourier series will be discontinuous at $x = 0$. An actual proof that this series is not uniformly convergent on $(0, \pi/2)$ is left as exercises 5.4.10 and 5.4.11.

Exercises

5.4.1. Determine whether or not each of the following series is uniformly convergent on the prescribed set S and justify your conclusion. If it is not uniformly convergent, prove that it is still convergent.

a. $\displaystyle\sum_{n=1}^{\infty} n^2 x^2 e^{-n^2 |x|}, \quad S = \mathbb{R}$

b. $\displaystyle\sum_{n=1}^{\infty} \frac{n^2}{\sqrt{n}} (x^n + x^{-n}), \quad S = [1/2, 2]$

c. $\displaystyle\sum_{n=1}^{\infty} 2^n \sin\left(\frac{1}{3^n x}\right), \quad S = (0, \infty)$

d. $\displaystyle\sum_{n=2}^{\infty} \ln\left(1 + \frac{x^2}{n \ln^2 n}\right), \quad S = (-a, a)$, where a is a positive constant.

e. $\displaystyle\sum_{n=1}^{\infty} \frac{\ln(1 + nx)}{n x^n}. \quad S = [2, \infty]$

f. $\displaystyle\sum_{n=1}^{\infty} \left(\frac{\pi}{2} - \arctan(n^2(1 + x^2))\right), \quad S = \mathbb{R}$

5.4.2. Let $a_0 + a_1 x + a_2 x^2 + \cdots$ be a power series with finite radius of convergence R. Prove that if the series of derivatives, $a_1 + 2a_2 x + 3a_3 x^2 + \cdots$ converges at $x = R$, then so does the original series.

5.4.3. Give an example of a power series with radius of convergence R that converges at $x = R$ but for which the series of derivatives does not converge at $x = R$.

5.4.4. Let $F(x) = f_1(x) + f_2(x) + f_3(x) + \cdots$ be a series that is uniformly convergent over (a, b) and for which each $f_k(x)$ is continuous on $[a, b]$. Prove that

$$F(b) = \lim_{x \to b^-} F(x).$$

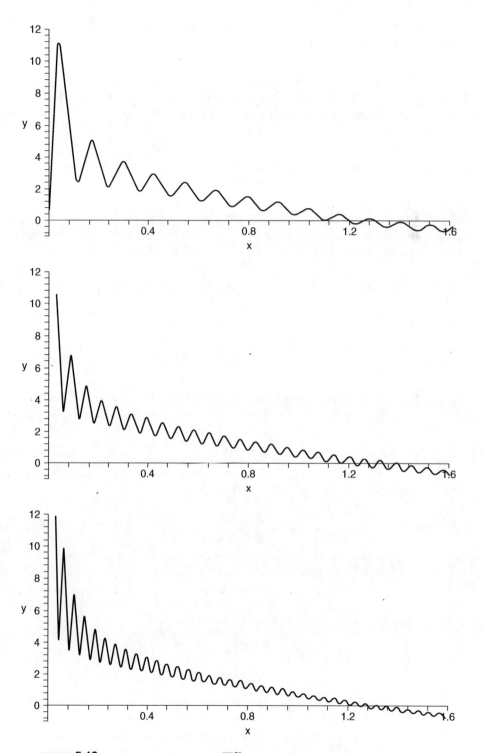

FIGURE 5.13. Partial sums of graphs of $\sum_{k=2}^{\infty} (\sin kx)/(\ln k)$ to 50, 100, and 150 terms.

5.4.5. Explain why it is that if a series converges on the closed interval $[a, b]$ and converges uniformly on the open interval (a, b), then it must converge uniformly on the closed interval $[a, b]$

5.4.6. Give an example of a series that is uniformly convergent over (a, b) and that is not uniformly convergent over $[a, b]$. Theorem 5.14 implies that such a series cannot be a power series.

5.4.7. Give an example of a series that converges at every point of (a, b), each summand is continuous at every point in $[a, b]$, but the series does not converge at every point of $[a, b]$.

5.4.8. If a power series has radius of convergence R and converges at $x = \pm R$, then we have shown that it converges uniformly over $[-R, 0]$ and over $[0, R]$. Prove that it converges uniformly over $[-R, R]$.

5.4.9. Prove Dirichlet's test for uniform convergence, Theorem 5.16.

5.4.10. Prove that

$$\sum_{k=2}^{\infty} \frac{\sin kx}{\ln k}$$

is discontinuous at $x = 0$ by proving that

$$\overline{\lim_{x \to 0+}} \sum_{k=2}^{\infty} \frac{\sin kx}{\ln k} = \infty.$$

5.4.11. Using the result from the previous exercise, prove that

$$\sum_{k=2}^{\infty} \frac{\sin kx}{\ln k}$$

is not uniformly convergent over $(0, \pi/2)$.

5.4.12. Use the Dirichlet test for uniform convergence to show that each of the following series converges uniformly on the indicated set S:

a. $\displaystyle\sum_{n=1}^{\infty} (-1)^{n+1} \frac{x^n}{n}, \quad S = [0, 1]$

b. $\displaystyle\sum_{n=1}^{\infty} \frac{\sin(nx)}{n}, \quad S = [\delta, 2\pi - \delta]$, for fixed $\delta, 0 < \delta < \pi$

c. $\displaystyle\sum_{n=1}^{\infty} \frac{\sin(n^2 x) \sin(nx)}{n + x^2} \quad S = \mathbb{R}$

d. $\displaystyle\sum_{n=1}^{\infty} \frac{\sin(nx)\,\arctan(nx)}{n}$, $S = [\delta, 2\pi - \delta]$, for fixed δ, $0 < \delta < \pi$

e. $\displaystyle\sum_{n=1}^{\infty} (-1)^{n+1}\, \frac{1}{n^x}$, $S = [a, \infty)$, for some constant $a > 0$

f. $\displaystyle\sum_{n=1}^{\infty} (-1)^{n+1}\, \frac{e^{-nx}}{\sqrt{n + x^2}}$, $S = [0, \infty)$

5.4.13. Let $F(x) = \sum_{k=1}^{\infty} f_k(x)$ be a series that is uniformly convergent over any closed interval $[c, d] \subseteq (a, b)$ where every $f_k(x)$ is continuous on $[a, b]$. Furthermore, assume that $\sum_{k=1}^{\infty} f_k(b)$ converges. Does this imply that

$$\lim_{x \to b^-} F(x) = F(b)?$$

6

Return to Fourier Series

In the spring of 1808, Siméon Denis Poisson wrote up the committee's report on Fourier's *Theory of the Propagation of Heat in Solid Bodies*. The conclusion was that it contained nothing that was new or interesting. Behind this opinion lay Lagrange's opposition to the admission of Fourier's trigonometric series and his conviction that they must not converge. In the following years, Fourier attempted to meet Lagrange's objections and to convince him that his series did in fact converge. In the meantime, he conducted experiments, comparing the predictions of his mathematical models with observed phenomena.

The problem of modeling the flow of heat was of concern to many scientists of the time. In 1811, the Institut de France announced a competition for the best explanation of heat diffusion. Fourier reworked his earlier manuscript and submitted it. Despite continuing objections from Lagrange, he was awarded the prize. Lagrange could not deny him the award, but he could postpone publication. Even after Lagrange died in 1813, Fourier's manuscript continued to languish in the Institut. Fourier began to prepare a book to disseminate his ideas.

After the second fall of Napoléon, Fourier came to Paris as the director of the Bureau of Statistics for the department of the Seine. He was back at the center of intellectual life. His book, *Théorie analytique de la chaleur* (*Analytic theory of heat*), appeared in 1822. That same year he was elected perpetual secretary of the Académie des Sciences, the highest of scientific honors. He used that position in succeeding years to encourage and promote the careers of emerging mathematicians. Gustav Dirichlet, Sophie Germain, Joseph Liouville, Claude Navier, and Charles Sturm were among those who received his assistance and would remember him fondly.

The problem of the convergence of Fourier series was given its first published treatment in 1820 in a paper by Poisson. His work suffers from the defect that in the course of proving the convergence of Fourier series he needed—in a subtle way—to assume that they converged. Fourier tried to supply a proof in his book. He did see the fundamental difficulty and so was able to show the way to an eventual proof, but he himself did not succeed. In 1826, Cauchy took up this problem and published what he considered to be a solution. There were flaws in his work.

In January of 1829, at the age of 23 and from his new professorship in Berlin, Gustav Lejeune Dirichlet submitted the paper "Sur la convergence des séries trigonométriques qui servent à représenter une fonction arbitraire entre des limites données." It begins with a critique of Cauchy's paper, pointing out Cauchy's mistaken assumption that if v_n approaches w_n as n approaches infinity, and if $w_1 + w_2 + w_3 + \cdots$ converges, then so must $v_1 + v_2 + v_3 + \cdots$. This was a critical assumption on Cauchy's part. Without it, his argument collapses. Dirichlet pointed out that if we define

$$ w_n = \frac{(-1)^n}{\sqrt{n}}, \qquad v_n = \frac{(-1)^n}{\sqrt{n}} + \frac{1}{n}, $$

then v_n approaches w_n and the series $\sum_{n=1}^{\infty} w_n$ converges, but the series $\sum_{n=1}^{\infty} v_n$ diverges.

After pointing out Cauchy's error, Dirichlet goes on to give the first substantially correct proof for the validity of Fourier series. It is this proof that we shall investigate.

Dirichlet's great interest in mathematics was always number theory. This is where he did most of his work. In 1829, Fourier's health was fading. He would die the following spring. For Dirichlet, this paper was more than an answer to an abstract question in mathematics; it was a tribute to a mentor and friend, a validation of the new and disturbing series that Joseph Fourier had introduced to the scientific community in 1807.

6.1 Dirichlet's Theorem

Until now, we have not done full justice to Fourier's series of trigonometric functions. In Chapter 1, we considered the expansion of an even function, a function for which $f(x) = f(-x)$. For such a function, we look for a cosine expansion

$$ f(x) = a_0 + a_1 \cos x + a_2 \cos 2x + a_3 \cos 3x + \cdots, $$

where f is defined over the interval $(-\pi, \pi)$. There is an analogous sine series for any odd function, $g(x) = -g(-x)$:

$$ g(x) = b_1 \sin x + b_2 \sin 2x + b_3 \sin 3x + \cdots. $$

An arbitrary function can be expressed uniquely as the sum of an even function and an odd function (see exercises 6.1.4 and 6.1.5 at the end of this section). For an arbitrary function defined over the interval $(-\pi, \pi)$, we try to represent it as the sum of a cosine series and a sine series:

$$ F(x) = a_0 + \sum_{k=1}^{\infty} \left[a_k \cos(kx) + b_k \sin(kx) \right]. \tag{6.1} $$

Fourier had considered such general series. The heuristic argument for finding the co-efficients in an arbitrary Fourier series rests on the observation that for integer values of k and m,

$$\int_{-\pi}^{\pi} \cos(kx)\cos(mx)\,dx = \begin{cases} 0 & \text{if } k \neq m, \\ 2\pi & \text{if } k = m = 0, \\ \pi & \text{if } k = m \neq 0, \end{cases} \tag{6.2}$$

$$\int_{-\pi}^{\pi} \sin(kx)\sin(mx)\,dx = \begin{cases} 0 & \text{if } k \neq m, \\ \pi & \text{if } k = m \neq 0, \end{cases} \tag{6.3}$$

$$\int_{-\pi}^{\pi} \sin(kx)\cos(mx)\,dx = 0. \tag{6.4}$$

If we assume that our function F has such a Fourier series expansion and that it is legal to interchange the summation and the integral, then the coefficients can be determined from the following formulæ:

$$a_0 = \frac{1}{2\pi} \int_{-\pi}^{\pi} F(x)\,dx, \tag{6.5}$$

$$a_k = \frac{1}{\pi} \int_{-\pi}^{\pi} F(x)\cos(kx)\,dx, \quad (k \geq 1), \tag{6.6}$$

$$b_k = \frac{1}{\pi} \int_{-\pi}^{\pi} F(x)\sin(kx)\,dx. \tag{6.7}$$

Fourier contended that if we define the coefficents a_k and b_k by equations (6.5–6.7), then $F(x)$ will equal the series in equation (6.1) when x lies between $-\pi$ and π.

The Nature of the Problem

The first problem Fourier encountered was that of defining what he meant by

$$\int_{-\pi}^{\pi} F(x)\cos(kx)\,dx.$$

In 1807, integration was defined and understood as the inverse process of differentiation, what some of today's textbooks call "antidifferentiation." The connection between integration and problems of areas and volumes was well understood, but that did not change the fact that one defined

$$\int F(x)\cos(kx)\,dx$$

as a function whose derivative was $F(x)\cos(kx)$.

This was a conceptual problem for many of those encountering Fourier series for the first time. It is not always true that $F(x)\cos(kx)$ can be expressed as the derivative of a known function. Fourier was responsible for changing the definition of an integral from an antiderivative to an area. It was his idea to put limits on the integration sign and to talk of a **definite integral** that was to be defined in terms of the area between $F(x)\cos(kx)$ and the x axis.

Dirichlet was the first to realize that not every function could be integrated. He mentions the function that takes on one value at every rational number and a different value at every irrational number, for example

$$f(x) = \begin{cases} 1, & \text{if } x \text{ is rational,} \\ 0, & \text{if } x \text{ is irrational.} \end{cases}$$

He thus highlighted the **first assumption** that we need to make about our function, that $F(x)$ **is integrable over** $[-\pi, \pi]$. Cauchy in the 1820s, Riemann in the 1850s, and Lebesgue in the 1900s were each to expand and clarify the meaning of integration. The Cauchy and Riemann integrals will be explained in the next two sections. A brief introduction to the Lebesgue integral can be found in the *Epilogue*, Chapter 7.

Poisson, Fourier, and Cauchy had concentrated their attention on proving that

$$a_0 + \sum_{k=1}^{\infty} [a_k \cos(kx) + b_k \sin(kx)]$$

always converges when a_k and b_k are defined by equations (6.5–6.7). When Dirichlet tackled the problem of Fourier series in 1829, he saw that the difficulties were greater than anyone else had imagined. It was not just a question of convergence. In many cases the convergence would not be absolute. This means that a rearrangement of the summands could lead to a different value. Even if the series was known to converge, there was a legitimate question of whether it converged to $F(x)$.

Specifically, the problem is as follows. We define coefficients a_k and b_k according to equations (6.5–6.7). We then form a partial sum:

$$\begin{aligned} F_n(x) &= a_0 + \sum_{k=1}^{n} [a_k \cos(kx) + b_k \sin(kx)] \\ &= \frac{1}{2\pi} \int_{-\pi}^{\pi} F(t)\, dt \\ &\quad + \sum_{k=1}^{n} \left[\left(\frac{1}{\pi} \int_{-\pi}^{\pi} F(t) \cos(kt)\, dt \right) \cos(kx) \right. \\ &\qquad \left. + \left(\frac{1}{\pi} \int_{-\pi}^{\pi} F(t) \sin(kt)\, dt \right) \sin(kx) \right]. \end{aligned} \qquad (6.8)$$

We must prove convergence to $F(x)$:

> *Given a positive error bound ϵ and a value for x in $(-\pi, \pi)$, we must be able to find a response N so that*
>
> $$n \geq N \quad \text{implies that} \quad |F(x) - F_n(x)| < \epsilon.$$

The proof that we give here is modeled on Dirichlet's original approach, but we shall incorporate some simplifications that were suggested by Ossian Bonnet in 1849 and Bernhard Riemann in 1854.

Simplifying $F_n(x)$

Since $F_n(x)$ involves finite sums, we are allowed to interchange the integrals and the summations. We can rewrite the partial sum as

$$
\begin{aligned}
F_n(x) &= \frac{1}{2\pi} \int_{-\pi}^{\pi} F(t)\, dt \\
&\quad + \frac{1}{\pi} \int_{-\pi}^{\pi} \left(\sum_{k=1}^{n} \cos(kt)\cos(kx) + \sin(kt)\sin(kx) \right) F(t)\, dt \\
&= \frac{1}{\pi} \int_{-\pi}^{\pi} \left(\frac{1}{2} + \sum_{k=1}^{n} \cos k(t - x) \right) F(t)\, dt.
\end{aligned}
\tag{6.9}
$$

In the last line, we used the trigonometric identity

$$
\cos(kt)\cos(kx) + \sin(kt)\sin(kx) = \cos k(t - x).
\tag{6.10}
$$

We now use another trigonometric identity (see exercise 6.1.16),

$$
\frac{1}{2} + \cos u + \cos 2u + \cdots + \cos nu = \frac{\sin[(2n + 1)u/2]}{2\sin[u/2]},
\tag{6.11}
$$

to simplify $F_n(x)$ further:

$$
F_n(x) = \frac{1}{\pi} \int_{-\pi}^{\pi} \frac{\sin[(2n + 1)(t - x)/2]}{2\sin[(t - x)/2]} F(t)\, dt.
\tag{6.12}
$$

We want to get our variable x out of the sine functions and back into the argument of F, and so we want to make a change of variable inside the integral. In order to simplify later calculations, it will be helpful if we first shift the entire interval of integration by distance x. In other words, we want to rewrite our last equation as

$$
F_n(x) = \frac{1}{\pi} \int_{-\pi+x}^{\pi+x} \frac{\sin[(2n + 1)(t - x)/2]}{2\sin[(t - x)/2]} F(t)\, dt.
\tag{6.13}
$$

This is legal as long as the integrand is periodic with period 2π. If it is periodic, then it does not matter which interval of length 2π we choose for our integration; the integral from $-\pi$ to π will be the same as the integral from $-\pi + x$ to $\pi + x$. There is no problem with the sine functions. Their ratio has period 2π. The only possible problem lies with the function F. But we have only specified the values of F for t between $-\pi$ and π. We are free to define F as we wish when t is outside this interval. In particular, we can choose to define F to be periodic with period 2π:

$$
F(t + 2\pi) = F(t).
\tag{6.14}
$$

This then is the **second assumption** about the function for which we seek a Fourier series representation: that it is **periodic with period** 2π. It is important to remember that this is not really a restriction since we began by only specifying the values of the function between $-\pi$ and π.

Splitting the Integral

We now split our integral into two pieces and use a different substitution on each. In the first piece, we replace t with $x - 2u$. In the second piece, t becomes $x + 2u$.

$$F_n(x) = \frac{1}{\pi} \int_{-\pi+x}^{x} \frac{\sin[(2n+1)(t-x)/2]}{2\sin[(t-x)/2]} F(t)\, dt$$
$$+ \frac{1}{\pi} \int_{x}^{\pi+x} \frac{\sin[(2n+1)(t-x)/2]}{2\sin[(t-x)/2]} F(t)\, dt$$
$$= \frac{1}{\pi} \int_{0}^{\pi/2} \frac{\sin[(2n+1)u]}{\sin u} F(x-2u)\, du$$
$$+ \frac{1}{\pi} \int_{0}^{\pi/2} \frac{\sin[(2n+1)u]}{\sin u} F(x+2u)\, du. \qquad (6.15)$$

This is essentially as far as Fourier went in his analysis, although he did give arguments why the sum of these integrals should approach $F(x)$. Before continuing with Dirichlet's paper, it is important to pause and look at what we have found. For convenience, we shall define $F_n^-(x)$ and $F_n^+(x)$ to be these two integrals,

$$F_n^-(x) = \frac{1}{\pi} \int_{0}^{\pi/2} \frac{\sin[(2n+1)u]}{\sin u} F(x-2u)\, du, \qquad (6.16)$$
$$F_n^+(x) = \frac{1}{\pi} \int_{0}^{\pi/2} \frac{\sin[(2n+1)u]}{\sin u} F(x+2u)\, du. \qquad (6.17)$$

We shall concentrate on F_n^+. Similar results apply to F_n^-.

Qualitative Analysis of F_n^+

The first thing that should strike you is that $F_n^+(x)$ depends not just on $F(x)$ but on the value of this function over the entire interval from x to $x + \pi$. In fact, the value of $F_n^+(x)$ is actually *independent* of the value of $F(x)$. If we leave this function the same at every point except x and change its value at x, then we do not change the value of the integral.

This is very discouraging news if we want to prove that

$$F(x) = \lim_{n \to \infty} [F_n^-(x) + F_n^+(x)]$$

since neither of these integrals depends on the value of F at x. It shows that not every function can have a Fourier series expansion. The value of F at x is going to have to be determined by its values at points to the left and right of x.

To see what this dependence is, we take a closer look at what we are integrating. The graphs of $y = \sin[(2n+1)u]/\sin u$ for $n = 4, 8$, and 12 are given in Figure 6.1. We can easily show that for each n the curve has a spike of height $2n + 1$ at the y axis. As n gets larger, the spike gets narrower. The graph hits the u axis for the first time at $u = \pi/(2n+1)$. We then get oscillations that damp down to a fairly constant amplitude as we move toward $\pi/2$. As n gets larger, these oscillations become tighter (see Figure 6.2), increasing in

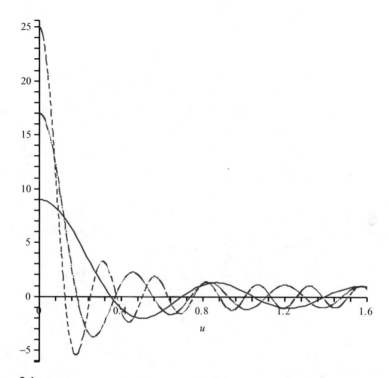

FIGURE 6.1. Graphs of $y = \sin[(2n+1)u]/\sin u$ for $n = 4$ (solid), 8 (dots), and 12 (dashes).

frequency. Because of its importance to the analysis of Fourier series, this function,

$$K_n(u) = \frac{\sin[(2n+1)u]}{\sin u},$$

has a name. It is called the **Dirichlet kernel**.

We take the Dirichlet kernel, multiply it by $F(x + 2u)$, and then integrate from $u = 0$ to $u = \pi/2$. Almost all of the area occurs inside the first spike. The value of the integral will be dominated by the values of $F(x + 2u)$ for $0 < u < \pi/(2n + 1)$. If F is continuous over this interval and n is large, then $F(x + 2u)$ will stay fairly constant, and this initial part of the integral will be approximately the value of

$$F\left(x + \frac{\pi}{2n+1}\right)$$

times the area under the spike. As we shall see, the area under the spike is approximately $\pi/2$.

After the spike, we are integrating $F(x + 2u)$ multiplied by a function that has tight oscillations. If $F(x + 2u)$ stays fairly constant over one complete oscillation, then the area above the u axis will approximate the area below the u axis for a net contribution that is close to zero. For a large value of n, we can expect the contribution from

$$\int_{\pi/(2n+1)}^{\pi/2} \frac{\sin[(2n+1)u]}{\sin u} F(x + 2u)\, du$$

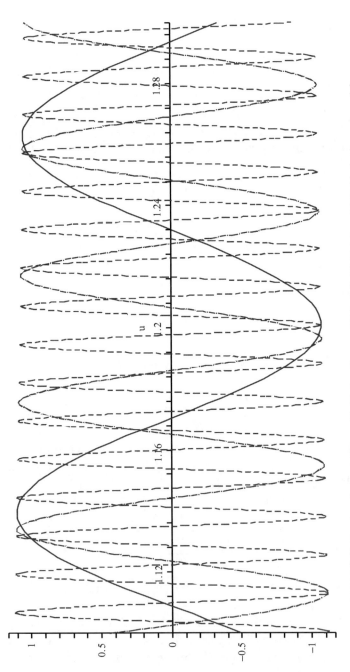

FIGURE 6.2. Graphs of $y = \sin[(2n+1)u]/\sin u$ for $n = 25$ (solid), 75 (dots), and 250 (dashes) near $u = 1.2$.

to be close to zero. While all of this needs to be made much more precise, you should be willing to believe that

$$
\begin{aligned}
F_n^+(x) &= \frac{1}{\pi} \int_0^{\pi/2} \frac{\sin[(2n+1)u]}{\sin u} F(x+2u)\, du \\
&\approx \frac{1}{\pi} \int_0^{\pi/(2n+1)} \frac{\sin[(2n+1)u]}{\sin u} F(x+2u)\, du \\
&\approx \frac{1}{\pi} \cdot \frac{\pi}{2} F\left(x + \frac{\pi}{2n+1}\right) = \frac{1}{2} F\left(x + \frac{\pi}{2n+1}\right).
\end{aligned} \tag{6.18}
$$

As n gets larger, this approaches $F(x)/2$ *provided F is continuous at x from the right.* Similarly, $F_n^-(x)$ approaches $F(x)/2$ if and only if F is continuous at x from the left. This suggests that F must be a continuous function at every value of x.

How to Avoid Continuity

Some of the most interesting and useful functions for which we want to find a Fourier series expansion are not continuous. One example is the series that we met in Chapter 1 that alternates between $+1$ and -1. Dirichlet was the first to see what it would mean to avoid continuity.

As n gets larger, $F_n^+(x)$ approaches the limit from the right of $F(x)/2$ (see page 92):

$$
\lim_{n \to \infty} F_n^+(x) = \frac{1}{2} \lim_{t \to x^+} F(t). \tag{6.19}
$$

Dirichlet invented a suggestive notation for this limit from the right:

$$
F(x+0) = \lim_{t \to x^+} F(t). \tag{6.20}
$$

Just as the "+" in an infinite summation is not really addition, so the "+" in $F(x+0)$ is not really addition. He similarly defined

$$
F(x-0) = \lim_{t \to x^-} F(t). \tag{6.21}
$$

We see that the best we can hope to prove is that

$$
\lim_{n \to \infty} [F_n^-(x) + F_n^+(x)] = \frac{F(x-0) + F(x+0)}{2}. \tag{6.22}
$$

If $F(x)$ is to have a Fourier series expansion, it does not have to be continuous, but it must be true that at every $x \in [-\pi, \pi]$ we have

$$
F(x) = \frac{F(x-0) + F(x+0)}{2}. \tag{6.23}
$$

This is our **third assumption**. It says that wherever F has a discontinuity, its **value must be the average of the limit from the left and the limit from the right**.

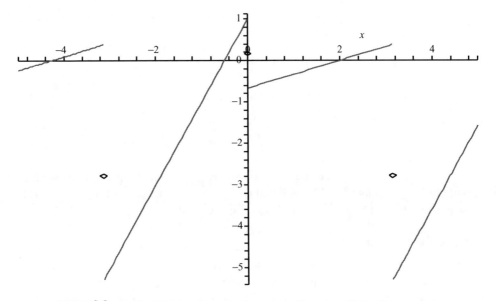

FIGURE 6.3. Graph of $F(x) = 2x + 1$, $(-\pi < x < 0)$, $(x - 2)/3$, $(0 < x < \pi)$.

For example, we can find a Fourier series expansion for the function

$$F(x) = \begin{cases} 2x + 1, & -\pi < x < 0, \\ (x - 2)/3, & 0 < x < \pi, \end{cases}$$

(see Figure 6.3). At $x = 0$, the limit from the left is 1, the limit from the right is $-2/3$. The Fourier series for this function will take the value

$$F(0) = \frac{1 - 2/3}{2} = \frac{1}{6}.$$

We also know that $F(x)$ is periodic of period 2π so that $F(-\pi) = F(\pi)$. To find the value at these endpoints, we find the limits from the right and the left:

$$F(-\pi + 0) = -2\pi + 1,$$
$$F(\pi - 0) = (\pi - 2)/3,$$
$$F(-\pi) = F(\pi) = \frac{-2\pi + 1 + (\pi - 2)/3}{2} = \frac{1 - 5\pi}{6}.$$

Dirichlet's Theorem

We have seen that proving that a function is equal to its Fourier expansion is equivalent to proving that $F(x - 0)/2 = \lim_{n \to \infty} F_n^-(x)$, and $F(x + 0)/2 = \lim_{n \to \infty} F_n^+(x)$ for all $x \in (-\pi, \pi)$. We have also seen that there are two pieces to the proof that $F(x + 0)/2 = \lim_{n \to \infty} F_n^+(x)$.

First, we must show that we can force

$$\frac{1}{\pi} \int_0^a \frac{\sin[(2n + 1)u]}{\sin u} F(x + 2u) \, du$$

to be as close as we want to $F(x+0)/2$ by taking an a close to 0 and an n that is sufficiently large. To do this, we shall have to have an interval of the form $(x, x+2a)$ where F is continuous. Even though we are allowing discontinuities, they must be separated by intervals on which F is continuous. This leads us to our **fourth assumption**, that F is **piecewise continuous**, which means that there are at most a finite number of values on $[-\pi, \pi]$ at which F is not continuous. As we shall see later, piecewise continuity implies integrability. Our first assumption has been subsumed under this stronger assumption.

Second, we must show that we can force

$$\int_a^{\pi/2} \frac{\sin[(2n+1)u]}{\sin u} F(x+2u)\,du$$

to be as close as we want to zero by taking an n that is sufficiently large. We need to use the frequent oscillations of $\sin[(2n+1)u]$ to get our cancellations. This means that we want to be able to control the oscillations of $F(x+2u)$. The **fifth and last assumption** is that F is **bounded and piecewise monotonic** (see page 87).

Dirichlet believed that this last assumption was not necessary, but he could not see how to prove his theorem without it. In fact, this last assumption can be weakened considerably. We shall content ourselves with the theorem as Dirichlet proved it.

Theorem 6.1 (Dirichlet's Theorem). *Let F be a bounded, piecewise continuous and piecewise monotonic function on $[-\pi, \pi]$. Furthermore, assume that F is periodic with period 2π and that*

$$F(x) = \frac{F(x+0) + F(x-0)}{2}$$

for every value of x. We define coefficients

$$a_0 = \frac{1}{2\pi} \int_{-\pi}^{\pi} F(x)\,dx,$$

$$a_k = \frac{1}{\pi} \int_{-\pi}^{\pi} F(x)\cos(kx)\,dx, \quad (k \geq 1),$$

$$b_k = \frac{1}{\pi} \int_{-\pi}^{\pi} F(x)\sin(kx)\,dx.$$

Then, at every value of x,

$$F(x) = a_0 + \sum_{k=1}^{\infty} \Big[a_k \cos(kx) + b_k \sin(kx) \Big]. \tag{6.24}$$

In fact, a piecewise monotonic function on a closed and bounded interval must be piecewise continuous. Since proving this would take us beyond the scope of this book, we leave both assumptions in the statement of Dirichlet's theorem.

Riemann's Lemma

We shall first show that

$$\int_a^{\pi/2} \frac{\sin[(2n+1)u]}{\sin u} F(x+2u)\, du, \quad 0 < a < \pi/2,$$

can be made arbitrarily close to zero by taking n sufficiently large. To do this, we shall concentrate on a single interval (a, b), $0 < a < b \le \pi/2$, where $F(x + 2u)/ \sin u$ is continuous and bounded. Since our whole interval from a to $\pi/2$ is made up of a finite number of such pieces, it is enough to show that the integral over each such piece can be forced to be arbitrarily close to zero.

Working with the integral over $[a, b]$, we do not change the value of the integral if we change the value of our function just at $x = a$ or b. It is convenient when working on this particular piece to redefine F, if necessary, so that $F(x + 2a) = F(x + 2a + 0)$ (the limit from the right) and $F(x + 2b) = F(x + 2b - 0)$ (the limit from the left) so that we can consider $F(x + 2u)/ \sin u$ to be continuous over the closed interval $[a, b]$. Since we are now working over a closed interval, the restriction that F be bounded is implied by the continuity (see Theorem 3.6). For simplicity of notation, we write $g(u)$ for $F(x + 2u)/ \sin u$. The only restriction on g is that it be continuous over $[a, b]$ where we mean continuous from the right at a, continuous from the left at b, and continuous in the usual sense at all points between a and b. We also write M in place of $2n + 1$. For this lemma, it is not necessary for this multiplier to be an odd integer.

Lemma 6.2 (Riemann's Lemma). *If $g(u)$ is a continuous function over the interval $[a, b]$, $0 < a < b \le \pi/2$, then*

$$\lim_{M \to \infty} \int_a^b \sin(Mu)\, g(u)\, du = 0. \tag{6.25}$$

Before we begin this proof, we need another result. We are going to need to know that given any $\epsilon > 0$, we can force $g(u)$ and $g(v)$ to be within ϵ of each other just by keeping u sufficiently close to v. This sounds like the definition of continuity, but it is not quite the same. Continuity is defined by being able to find a response to ϵ at a specific value of u. We are going to need a response that works for *all* u in $[a, b]$. This is more than continuity. This is **uniform continuity**.

Definition: uniform continuity

We say that f is **uniformly continuous over the interval** I if given any positive error bound ϵ, we can always reply with a tolerance δ such that for *any* points a and x in I, if x is within δ of a, then $f(x)$ is within ϵ of $f(a)$:

$$|x - a| < \delta \quad \text{implies that} \quad |f(x) - f(a)| < \epsilon.$$

Uniform continuity does not have to be defined over an interval. It can equally well be defined over any set. This should be compared with the definition of continuity given on page 81. With uniform continuity, the value of δ does not depend on a.

The hypothesis of Riemann's lemma only called for continuity. Fortunately, the next lemma tells us that since we are working over a closed and bounded interval, this is enough.

Lemma 6.3 (Continuity on $[a, b] \implies$ Unif. Continuity). *If f is continuous over the closed and bounded interval $[a, b]$, then it is uniformly continuous over this interval.*

Proof: We shall prove the contra-positive: if f is not uniformly continuous, then there is at least one point in $[a, b]$ at which f is not continuous.

To say that f is not uniformly continuous over $[a, b]$ means that there is some $\epsilon > 0$ for which there is no uniform response, no single δ that works at every point x in $[a, b]$. This, in turn, means that given any $\delta > 0$, we can always find an $x \in [a, b]$ and another point $y \in [a, b]$ such that $0 < |x - y| < \delta$ but $|f(x) - f(y)| \geq \epsilon$.

We choose an $\epsilon > 0$ for which there is no uniform response and, for each $n \in \mathbb{N}$, choose x_n, y_n in $[a, b]$ such that $|x_n - y_n| < 1/n$ and $|f(x_n) - f(y_n)| \geq \epsilon$. Let $x = \overline{\lim}_{n \to \infty} x_n$ and $y = \overline{\lim}_{n \to \infty} y_n$. These exist because both sequences are bounded. They must be equal because

$$|x - y| \leq |x - x_n| + |x_n - y_n| + |y_n - y|,$$

and each of the pieces on the right can be made as small as desired by taking n sufficiently large.

Since $x = y$, the function f cannot be continuous at this common upper limit. We can find pairs (x_n, y_n) as close to $x = y$ as we wish, but the values of $F(x_n)$ and $F(y_n)$ will always stay at least ϵ apart.

Q.E.D.

Proof: (Riemann's Lemma) We must show that for any specified error bound ϵ, there is a response M such that $N \geq M$ implies that

$$\left| \int_a^b \sin(Nu)\, g(u)\, du \right| < \epsilon.$$

The key is to partition our interval $[a, b]$ into m equal subintervals:

$$a = u_0 < u_1 < u_2 < \cdots < u_m = b,$$

$$u_k - u_{k-1} = \frac{b - a}{m},$$

where we choose m so that if $|u - v| \leq (b - a)/m$, then $|g(u) - g(v)| < \epsilon/2(b - a)$. Since g is uniformly continuous over $[a, b]$, we can always find such an m.

The proof proceeds by approximating our function g by the constant $g(u_{k-1})$ on the kth interval. As M increases, the integral of

$$\sin(Mu)\, g(u_{k-1})$$

can be made arbitrarily small. Uniform continuity enables us to control the size of the error introduced when we substitute $g(u_{k-1})$ for $g(u)$.

We break our integral up into a sum of integrals over these subintervals:

$$\left| \int_a^b \sin(Mu)\, g(u)\, du \right| = \left| \sum_{k=1}^m \int_{u_{k-1}}^{u_k} \sin(Mu)\, g(u)\, du \right|$$

$$= \left| \sum_{k=1}^m \int_{u_{k-1}}^{u_k} \sin(Mu)\, [g(u_{k-1}) + g(u) - g(u_{k-1})]\, du \right|$$

$$\leq \left| \sum_{k=1}^m \int_{u_{k-1}}^{u_k} \sin(Mu)\, g(u_{k-1})\, du \right|$$

$$+ \left| \sum_{k=1}^m \int_{u_{k-1}}^{u_k} \sin(Mu)\, [g(u) - g(u_{k-1})]\, du \right|$$

$$\leq \sum_{k=1}^m \left| \int_{u_{k-1}}^{u_k} \sin(Mu)\, g(u_{k-1})\, du \right|$$

$$+ \sum_{k=1}^m \int_{u_{k-1}}^{u_k} \left| \sin(Mu)\, [g(u) - g(u_{k-1})] \right|\, du. \tag{6.26}$$

Since g is continuous on this closed interval, it must be bounded. Let K be an upper bound for $|g|$:

$$|g(u)| \leq K \qquad \text{for all } u \in [a, b].$$

In our first integral, $g(u_{k-1})$ is independent of u, and so we can pull it outside the integral and then replace $|g(u_{k-1})|$ by the upper bound K. In the second integral, we use the fact that $|\sin(Mu)| \leq 1$ and

$$|g(u) - g(u_{k-1})| \leq \frac{\epsilon}{2(b-a)}.$$

Using these results, we simplify our inequality,

$$\left| \int_a^b \sin(Mu)\, g(u)\, du \right| \leq \sum_{k=1}^m \left| \int_{u_{k-1}}^{u_k} \sin(Mu)\, g(u_{k-1})\, du \right|$$

$$+ \sum_{k=1}^m \int_{u_{k-1}}^{u_k} \left| \sin(Mu)\, [g(u) - g(u_{k-1})] \right|\, du$$

$$\leq \sum_{k=1}^m K \left| \int_{u_{k-1}}^{u_k} \sin(Mu)\, du \right| + \sum_{k=1}^m \int_{u_{k-1}}^{u_k} \frac{\epsilon}{2(b-a)}\, du$$

$$= \sum_{k=1}^m K \frac{|-\cos(Mu_k) + \cos(Mu_{k-1})|}{M} + \frac{\epsilon}{2}. \tag{6.27}$$

Since $|-\cos(Mu_k) + \cos(Mu_{k-1})| \leq 2$, the first part of this bound can be bounded by $2mK/M$. Our upper bound can be simplified to

$$\left| \int_a^b \sin(Mu)\, g(u)\, du \right| \leq \frac{2mK}{M} + \frac{\epsilon}{2}. \tag{6.28}$$

We have to careful. The value of m has been forced by our choice of ϵ and the values of a, b and K are outside our control. But once we have chosen ϵ (thus m), we are still free to choose M as large as we wish. We find an M so that

$$\frac{2mK}{M} < \frac{\epsilon}{2}.$$

If $N \geq M$, then the absolute value of the integral is strictly less than ϵ.

Q.E.D.

The Integral of the Dirichlet Kernel

> **Lemma 6.4 (Integral of Dirichlet Kernel).** *For any positive integer n,*
> $$\int_0^{\pi/2} \frac{\sin[(2n+1)u]}{\sin u}\, du = \frac{\pi}{2}.$$

Proof: From equation (6.11), we know that

$$\frac{\sin[(2n+1)u]}{\sin u} = 1 + 2\cos(2u) + 2\cos(4u) + \cdots + 2\cos(2nu).$$

Substituting this into our integral and integrating each summand, we get exactly $\pi/2$.

Q.E.D.

Bonnet's Mean Value Theorem

The final lemma that we need is Bonnet's form of the mean value theorem, a version that he discovered and proved in 1849 specifically to simplify the proof of Dirichlet's theorem. As he pointed out, it also has many other applications. We shall postpone the proof of this lemma until the next section. Here, for the first time, we shall need to be very careful about exactly what we mean by an integral.

> **Lemma 6.5 (Bonnet's Mean Value Theorem).** *Let f be integrable and let g be a nonnegative, increasing function on $[\alpha, \beta]$. There is at least one value ζ strictly between α and β for which*
> $$\int_\alpha^\beta f(t)g(t)\, dt = g(\beta) \int_\zeta^\beta f(t)\, dt. \tag{6.29}$$

As an example, let $f(t) = \sin t$ and $g(t) = t^2$ on the interval $[0, 2\pi]$. This lemma promises us a number ζ for which

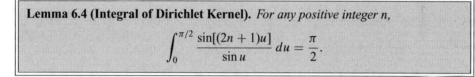

$$\int_0^{2\pi} t^2 \sin t\, dt = 4\pi^2 \int_\zeta^{2\pi} \sin t\, dt.$$

Conclusion to the Proof of Dirichlet's Theorem

As we have seen, we need to show that we can force

$$\left| \frac{F(x+0)}{2} - \frac{1}{\pi} \int_0^{\pi/2} \frac{\sin[(2n+1)u]}{\sin u} F(x+2u)\,du \right|$$

and

$$\left| \frac{F(x-0)}{2} - \frac{1}{\pi} \int_0^{\pi/2} \frac{\sin[(2n+1)u]}{\sin u} F(x-2u)\,du \right|$$

to each be arbitrarily small by taking n sufficiently large. Any argument that works for one of these differences will work for the other. Given a positive error bound ϵ, our problem is to find a response N for which $n \geq N$ implies that

$$\left| \frac{F(x+0)}{2} - \frac{1}{\pi} \int_0^{\pi/2} \frac{\sin[(2n+1)u]}{\sin u} F(x+2u)\,du \right| < \epsilon. \tag{6.30}$$

Lemma 6.4 implies that

$$\frac{F(x+0)}{2} = \frac{1}{\pi} \int_0^{\pi/2} \frac{\sin[(2n+1)u]}{\sin u} F(x+0)\,du. \tag{6.31}$$

Making this substitution, we can rewrite the left side of equation (6.30) as

$$\left| \frac{1}{\pi} \int_0^{\pi/2} \frac{\sin[(2n+1)u]}{\sin u} F(x+0)\,du \ - \ \frac{1}{\pi} \int_0^{\pi/2} \frac{\sin[(2n+1)u]}{\sin u} F(x+2u)\,du \right|$$

$$= \left| \frac{1}{\pi} \int_0^{\pi/2} \frac{\sin[(2n+1)u]}{\sin u} [F(x+0) - F(x+2u)]\,du \right|$$

$$\leq \frac{1}{\pi} \left| \int_0^a \frac{\sin[(2n+1)u]}{\sin u} [F(x+0) - F(x+2u)]\,du \right|$$

$$+ \frac{1}{\pi} \left| \int_a^{\pi/2} \frac{\sin[(2n+1)u]}{\sin u} [F(x+0) - F(x+2u)]\,du \right|, \tag{6.32}$$

where a is some point between 0 and $\pi/2$ whose exact position we shall determine later. Lemma 6.2 tells us that once we have chosen a value for a, we can make the second term as small as we want by taking a sufficiently large value for n.

The idea at this point is to choose our a small enough so that

$$|F(x+0) - F(x+2a)|$$

is very small when $0 < u < a$. We can do this because $F(x+0)$ is the limit from the right of $F(x+2u)$. There is one potential problem. We want to make

$$\left| \int_0^a \frac{\sin[(2n+1)u]}{\sin u} [F(x+0) - F(x+2u)]\,du \right|$$

very small. How do we know that the choice of a does not depend on the choice of n? This is a real danger. If it does, then our argument collapses: in the second integral the choice of n depends on a and in the first integral the choice of a depends on n. We might find ourselves in exactly the kind of trap that Cauchy fell into when he proved that every series of continuous functions is continuous.

Bonnet's mean value theorem, Lemma 6.5, comes to our rescue. We insist that a be close enough to 0 that $F(x + 2u)$ is continuous and monotonic on $(0, a]$. We define

$$g(u) = \begin{cases} |F(x + 0) - F(x + 2a)|, & 0 < u \le a, \\ 0, & u = 0. \end{cases}$$

Since $F(x + 0) - F(x + 2u)$ is monotonic on $(0, a]$ and approaches 0 as u approaches 0 from the right, either it is ≥ 0 for all $u \in [0, a]$ or it is ≤ 0 for all $u \in [0, a]$. In either case, $g(u)$ will be nonnegative and increasing on $[0, a]$. Since $F(x + 0) - F(x + 2u)$ does not change sign on this interval,

$$\left| \int_0^a \frac{\sin[(2n + 1)u]}{\sin u} [F(x + 0) - F(x + 2u)] \, du \right| = \left| \int_0^a \frac{\sin[(2n + 1)u]}{\sin u} g(u) \, du \right|.$$

$$(6.33)$$

We can now apply Lemma 6.5:

$$\left| \int_0^a \frac{\sin[(2n + 1)u]}{\sin u} [F(x + 0) - F(x + 2u)] \, du \right| = g(a) \left| \int_\zeta^a \frac{\sin[(2n + 1)u]}{\sin u} \, du \right|$$

$$= |F(x + 0) - F(x + 2a)| \left| \int_\zeta^a \frac{\sin[(2n + 1)u]}{\sin u} \, du \right|. \qquad (6.34)$$

Lemmas 6.2 and 6.4 imply that the integral

$$\left| \int_\zeta^a \frac{\sin[(2n + 1)u]}{\sin u} \, du \right|$$

is bounded as n increases, regardless of the value of ζ. It is not difficult to see that it is bounded by π (exercises 6.1.12 and 6.1.15). We need to choose an a that is close enough to zero so that

$$|F(x + 0) - F(x + 2a)| < \epsilon/2. \qquad (6.35)$$

It follows that

$$\frac{1}{\pi} \left| \int_0^a \frac{\sin[(2n + 1)u]}{\sin u} [F(x + 0) - F(x + 2u)] \, du \right|$$

$$= \frac{1}{\pi} \left| F(x + 0) - F(x + 2a) \right| \left| \int_\zeta^a \frac{\sin[(2n + 1)u]}{\sin u} \, du \right|$$

$$< \frac{1}{\pi} \cdot \frac{\epsilon}{2} \cdot \pi = \frac{\epsilon}{2}. \qquad (6.36)$$

Thanks to Bonnet's mean value theorem, we have been able to find a value of a that makes the first piece of our integral less than $\epsilon/2$ regardless of the choice of n. We are now free to choose an N that depends on a. We respond with any N for which $n \ge N$ implies that

$$\frac{1}{\pi} \left| \int_a^{\pi/2} \frac{\sin[(2n + 1)u]}{\sin u} [F(x + 0) - F(x + 2u)] \, du \right| < \frac{\epsilon}{2}. \qquad (6.37)$$

Q.E.D.

Exercises

The symbol $\boxed{\textbf{M\&M}}$ indicates that *Maple* and *Mathematica* codes for this problem are available in the **Web Resources** at **www.macalester.edu/aratra**.

6.1.1. Prove that

$$\sum_{k=1}^{\infty} \frac{(-1)^k}{\sqrt{k}}$$

converges but

$$\sum_{k=1}^{\infty} \left(\frac{(-1)^k}{\sqrt{k}} + \frac{1}{k} \right)$$

diverges.

6.1.2. Let $\sum_{k=1}^{\infty} w_k$ be any convergent series. Prove the divergence of

$$\sum_{k=1}^{\infty} \left(w_k + \frac{1}{k} \right).$$

6.1.3. Find two other examples of infinite series, $\sum_{k=1}^{\infty} a_k$ and $\sum_{k=1}^{\infty} b_k$, for which the first converges, the second diverges, but

$$\lim_{k \to \infty} (a_k - b_k) = 0.$$

6.1.4. Let F be an arbitrary function defined for all real values of x. Let

$$f(x) = \frac{F(x) + F(-x)}{2}, \tag{6.38}$$

$$g(x) = \frac{F(x) - F(-x)}{2}. \tag{6.39}$$

Prove that f is an even function and that g is an odd function and that

$$F(x) = f(x) + g(x).$$

6.1.5. Prove that if F can be written as the sum of an even function and an odd function, $F = f + g$, then f and g *must* satisfy equations (6.38) and (6.39).

6.1.6. We have seen that if $\sum_{k=1}^{\infty} a_k$ and $\sum_{k=1}^{\infty} b_k$ each converge, then the series $\sum_{k=1}^{\infty} (a_k + b_k)$ must converge. It does not necessarily work the other way. For example,

$$\sum_{k=1}^{\infty} \left(\frac{1}{k} - \frac{1}{k+1} \right) = \sum_{k=1}^{\infty} \frac{1}{k(k+1)}$$

converges, but neither

$$\sum_{k=1}^{\infty} \frac{1}{k} \quad \text{nor} \quad \sum_{k=1}^{\infty} \frac{-1}{k+1}$$

converges. Discuss whether or not it is possible to have a Fourier series,

$$a_0 + \sum_{k=1}^{\infty} [a_k \cos(kx) + b_k \sin(kx)]$$

converge for all x without either

$$a_0 + \sum_{k=1}^{\infty} a_k \cos(kx) \quad \text{or} \quad \sum_{k=1}^{\infty} b_k \sin(kx)$$

converging.

6.1.7. It is often convenient to work with the Fourier series of a function F with period 2 that is defined on $[-1, 1]$. Show that, in this case, the Fourier series is given by

$$F(x) = a_0 + \sum_{k=1}^{\infty} [a_k \cos(k\pi x) + b_k \sin(k\pi x)],$$

where

$$a_0 = \frac{1}{2} \int_{-1}^{1} F(x) \, dx,$$

$$a_k = \int_{-1}^{1} F(x) \cos(k\pi x) \, dx, \quad (k \geq 1),$$

$$b_k = \int_{-1}^{1} F(x) \sin(k\pi x) \, dx.$$

6.1.8. **M&M** Find the Fourier series expansions for each of the following functions that are defined on $(-1, 1)$ and have period 2: Find the value of this Fourier series at $x = 1$.

a. $f(x) = x^2$
b. $f(x) = \cos(3\pi x)$
c. $f(x) = \sin x$
d. $f(x) = e^x$

6.1.9. Find the Fourier series expansion for the function with period 2 that is equal to x^2 on the interval $(1, 3)$. What is the value of this Fourier series at $x = 1$?

6.1.10. Assume that

$$F(x) = a_0 + \sum_{k=1}^{\infty} [a_k \cos(kx) + b_k \sin(kx)]$$

converges uniformly. Use equations (6.2) and (6.3) to prove that

$$a_0 = \frac{1}{2\pi} \int_{-\pi}^{\pi} F(x) \, dx,$$

$$a_k = \frac{1}{\pi} \int_{-\pi}^{\pi} F(x) \cos(kx) \, dx, \quad (k \geq 1),$$

$$b_k = \frac{1}{\pi} \int_{-\pi}^{\pi} F(x) \sin(kx) \, dx.$$

6.1.11. **M&M** We know that

$$\cos x - \frac{1}{3} \cos 3x + \frac{1}{5} \cos 5x - \frac{1}{7} \cos 7x + \cdots$$

converges to $\pi/4$ when $-\pi/2 < x < \pi/2$, but that it does not converge absolutely. Choose at least four different values of x between 0 and $\pi/2$. For each value of x, apply the Riemann algorithm as described on page 177 to find the first twenty terms of the rearrangement of this series that converges to 1. Does the same rearrangement work for every value of x?

6.1.12. Prove equation (6.11):

$$\frac{1}{2} + \cos u + \cos 2u + \cdots + \cos nu = \frac{\sin[(2n+1)u/2]}{2\sin[u/2]}.$$

6.1.13. (**M&M**) Approximate the values of

$$\frac{1}{\pi} \int_0^{\pi/2} \frac{\sin[(2n+1)u]}{\sin u} \sqrt{9+2u}\, du$$

for various values of n. Describe what happens as n increases. What value do you expect it to approach?

6.1.14. Find a value of ζ for which

$$\int_0^{2\pi} t \sin t\, dt = 2\pi \int_\zeta^{2\pi} \sin t\, dt.$$

6.1.15. Prove that

$$\int_0^{\pi/(2n+1)} \frac{\sin[(2n+1)u]}{\sin u}\, du < \pi.$$

6.1.16. Justify the statement that if $0 \le \zeta < a \le \pi/2$, then

$$\left| \int_\zeta^a \frac{\sin[(2n+1)u]}{\sin u}\, du \right| \le \int_0^{\pi/(2n+1)} \frac{\sin[(2n+1)u]}{\sin u}\, du.$$

6.1.17. (**M&M**) Find the coefficients of the Fourier series expansion of

$$F(x) = \begin{cases} 2x+1, & -\pi < x < 0, \\ (x-2)/3, & 0 < x < \pi. \end{cases}$$

Evaluate partial sums for $x = 0$ and $x = \pi$. Do they approach the expected values?

6.1.18. Show that Bonnet's mean value theorem is equivalent to the statement that if f is integrable and g is a nonnegative, decreasing function on $[\alpha, \beta]$, then there is at least one value ζ strictly between α and β for which

$$\int_\alpha^\beta f(t)g(t)\, dt = g(\alpha) \int_\alpha^\zeta f(t)\, dt. \tag{6.40}$$

6.2   The Cauchy Integral

If we have waited this long before defining integration, it is because we have not needed a careful definition. For more than a hundred years, it was enough to define integration as the inverse process to differentiation. As we saw in the last section, this is no longer sufficient

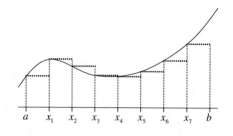

FIGURE 6.4. $\int_a^b f(x)\,dx \approx \sum_{j=1}^n f(x_{j-1})(x_j - x_{j-1})$.

when we start using Fourier series. We need a broader and clearer definition. Fourier's solution, to define the definite integral in terms of area, raises the question: what do we mean by "area"? As the nineteenth century progressed, it became increasingly evident that the problem of defining areas was equivalent to the problem of defining integrals. If we wanted a meaningful definition of either, then we needed to look elsewhere.

It was Cauchy who first proposed the modern solution to this problem. He defined the integral as the limit of approximating sums. Ever since the invention of the calculus, those who used it knew that integrals were limits of these sums. We have seen that Archimedes calculated areas by using approximating sums. But when pushed to define integration, they chose to define it as the inverse process of differentiation. The fact that this inverse process could yield the limits of these approximating sums was the key theorem of calculus that made it such a powerful tool for calculation. It never occurred to them to define the integral as the limit of these sums.

The reason that no one used this definition before Cauchy is that it is ungainly. For the functions that were studied before the 19th century, it is much easier to define integration as anti-differentiation. Cauchy was on the first wave of the mathematical realization that the existing concept of function, something that could be expressed by an algebraic formula involving a small family of common functions, was far too restrictive. Cauchy needed a definition of integration that would enable him to establish that *any* continuous function is integrable.

Following Cauchy, we shall assume that we are working with a continuous function f on a closed and bounded interval $[a, b]$. We choose a positive integer n and an arbitrary partition of $[a, b]$ into n subintervals:

$$a = x_0 < x_1 < x_2 < \cdots < x_n = b.$$

These subintervals do not have to be of equal length. We form a sum that approximates the value of the definite integral of f from a to b:

$$\int_a^b f(x)\,dx \approx \sum_{j=1}^n f(x_{j-1})(x_j - x_{j-1}). \tag{6.41}$$

In terms of area, this is an approximation by rectangles (see Figure 6.4). The jth rectangle sits over the interval $[x_{j-1}, x_j]$ and has height $f(x_{j-1})$, the value of the function at the left-hand edge of the interval. Cauchy now defines the value of the definite integral to be the limit of all such sums as the lengths of the subintervals approach zero.

It is significant that he does not merely take the limit as n approaches infinity. As is shown in the exercises, increasing only the number of subintervals is not enough to give us convergence to the desired value. The lengths of these subintervals must all shrink. For precision, we state Cauchy's definition of the definite integral in the language of the ϵ–δ game.

Definition: integration (Cauchy)

A function f is said to be **integrable** over the interval $[a, b]$ and its integral has the value V provided that the following condition is satisfied. Given any specified error bound ϵ, there must be a response δ such that for any partition

$$a = x_0 < x_1 < x_2 < \cdots < x_n = b$$

where all of the subintervals have length less than δ,

$$|x_j - x_{j-1}| < \delta, \quad \text{for all } j,$$

the corresponding approximating sum will lie within ϵ of V,

$$\left| \sum_{j=1}^{n} f(x_{j-1})(x_j - x_{j-1}) - V \right| < \epsilon.$$

The value of the integral is denoted by $V = \int_a^b f(x)\, dx$.

An Example

To see how cumbersome this definition is, we shall use it to justify the simple integral evaluation

$$\int_1^4 x\, dx = \frac{16 - 1}{2} = \frac{15}{2}. \tag{6.42}$$

Given any positive error bound ϵ, we must show how to respond with a tolerance δ so that for any partition with subintervals of length less than δ, the approximating sum will lie within ϵ of 15/2.

The approximating sum is

$$\sum_{j=1}^{n} x_{j-1}(x_j - x_{j-1}).$$

We have approximated the area under $y = x$ in Figure 6.5 by the sum of the areas of the rectangles with height x_{j-1} and width $x_j - x_{j-1}$. The correct area for each trapezoid is $(x_j^2 - x_{j-1}^2)/2$. Our approximation is too low by precisely

$$\frac{x_j^2 - x_{j-1}^2}{2} - x_{j-1}(x_j - x_{j-1}) = \frac{x_j^2 - 2x_j x_{j-1} + x_{j-1}^2}{2} = \frac{(x_j - x_{j-1})^2}{2}.$$

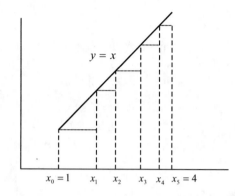

FIGURE 6.5. The area under $y = x$ approximated by rectangles.

If we replace each summand in the approximating sum by the correct value minus the error, we see that our approximating sum is

$$\sum_{j=1}^{n} x_{j-1}(x_j - x_{j-1}) = \sum_{j=1}^{n} \left(\frac{x_j^2 - x_{j-1}^2}{2} - \frac{(x_j - x_{j-1})^2}{2} \right)$$

$$= \sum_{j=1}^{n} \frac{x_j^2 - x_{j-1}^2}{2} - \sum_{j=1}^{n} \frac{(x_j - x_{j-1})^2}{2}$$

$$= \frac{x_n^2 - x_0^2}{2} - \sum_{j=1}^{n} \frac{(x_j - x_{j-1})^2}{2}$$

$$= \frac{15}{2} - \sum_{j=1}^{n} \frac{(x_j - x_{j-1})^2}{2}. \tag{6.43}$$

If each subinterval has length $x_j - x_{j-1} < \delta$, then the total error is less than

$$\sum_{j=1}^{n} \frac{(x_j - x_{j-1})^2}{2} < \frac{\delta}{2} \sum_{j=1}^{n} (x_j - x_{j-1}) = \frac{3\delta}{2}. \tag{6.44}$$

We can guarantee that this error is less than ϵ if we choose $\delta = 2\epsilon/3$.

The Cauchy Criterion

If the target value V is not known, there is a corresponding Cauchy criterion (p. 240).

This definition is slippery because the two partitions may have few or no interior points in common. Cauchy used this definition to prove the integrability of any continuous function. He recognized that to compare these sums, we must have common points in the partitions. His solution is the one we use today. We look for a common **refinement**, a partition that combines the break points of both of the original partitions. If we define a partition by its set of break points,

$$P_1 = \{a, x_1, x_2, \ldots, x_{n-1}, b\},$$
$$P_2 = \{a, x_1', x_2', \ldots, x_{m-1}', b\},$$

Definition: Cauchy criterion for integration

A function f is said to be **integrable** over the interval $[a, b]$ provided that the following condition is satisfied. Given any specified error bound ϵ, there must be a response δ such that for any pair of partitions of the interval $[a, b]$:

$$a = x_0 < x_1 < x_2 < \cdots < x_n = b,$$
$$a = x'_0 < x'_1 < x'_2 < \cdots < x'_m = b,$$

where all of the subintervals have length less than δ,

$$|x_j - x_{j-1}| < \delta, \quad |x'_j - x'_{j-1}| < \delta \quad \text{for all } j,$$

the corresponding approximating sums will lie within ϵ of each other,

$$\left| \sum_{j=1}^{n} f(x_{j-1})(x_j - x_{j-1}) - \sum_{j=1}^{m} f(x'_{j-1})(x'_j - x'_{j-1}) \right| < \epsilon.$$

then a refinement of P_1 is any partition of $[a, b]$ that contains P_1. The smallest common refinement of P_1 and P_2 is the union of these sets,

$$P = P_1 \bigcup P_2.$$

To prove the integrability of a function, we need only find a response δ so that if we start with a partition whose subintervals have length less than δ and refine this partition, then we change the value of the approximating sum by at most $\epsilon/2$. Let $P = \{t_0, t_1, t_2, \ldots, t_r\}$ be a common refinement of P_1 and P_2. If

$$\left| \sum_{j=1}^{n} f(x_{j-1})(x_j - x_{j-1}) - \sum_{j=1}^{r} f(t_{j-1})(t_j - t_{j-1}) \right| < \epsilon/2$$

and

$$\left| \sum_{j=1}^{n} f(t_{j-1})(t_j - t_{j-1}) - \sum_{j=1}^{m} f(x'_{j-1})(x'_j - x'_{j-1}) \right| < \epsilon/2,$$

then

$$\left| \sum_{j=1}^{n} f(x_{j-1})(x_j - x_{j-1}) - \sum_{j=1}^{m} f(x'_{j-1})(x'_j - x'_{j-1}) \right| < \epsilon.$$

Continuity Implies Integrability

Cauchy's definition of integrability may be cumbersome, but it accomplished the task that no previous definition had been able to do. It made it possible to prove that any continuous function is integrable.

> **Theorem 6.6 (Continuous $\Longrightarrow$ Integrable).** *If f is a continuous function on the closed, bounded interval $[a, b]$, then f is integrable over $[a, b]$.*

Proof: We have outlined what needs to be done. If we are given an error bound ϵ, we must be able to construct a response δ with the property that if

$$P_1 = \{a, x_1, x_2, \ldots, x_{n-1}, b\}$$

is any partition of $[a, b]$ with interval lengths less than δ and if P_2 is any refinement of P_1 ($P_1 \subseteq P_2$), then the difference between the corresponding approximating sums is less than $\epsilon/2$. We take each interval $[x_{j-1}, x_j]$ whose endpoints are consecutive points of P_1 and denote its partition in P_2 by

$$P_{1,j} = \{x_{j-1} = x_{j0}, x_{j1}, x_{j2}, \ldots, x_{jd_j} = x_j\}.$$

The partition P_2 is the union of these partitions of the subintervals,

$$P_2 = \bigcup_{j=1}^{n} P_{1,j}.$$

We must show that

$$\left| \sum_{j=1}^{n} f(x_{j-1})(x_j - x_{j-1}) - \sum_{j=1}^{n} \sum_{k=1}^{d_j} f(x_{jk-1})(x_{jk} - x_{jk-1}) \right| < \epsilon/2.$$

Let us consider just the jth subinterval of our original partition. This subinterval contributes

$$f(x_{j-1})(x_j - x_{j-1}) - \sum_{k=1}^{d_j} f(x_{jk-1})(x_{jk} - x_{jk-1})$$

to the difference of the sums. Let M_j be the maximum value of f over the interval $[x_{j-1}, x_j]$ and let m_j be the minimum value over the same interval. Since every x_{jk-1} is contained in this interval, we have that

$$m_j \leq f(x_{jk-1}) \leq M_j.$$

We see that

$$\sum_{k=1}^{d_j} f(x_{jk-1})(x_{jk} - x_{jk-1}) \leq M_j \sum_{k=1}^{d_j} (x_{jk-1} - x_{jk})$$
$$= M_j (x_j - x_{j-1}) \tag{6.45}$$

and

$$\sum_{k=1}^{d_j} f(x_{jk-1})(x_{jk-1} - x_{jk}) \geq m_j \sum_{k=1}^{d_j} (x_{jk} - x_{jk-1})$$
$$= m_j (x_j - x_{j-1}). \tag{6.46}$$

We now invoke the intermediate value theorem. Since

$$\frac{\sum_{k=1}^{d_j} f(x_{jk-1})(x_{jk} - x_{jk-1})}{x_j - x_{j-1}}$$

is a constant sitting somewhere between the minimal and maximal values of f over $[x_{j-1}, x_j]$, it must actually equal $f(c_j)$ at some point c_j in this interval. We have proven that, for some $c_j \in [x_{j-1}, x_j]$,

$$\sum_{k=1}^{d_j} f(x_{jk-1})(x_{jk} - x_{jk-1}) = f(c_j)(x_j - x_{j-1}). \tag{6.47}$$

We use this to simplify the contribution from the jth subinterval:

$$\left| f(x_{j-1})(x_j - x_{j-1}) - \sum_{k=1}^{d_j} f(x_{jk-1})(x_{jk} - x_{jk-1}) \right|$$

$$= |f(x_{j-1}) - f(c_j)| (x_j - x_{j-1}). \tag{6.48}$$

Both x_{j-1} and c_j lie in the same subinterval of the original partition. Using the continuity of f, *we choose our δ so that if $|x_{j-1} - c_j| < \delta$, then*

$$|f(x_{j-1}) - f(c_j)| < \frac{\epsilon}{2(b-a)}. \tag{6.49}$$

This is our response. It only remains to verify that the difference of the two sums is within the allowed error:

$$\left| \sum_{j=1}^{n} f(x_{j-1})(x_j - x_{j-1}) - \sum_{j=1}^{n} \sum_{k=1}^{d_j} f(x_{jk-1})(x_{jk} - x_{jk-1}) \right|$$

$$= \left| \sum_{j=1}^{n} f(x_{j-1})(x_j - x_{j-1}) - \sum_{j=1}^{n} f(c_j)(x_j - x_{j-1}) \right|$$

$$\leq \sum_{j=1}^{n} |f(x_{j-1}) - f(c_j)| (x_j - x_{j-1})$$

$$< \frac{\epsilon}{2(b-a)} \sum_{j=1}^{n} (x_j - x_{j-1}) \quad = \quad \frac{\epsilon}{2}. \tag{6.50}$$

Q.E.D.

The proof we have just seen is a carefully stated version of Cauchy's proof. There are no mistakes in it, but it does reflect an oversight by Cauchy. The problem comes in the italicized portion of the last paragraph, the place where we choose our δ. Notice that we need uniform continuity in exactly the same way that we needed uniform continuity to prove Riemann's lemma. We *have* uniform continuity because we are working with a continuous function on a closed and bounded interval, but Cauchy never explicitly recognized this need. This should be reminiscent of the earlier flaw in Cauchy's reasoning with regard to the continuity of infinite series. There he missed the need for uniform convergence.

A Mean Value Theorem Implying Continuity

There is a version of the mean value theorem that deals with integrals rather than derivatives. We observe that if f is bounded below by A and above by B as x ranges over the interval $[a, b]$, then we have for any partition of this interval,

$$A \sum_{j=1}^{n} (x_j - x_{j-1}) \leq \sum_{j=1}^{n} f(x_{j-1})(x_j - x_{j-1}) \leq B \sum_{j=1}^{n} (x_j - x_{j-1}),$$

$$A(b-a) \leq \sum_{j=1}^{n} f(x_{j-1})(x_j - x_{j-1}) \leq B(b-a),$$

$$A \leq \frac{\sum_{j=1}^{n} f(x_{j-1})(x_j - x_{j-1})}{b-a} \leq B.$$

If these bounds hold for *every* approximating sum, then they also must hold for the integral:

$$A \leq \frac{\int_a^b f(x)\,dx}{b-a} \leq B. \tag{6.51}$$

If f is continuous over the interval $[a, b]$, then the intermediate value theorem tells us that it must actually equal this ratio at some point strictly between a and b.

Theorem 6.7 (Integral Form of the Mean Value Theorem). *If f is continuous on the interval $[a, b]$, then there is at least one point $c \in (a, b)$ for which*

$$\int_a^b f(x)\,dx = f(c)(b-a). \tag{6.52}$$

Theorem 6.7 says something important about the continuity of the integral regardless of whether f is continuous or not.

Corollary 6.8 (Continuity of Integral). *Let f be a bounded integrable function on $[a, b]$ and define F for x in $[a, b]$ by*

$$F(x) = \int_a^x f(t)\,dt.$$

Then F is continuous at every point between a and b.

Proof: We choose a point c between a and b. We must show that for any error bound ϵ supplied by our opponent, we always have a response δ for which

$$|t - c| < \delta \quad \text{implies that} \quad \left| \int_a^t f(x)\,dx - \int_a^c f(x)\,dx \right| < \epsilon.$$

We choose a bound M so that $|f(x)| \leq M$ for all x in $[a, b]$. It follows that the absolute value of the difference of the integrals is bounded by

$$\left| \int_a^t f(x)\, dx - \int_a^c f(x)\, dx \right| = \left| \int_c^t f(x)\, dx \right| \leq M\,|t - c|. \tag{6.53}$$

We respond with $\delta = \epsilon / M$.

Q.E.D.

Proof of Bonnet's Mean Value Theorem

Cauchy's definition of the integral enables us to prove results such as Lemma 6.5 on page 231, Bonnet's mean value theorem. The result that Bonnet actually proved was the following lemma.

Lemma 6.9 (Bonnet's Lemma). *Let f be integrable and let g be a nonnegative, increasing function on $[\alpha, \beta]$. If the integral of f from x to β has least upper bound B and greatest lower bound A as x ranges over the values in $[\alpha, \beta]$,*

$$A \leq \int_x^\beta f(t)\, dt \leq B,$$

then

$$A\, g(\beta) \leq \int_\alpha^\beta f(t)\, g(t)\, dt \leq B\, g(\beta). \tag{6.54}$$

To see that this implies Lemma 6.5, we observe that, as a function of x,

$$g(\beta) \int_x^\beta f(t)\, dt$$

is continuous and so achieves the values of its least upper and greatest lower bounds: $A\, g(\beta)$ and $B\, g(\beta)$. The intermediate value theorem implies that there is some ζ between α and β for which

$$g(\beta) \int_\zeta^\beta f(t)\, dt = \int_\alpha^\beta f(t)\, g(t)\, dt.$$

Proof: Bonnet refers to Lemma 6.9 as a particular case of Abel's lemma which we have seen as Theorem 4.16 on page 161. We use the definition of the integral to work with the approximating sum and do exactly the same manipulation on this sum that Abel performed in obtaining his lemma. The proof is complicated by the fact that we have to keep track of the tolerances involved, but Bonnet is correct that the basic idea is contained in Abel's lemma.

Let us forget tolerances for a moment and concentrate on the summations. We choose a partition of $[\alpha, \beta]$, $P = \{\alpha = x_0, x_1, \ldots, x_n = \beta\}$, and look at the sum that approximates $\int_\alpha^\beta f(t)\, g(t)\, dt$ by evaluating the function at the right-hand edge of each interval:

$$S_P = \sum_{j=1}^n f(x_j)\, g(x_j)\, (x_j - x_{j-1}). \tag{6.55}$$

For $k = 1, 2, \ldots, n$, we define

$$S_{P,k} = \sum_{j=k}^{n} f(x_j)(x_j - x_{j-1}).\qquad (6.56)$$

The partial sum $S_{P,k}$ will be a good approximation to $\int_{x_{k-1}}^{\beta} f(t)\,dt$ which we know lies between A and B.

We see that

$$f(x_j)(x_j - x_{j-1}) = S_{P,j} - S_{P,j+1}.\qquad (6.57)$$

If we define $S_{P,n+1} = 0$, then S_P can be rewritten as

$$
\begin{aligned}
S_P &= \sum_{j=1}^{n} g(x_j)(S_{P,j} - S_{P,j+1})\\
&= g(x_1)(S_{P,1} - S_{P,2}) + g(x_2)(S_{P,2} - S_{P,3}) + \cdots + g(x_n)(S_{P,n} - S_{P,n+1})\\
&= g(x_1) S_{P,1} + [g(x_2) - g(x_1)] S_{P,2} + \cdots + [g(x_n) - g(x_{n-1})] S_{P,n}.\qquad (6.58)
\end{aligned}
$$

Since g is nonnegative and increasing, each coefficient of $S_{P,k}$ is greater than or equal to 0. We let A_P and B_P be, respectively, lower and upper bounds on the set of values of $S_{P,k}$,

$$A_P \le S_{P,k} \le B_P \quad \text{for } 1 \le k \le n.$$

We can now bound S_P:

$$
\begin{aligned}
S_P &\ge A_P\,(g(x_1) + g(x_2) - g(x_1) + \cdots + g(x_n) - g(x_{n-1}))\\
&= A_P\, g(x_n) \quad = \quad A_P\, g(\beta),\qquad (6.59)\\
S_P &\le B_P\,(g(x_1) + g(x_2) - g(x_1) + \cdots + g(x_n) - g(x_{n-1}))\\
&= B_P\, g(x_n) \quad = \quad B_P\, g(\beta).\qquad (6.60)
\end{aligned}
$$

The idea at this point is to argue that as the partition becomes finer, S_P approaches $\int_{\alpha}^{\beta} f(t)g(t)\,dt$ and the upper and lower bounds, B_P and A_P, approach B and A, respectively. We use the definition of the integral to make this part of the argument more precise.

We choose an error bound ϵ and find a response δ so that for any partition of $[\alpha, \beta]$ with subintervals of length less than δ, we have that

$$\sum_{j=1}^{n} f(x_j)\, g(x_j)(x_j - x_{j-1})$$

is within ϵ of $\int_{\alpha}^{\beta} f(t)\, g(t)\,dt$ and *at the same time*

$$\sum_{j=1}^{n} f(x_j)(x_j - x_{j-1})$$

is within ϵ of $\int_{\alpha}^{\beta} f(t)\,dt$. There is a δ_1 that works in the first case and a δ_2 that works in the second. We choose whichever is smaller.

Once we have chosen our δ, any partition with subintervals of length less than δ will yield an approximating sum within the allowable error ϵ. For any k, $1 \le k \le n$,

$$\sum_{j=k}^{n} f(x_j)(x_j - x_{j-1})$$

must also be within ϵ of $\int_{x_{k-1}}^{\beta} f(t)\,dt$. This is because we are allowed to choose as fine a partition as we might wish over $[\alpha, x_{k-1}]$ so that the missing summands add to an amount arbitrarily close to $\int_{\alpha}^{x_{k-1}} f(t)\,dt$. It follows that $A_P > A - \epsilon$ and $B_P < B + \epsilon$, and therefore

$$(A - \epsilon)g(\beta) < A_P\,g(\beta) \le S_P \le B_P\,g(\beta) < (B + \epsilon)g(\beta). \tag{6.61}$$

Combining this with the inequality

$$S_P - \epsilon < \int_{\alpha}^{\beta} f(t)\,g(t)\,dt < S_P + \epsilon \tag{6.62}$$

yields

$$A\,g(\beta) - \epsilon[1 + g(\beta)] < \int_{\alpha}^{\beta} f(t)\,g(t)\,dt < B\,g(\beta) + \epsilon[1 + g(\beta)]. \tag{6.63}$$

Since this is true for *every* positive ϵ, we have proved the desired inequalities:

$$A\,g(\beta) \le \int_{\alpha}^{\beta} f(t)\,g(t)\,dt \le B\,g(\beta).$$

Q.E.D.

Exercises

The symbol $(\mathbf{M\&M})$ indicates that *Maple* and *Mathematica* codes for this problem are available in the **Web Resources** at **www.macalester.edu/aratra**.

6.2.1. $(\mathbf{M\&M})$ If we use n subintervals of equal length, then $\int_0^1 (x^3 - 2x^2 + x)\,dx$ is approximated by the sum

$$\sum_{j=0}^{n-1} \left(\frac{j^3}{n^3} - 2\frac{j^2}{n^2} + \frac{j}{n} \right) \frac{1}{n}.$$

Evaluate these approximations for $1 \le n \le 20$. How large must n be before you are within 0.001 of the correct value?

6.2.2. $(\mathbf{M\&M})$ Experiment with the value of

$$\sum_{j=1}^{n} (x_{j-1}^3 - 2x_{j-1}^2 + x_{j-1})(x_j - x_{j-1})$$

for different partitions of $[0, 1]$ into intervals of length at most 0.25. Find a partition for which the difference between this sum and $\int_0^1 (x^3 - 2x^2 + x)\, dx$ is as large as possible.

6.2.3. **M&M** Define $\sin(1/0) = 0$. To see whether

$$\int_0^1 \sin(1/x)\, dx$$

exists, we can look at the approximating sums and see if they seem to converge. Does it appear that

$$\left(\sum_{j=1}^{n-1} \sin(n/j) \right) \frac{1}{n}$$

converges as n increases? If so, what does the value appear to be?

6.2.4. Determine whether or not $\sin(1/x)$ is integrable over $[0, 1]$ and prove your assertion.

6.2.5. What is the value of

$$\int_0^1 \cos^2(100\pi x)\, dx?$$

Consider the approximating sum

$$\frac{1}{n} \sum_{j=0}^{n-1} \cos^2(100\pi j/n).$$

Describe what happens as n increases. Does increasing the value of n always bring you closer to the actual value of the integral?

6.2.6. Prove that if f is differentiable and the derivative is bounded on the interval I, then f is uniformly continuous on I.

6.2.7. Give an example of a function f and an interval I such that f is differentiable at every point of I, the derivative of f over I is *not* bounded, but f *is* uniformly continuous over I.

6.2.8. Is $\sin(1/x)$ uniformly continuous on $(0, 1)$? Justify your answer.

6.2.9. Is $\sin(1/x)$ uniformly continuous on $(1, \infty)$? Justify your answer.

6.2.10. Prove that if *every* approximating sum for the integral $\int_a^b f(x)\, dx$ is bounded below by m and above by M,

$$m \leq \sum_{j=1}^{n} f(x_{j-1})(x_j - x_{j-1}) \leq M,$$

then the integral must also be bounded below by m and above by M.

6.2.11. Use the integral form of the mean value theorem to prove that if f is continuous over $[a, b]$, then

$$\frac{d}{dx} \int_a^x f(t)\, dt = f(x) \tag{6.64}$$

for every $x \in [a, b]$.

6.2.12. Discuss whether or not equation (6.64) is valid at a point where f is *not* continuous.

6.3   The Riemann Integral

A more useful definition of integration was given by Bernhard Riemann in "Über die Darstellbarkeit einer Function durch eine trigonometrische Reihe." As mentioned in section 5.1, this was written after the summer of 1852 when Riemann had discussed questions of Fourier series with Dirichlet. Its purpose was nothing less than to find necessary and sufficient conditions for a function to have a representation as a trigonometric series. Riemann never published it, probably because it raised many new questions that he was hoping to answer. It appeared in 1867, after his death.

Riemann begins with a summary of the history of the subject, describing the contributions from d'Alembert, Euler, Bernoulli, and Lagrange and the questions that arose concerning the validity of a trigonometric expansion for arbitrary functions. He discusses Fourier's contributions and Dirichlet's proof, emphasizing Dirichlet's recognition of the distinction between absolute and conditional convergence. This is where the Riemann rearrangement theorem is stated, not as a theorem but as an observation. He points out the difficulty with Fourier series: that in general the convergence will not be absolute.

This is followed by a list of the assumptions that Dirichlet needed to impose on a function in order to prove that it did have representation as a trigonometric series:

    I. it must be integrable,
   II. at each point of discontinuity, its value must be the average of the limit from the left and the limit from the right,
 III. it must be piecewise continuous, bounded, and piecewise monotonic.

The second condition is essential. We have seen that the Fourier series cannot equal the original function at any point where this is not true. The third assumption is not as clearly necessary. Most of Riemann's work involved probing how far the third assumption could be weakened.

The Riemann Integral

The first task is to clarify the meaning of the integral. Cauchy's definition was adequate for proving that any bounded continuous function is integrable. It is also sufficient for a demonstration that any bounded piecewise continuous function is integrable. Riemann wished to consider even more general functions, functions with infinitely many discontinuities within any finite interval. His definition is very similar to Cauchy's. Like Cauchy, he uses approximating sums:

$$\int_a^b f(x)\,dt \approx \sum_{j=1}^n f(x_{j-1}^*)(x_j - x_{j-1}).$$

Unlike Cauchy who evaluated the function f at the left-hand endpoint of each interval, Riemann allows approximating sums in which x_{j-1}^* can be any point in the interval $[x_{j-1}, x_j]$. Because of this extra freedom, it appears more difficult to guarantee convergence of these series. In fact, for bounded functions Riemann's definition is equivalent to Cauchy's. Cauchy wanted to be able to prove that any continuous function is integrable. Riemann was interested in seeing how discontinuous a function could be and still remain integrable. As he realized, to be tied to the left-hand endpoints obscures what is happening in general. Riemann's definition—in the language of the ϵ–δ game—is the following.

Definition: integration (Riemann)

A function f is said to be **Riemann integrable** over the interval $[a, b]$ and its integral has the value V provided that the following condition is satisfied. Given any specified error bound ϵ, there must be a response δ such that for any partition

$$a = x_0 < x_1 < x_2 < \cdots < x_n = b$$

where each of the subintervals has length less than δ,

$$|x_j - x_{j-1}| < \delta, \quad \text{for all } j,$$

and for any set of values $x_0^* \in [x_0, x_1]$, $x_1^* \in [x_1, x_2]$, $\ldots$, $x_{n-1}^* \in [x_{n-1}, x_n]$, the corresponding approximating sum will lie within ϵ of the value V,

$$\left| \sum_{j=1}^n f(x_{j-1}^*)(x_j - x_{j-1}) - V \right| < \epsilon.$$

The value of the integral is denoted by

$$V = \int_a^b f(x)\,dx.$$

What Riemann gains in allowing x_{j-1}^* to take on any value in $[x_{j-1}, x_j]$ is greater flexibility. In particular, it enables him to establish necessary and sufficient conditions for the existence of the integral.

Necessary and Sufficient Conditions

If we want to prove that the integral exists without knowing its value, then we are again thrown back on the Cauchy criterion. Given an error bound ϵ, we must show that we have a response δ with the property that any two approximating sums with subintervals of length less than δ must differ from each other by less than ϵ. As Cauchy did in proving the integrability of continuous functions, it is enough if each sum can be brought within $\epsilon/2$ of any approximating sum that uses the common refinement.

Let $P_1 = \{a, x_1, x_2, \ldots, x_{n-1}, b\}$ be a partition of $[a, b]$ and let P_2 be a refinement of P_1. As before, we take each interval $[x_{j-1}, x_j]$ whose endpoints are consecutive points of P_1 and denote its partition in P_2 by

$$P_{1,j} = \{x_{j-1} = x_{j0}, x_{j1}, x_{j2}, \ldots, x_{jd_j} = x_j\}.$$

The partition P_2 is the union of these partitions of the subintervals,

$$P_2 = \bigcup_{j=1}^{n} P_{1,j}.$$

We need to show that we can force

$$\left| \sum_{j=1}^{n} f(x_{j-1}^*)(x_j - x_{j-1}) - \sum_{j=1}^{n} \sum_{k=1}^{d_j} f(x_{jk-1}^{**})(x_{jk} - x_{jk-1}) \right| < \epsilon/2$$

by establishing a bound on $x_j - x_{j-1}$. Beyond the fact that $x_{j-1}^* \in [x_{j-1}, x_j]$ and $x_{jk-1}^{**} \in [x_{jk-1}, x_{jk}] \subseteq [x_{j-1}, x_j]$, there is no necessary relationship between x_{j-1}^* and x_{jk-1}^{**}.

We assume that f is bounded on $[a, b]$, and we let M_j be the least upper bound of the values of $f(x)$ for $x \in [x_{j-1}, x_j]$, m_j be the greatest lower bound. We define the **variation** of f on $[x_{j-1}, x_j]$ to be

$$D_j = M_j - m_j.$$

It follows that

$$\left| f(x_{j-1}^*)(x_j - x_{j-1}) - \sum_{k=1}^{d_j} f(x_{jk-1}^{**})(x_{jk} - x_{jk-1}) \right|$$

$$= \left| \sum_{k=1}^{d_j} f(x_{j-1}^*)(x_{jk} - x_{jk-1}) - \sum_{k=1}^{d_j} f(x_{jk-1}^{**})(x_{jk} - x_{jk-1}) \right|$$

$$\leq \sum_{k=1}^{d_j} |f(x_{j-1}^*) - f(x_{jk-1}^{**})|(x_{jk} - x_{jk-1})$$

$$\leq D_j \sum_{k=1}^{d_j} (x_{jk} - x_{jk-1})$$

$$= D_j(x_j - x_{j-1}). \tag{6.65}$$

This is an upper bound that we can approach as closely as we please by taking the refinement to be the original partition, $P_2 = P_1$, choosing x_{j-1}^* so that $f(x_{j-1}^*)$ is close to M_j, and choosing x_{j0}^{**} so that $f(x_{j0}^{**})$ is close to m_j.

This means that we have integrability if and only if there is a response δ such that for any partition with subintervals of length less than δ,

$$D_1(x_1 - x_0) + D_2(x_2 - x_1) + \cdots + D_n(x_n - x_{n-1}) < \epsilon/2, \tag{6.66}$$

where D_j is the variation of f on the interval $[x_{j-1}, x_j]$. From here, Riemann derived the following theorem.

> **Theorem 6.10 (Conditions for Riemann Integrability).** *Let f be a bounded function on $[a, b]$. This function is integrable over $[a, b]$ if and only if for any pair of positive numbers (v, σ), there is a δ such that for any partition of $[a, b]$ with subintervals of length less than δ, the subintervals on which the variation is $\geq \sigma$ have a combined length that is $< v$.*

For example, if $f(x) = x^2$ on $[0, 1]$ and we are asked to respond to the challenge $v = 0.3$, $\sigma = 0.23$, we must find a δ such that any partition of $[0, 1]$ into subintervals of length less than δ results in subintervals of combined length less than 0.3 on which the variation is greater than 0.23. The response $\delta = 0.2$ will not work. For the partition $\{0, 0.2, 0.4, 0.6, 0.8, 1\}$, there are two subintervals on which the variation exceeds 0.23. On $[0.6, 0.8]$ the variation is $0.8^2 - 0.6^2 = 0.28$, and on $[0.8, 1]$ the variation is $1^2 - 0.8^2 = 0.36$. The combined length of these intervals is 0.4, and this is larger than $v = 0.3$. There is a response, however, and it is left as an exercise to find one.

This theorem implies that Dirichlet's function

$$f(x) = \begin{cases} 1, & \text{if } x \text{ is rational,} \\ 0, & \text{if } x \text{ is irrational} \end{cases}$$

is not Riemann integrable over $[0, 2]$. Every subinterval contains both rational and irrational numbers. If we are challenged with $v = 1/2$, $\sigma = 1/3$, then every subinterval has variation equal to 1 no matter how short it might be. The sum of the lengths of the subintervals with variation larger than $1/3$ is 2 which is larger than $1/2$.

On the other hand, $\int_0^1 \sin(1/x)\, dx$ does exist. Once we are given v and σ, we choose some point α between 0 and v. The function $\sin(1/x)$ is uniformly continuous on the interval $[\alpha, 1]$, and so we can choose a δ so that on each subinterval of $[\alpha, 1]$ the variation is less than σ. It follows that all of the subintervals with variation larger than σ lie inside $[0, \alpha + \delta)$, and so the sum of their lengths is less than $\alpha + \delta$. If we also restrict δ so that $\alpha + \delta$ is less than v, then the sum of the lengths of the subintervals with variation larger than σ will be less than v.

Proof: We assume that f is integrable and so $D_1(x_1 - x_0) + D_2(x_2 - x_1) + \cdots + D_n(x_n - x_{n-1})$ can be made arbitrarily small by taking a partition with sufficiently short subintervals. Given v and σ and a partition of $[a, b]$, let s be the sum of the lengths of the subintervals on which the variation is larger than σ. We have that

$$\sigma s < D_1(x_1 - x_0) + D_2(x_2 - x_1) + \cdots + D_n(x_n - x_{n-1}). \tag{6.67}$$

We choose our δ so that the right side of this inequality is less than σv. This implies that $\sigma s < \sigma v$, and so s is less than v.

In the other direction, we assume that there is such a δ for any choice of v and σ. Let D be the variation of f over the entire interval $[a, b]$ so that each D_j is less than or equal to D. Those subintervals with variation greater than σ contribute at most Dv, while the subintervals with variation less than σ contribute at most $\sigma(b - a)$. It follows that

$$D_1(x_1 - x_0) + D_2(x_2 - x_1) + \cdots + D_n(x_n - x_{n-1}) < Dv + (b - a)\sigma.$$

If we choose $v = \epsilon/4D$ and $\sigma = \epsilon/4(b-a)$, then

$$D_1(x_1 - x_0) + D_2(x_2 - x_1) + \cdots + D_n(x_n - x_{n-1}) < \epsilon/2.$$

Q.E.D.

Improper Integrals

> **Definition: improper integral** An integral is **improper** if either the function that is being integrated or the interval over which the function is integrated is unbounded. Strictly speaking, the Riemann integral does not exist in either case. However, there may be a value that can be assigned to such an integral by taking a limit of integrals that are Riemann integrable.

Riemann's definition only applies to bounded functions on closed, bounded intervals. Cauchy had shown how to integrate unbounded functions. Riemann's treatment is exactly the same. If $f(x)$ is unbounded as x approaches c for some $c \in [a, b]$, then he defines

$$\int_a^b f(x)\,dx = \lim_{\epsilon_1 \to 0^-} \left(\int_a^{c+\epsilon_1} f(x)\,dx \right) + \lim_{\epsilon_2 \to 0^+} \left(\int_{c+\epsilon_2}^b f(x)\,dx \right). \tag{6.68}$$

Both limits must exist independently.

For example,

$$\int_{-1}^1 \frac{dx}{x} = \lim_{\epsilon_1 \to 0^-} \left(\int_{-1}^{\epsilon_1} \frac{dx}{x} \right) + \lim_{\epsilon_2 \to 0^+} \left(\int_{\epsilon_2}^1 \frac{dx}{x} \right)$$
$$= \lim_{\epsilon_1 \to 0^-} (\ln|\epsilon_1|) - \lim_{\epsilon_2 \to 0^+} (\ln \epsilon_2),$$

which does not exist. On the other hand,

$$\int_0^1 \frac{dx}{\sqrt{x}} = \lim_{\epsilon \to 0^+} 2\sqrt{x}\,\Big|_\epsilon^1 = \lim_{\epsilon \to 0^+} (2 - 2\sqrt{\epsilon}) = 2.$$

See page 138 for a discussion of integration over an unbounded domain.

Integrability with Infinitely Many Discontinuities

One of the surprising results that Riemann produces is an example of an integrable function with infinitely many discontinuities between 0 and 1. He defines the function

$$((x)) = \begin{cases} x - \lfloor x \rfloor, & \lfloor x \rfloor \le x < \lfloor x \rfloor + 1/2, \\ 0, & x = \lfloor x \rfloor + 1/2, \\ x - \lfloor x \rfloor - 1, & \lfloor x \rfloor + 1/2 < x < \lfloor x \rfloor + 1, \end{cases} \tag{6.69}$$

(see Figure 6.6). He then defines

$$f(x) = \sum_{n=1}^\infty \frac{((nx))}{n^2}. \tag{6.70}$$

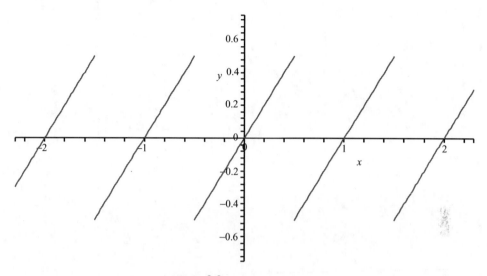

FIGURE 6.6. Graph of $y = ((x))$.

Since $|((nx))| < 1/2$, this series converges for all x. It has a discontinuity whenever nx is half of an odd integer, and that will happen for every x that is a rational number with an even denominator (see Figure 6.7).

Specifically, if $x = a/2b$ where a is odd and a and b are relatively prime, and if n is an odd multiple of b, then

$$((na/2b + 0)) - ((na/2b)) = -1/2 \qquad \text{and} \qquad ((na/2b - 0)) - ((na/2b)) = 1/2.$$

We want to be able to assert that

$$f\left(\frac{a}{2b} + 0\right) - f\left(\frac{a}{2b}\right) = \sum_{\substack{m=1 \\ m \text{ odd}}}^{\infty} \frac{-1/2}{(mb)^2}$$

$$= \frac{-1}{2b^2}\left(1 + \frac{1}{9} + \frac{1}{25} + \cdots\right)$$

$$= \frac{-\pi^2}{16b^2}, \tag{6.71}$$

$$f\left(\frac{a}{2b} - 0\right) - f\left(\frac{a}{2b}\right) = \sum_{\substack{m=1 \\ m \text{ odd}}}^{\infty} \frac{1/2}{(mb)^2}$$

$$= \frac{1}{2b^2}\left(1 + \frac{1}{9} + \frac{1}{25} + \cdots\right)$$

$$= \frac{\pi^2}{16b^2}. \tag{6.72}$$

The first line of these equalities assumes that we can interchange limits, that

$$f(x + 0) - f(x) = \lim_{v \to 0^+}\left(\sum_{n=1}^{\infty} \frac{((nx + nv)) - ((nx))}{n^2}\right)$$

$$= \sum_{n=1}^{\infty}\left(\lim_{v \to 0^+} \frac{((nx + nv)) - ((nx))}{n^2}\right). \tag{6.73}$$

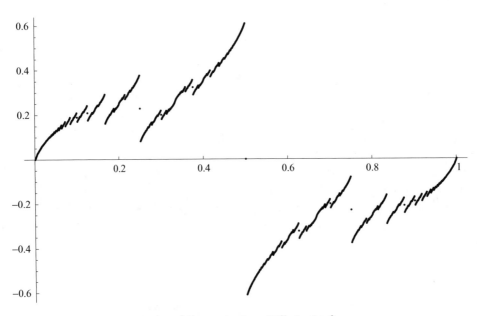

FIGURE 6.7. Graph of $y = \sum_{n=1}^{\infty} ((nx))/n^2$.

The justification of this interchange rests on the uniform convergence of our series over the set of all x and is left as exercise 6.3.16.

Our function f has a discontinuity at every rational number with an even denominator, but it is integrable. Given ν and σ, there are only finitely many rational numbers between 0 and 1 at which the variation is larger than σ. If the variation is larger than σ at $x = a/2b$, then b must satisfy

$$\frac{\pi^2}{8b^2} > \sigma$$

which means that b is a positive integer less than $\pi/\sqrt{8\sigma}$. If there are N such rational numbers, then we choose our response δ so that $N\delta$ is less than ν and so that the variation is less than σ on every other subinterval.

Fourier Series

Now that he has settled the problem of integrability, Riemann moves to the main theme of his paper. He points out that previous work had focused on the question: when is a function representable by a trigonometric series? Characteristic of his insightfulness, he realizes that the question needs to be reversed. "We must proceed from the inverse question: if a function is representable by a trigonometric series, what consequences does this have for its behavior, for the variation of its value with the continuous variation of the argument?"

This shift of focus enabled him to find necessary and sufficient conditions for a function to be representable as a trigonometric series. If f is a convergent trigonometric series, then there exists a function F for which f is the second derivative of F. Furthermore, for arbitrary constants a, b, and c and any function λ that is continuous on $[b, c]$ and zero at b

and at c, whose derivative is continuous on $[b, c]$ and zero at b and at c, and whose second derivative is piecewise monotonic on $[b, c]$, we must have that

$$\lim_{\mu \to 0} \mu^2 \int_b^c F(x) \cos \mu (x - a) \lambda(x) \, dx = 0. \tag{6.74}$$

An example of a function that fails these conditions and thus does not have a Fourier series representation is

$$f(x) = \frac{d}{dx}[x^\nu \cos(1/x)] = \nu x^{\nu-1} \cos(1/x) + x^{\nu-2} \sin(1/x),$$

where ν is any constant between 0 and 1/2.

Riemann then proves that these conditions are not only necessary, they are sufficient. If there exists a function F satisfying equation (6.74) with the conditions described above, then f has a representation as a trigonometric series.

Riemann opened new worlds of possibilities: integrable functions whose Fourier series do not converge, convergent trigonometric series whose sum is not integrable, trigonometric series that converge only at rational values of x or are unbounded in any open interval, functions that are continuous at every point but that lack a derivative at any point.

Exercises

The symbol (**M&M**) indicates that *Maple* and *Mathematica* codes for this problem are available in the **Web Resources** at **www.macalester.edu/aratra**.

6.3.1. While the Riemann and Cauchy definitions of integration are equivalent for bounded functions, they are not entirely equivalent for unbounded functions. Show that

$$f(x) = \begin{cases} 1/\sqrt{|x|}, & -1 \leq x < 0, \\ 0, & x = 0, \end{cases}$$

is integrable over $[-1, 0]$ in the Cauchy sense but not in the Riemann sense.

6.3.2. Prove that if f is continuous on the closed and bounded interval $[a, b]$, then f is Riemann integrable over $[a, b]$.

6.3.3. Prove that if f is Riemann integrable over $[a, b]$, then it also satisfies Cauchy's definition of integrability.

6.3.4. When looking for a response to a (ν, σ) challenge—to find a δ so that for any partition of $[a, b]$ with subintervals of length less than δ, ν is larger than the sum of the lengths of the subintervals with variation larger than σ—it is important to realize that shifting the partition can affect the sum of the lengths of the intervals on which the variation exceeds σ. For the example given on page 251, $f(x) = x^2$ on $[0, 1]$, $\nu = 0.3$, $\sigma = 0.23$, we saw that if the partition is

$$0 < 0.2 < 0.4 < 0.6 < 0.8 < 1,$$

then the sum of the lengths is 0.4. Find the sum of the lengths of the subintervals on which the variation exceeds 0.23 for each of the following partitions in which the subintervals still have length ≤ 0.2:

$$0 < 0.1 < 0.3 < 0.5 < 0.7 < 0.9 < 1,$$
$$0 < 0.15 < 0.35 < 0.55 < 0.73 < 0.88 < 1,$$
$$0 < 0.13 < 0.33 < 0.53 < 0.72 < 0.87 < 1.$$

6.3.5. Continuing the previous exercise, find a partition of $[0, 1]$ into subintervals of length less than or equal to 0.2 that maximizes the sum of the lengths of the subintervals on which the variation equals or exceeds 0.23.

6.3.6. Continuing the previous two exercises, find a response δ to the challenge $\nu = 0.3$, $\sigma = 0.23$. Prove that you have satisfied the challenge for *any* partition of $[0, 1]$ into subintervals of length less than δ.

6.3.7. Prove or disprove that the function f defined by

$$f(x) = \begin{cases} 1/q, & \text{if } x = p/q \text{ is rational,} \\ 0, & \text{if } x \text{ is irrational,} \end{cases}$$

is Riemann integrable over $[0, 1]$. We define $f(0) = f(1) = 1$.

6.3.8. Prove or disprove that the function g defined by

$$g(x) = \frac{1}{x} - \left\lfloor \frac{1}{x} \right\rfloor$$

is Riemann integrable over $[0, 1]$.

6.3.9. Let h be defined by

$$h(x) = \begin{cases} 1, & x = 1/n, \ n \in \mathbb{N}, \\ 0, & \text{otherwise.} \end{cases}$$

Prove that f is Riemann integrable over $[0, 1]$ and that $\int_0^1 f(x)\,dx = 0$.

6.3.10. Using Riemann integrals of suitably chosen functions, find the following limits.

a. $\displaystyle \lim_{n \to \infty} \frac{1}{n} \left(e^{1/n} + e^{2/n} + e^{3/n} + \cdots + e^{n/n} \right)$

b. $\displaystyle \lim_{n \to \infty} \frac{1}{n^3} \left(1^2 + 2^2 + 3^2 + \cdots + n^2 \right)$

c. $\displaystyle \lim_{n \to \infty} \frac{1}{n^{k+1}} \left(1^k + 2^k + 3^k + \cdots + n^k \right), \quad k \geq 0$

d. $\displaystyle \lim_{n \to \infty} \left(\frac{1}{n+1} + \frac{1}{n+2} + \cdots + \frac{1}{3n} \right)$

e. $\displaystyle \lim_{n \to \infty} n^2 \left(\frac{1}{n^3 + 1^3} + \frac{1}{n^3 + 2^3} + \cdots + \frac{1}{n^3 + n^3} \right)$

f. $\displaystyle\lim_{n\to\infty} \frac{1}{n}\sqrt[n]{(n+1)(n+2)\cdots(n+n)}$

6.3.11. Prove that if f is Riemann integrable on $[0, 1]$ and $|q| < 1$, then

$$\lim_{q\to 1^-} (1-q)\sum_{n=1}^{\infty} q^n f(q^n) = \int_0^1 f(x)\,dx. \qquad (6.75)$$

6.3.12. Find a bound (in terms of $\alpha > 0$) on the size of

$$\frac{d}{dx}\sin(1/x), \quad x \in [\alpha, 1].$$

For the function $\sin(1/x)$ on the interval $[0, 1]$, find a response δ to the challenge $\nu = 0.3$, $\sigma = 0.1$.

6.3.13. Use the fact that

$$1 + \frac{1}{4} + \frac{1}{9} + \frac{1}{16} + \frac{1}{25} + \cdots = \frac{\pi^2}{6}$$

(appendix A.3) to prove that

$$1 + \frac{1}{9} + \frac{1}{25} + \frac{1}{49} + \cdots = \frac{\pi^2}{8}.$$

6.3.14. For the function

$$f(x) = \sum_{n=1}^{\infty} \frac{((nx))}{n^2}$$

defined on page 252, find a response δ to the challenge $\nu = 0.2$, $\sigma = 0.1$.

6.3.15. (**M&M**) Graph the partial sums

$$f_N(x) = \sum_{n=1}^{N} \frac{((nx))}{n^2}$$

over $[0, 1]$ for $N = 10, 100$, and 1000.

6.3.16. Prove that $f(x) = \sum_{n=1}^{\infty}((nx))/n^2$ converges uniformly. Prove that the interchange of limits in equation (6.73) is allowable.

6.3.17. To find the Fourier expansion of $((nx))$ over $[-1, 1]$, we observe that this function is odd and so $a_k = 0$ for all k. Using exercise 6.1.7 from section 6.1, show that

$$b_k = \begin{cases} 0, & \text{if } 2n \text{ does not divide } k, \\ -(-1)^{k/2n}(2n/k\pi), & \text{if } 2n \text{ does divide } k. \end{cases} \qquad (6.76)$$

6.3.18. Use the results from exercises 6.3.16 and 6.3.17 to prove that

$$\sum_{n=1}^{\infty} \frac{((nx))}{n^2} = \sum_{k=1}^{\infty} \frac{-\psi(k)}{k^2 \pi} \sin(2k\pi x), \tag{6.77}$$

where

$$\psi(k) = \sum_{d|k} (-1)^d d, \tag{6.78}$$

the sum being over all positive integers d that divide evenly into k.

6.3.19. (**M&M**) Investigate the function ψ defined in equation (6.78). Calculate its values up to at least $k = 100$. How fast does its absolute value grow? When is it positive? What else can you say about ψ?

6.3.20. Show that if

$$g(x) = \sum_{n=1}^{\infty} \frac{((nx))}{n}, \tag{6.79}$$

then

$$g(1/4) = \frac{1}{4} \left(1 - \frac{1}{3} + \frac{1}{5} - \frac{1}{7} + \cdots \right) = \frac{\pi}{16}.$$

Find an approximate value for $g(1/5)$. Prove that this series converges when $x = 1/5$.

6.3.21. Prove that the series g of equation (6.79) converges at every rational value of x. Discuss what you think happens at irrational values of x.

6.4 Continuity without Differentiability

Few mathematical feats have been as surprising as the exhibition of a function that is continuous at every value and differentiable at none. It illustrates that confusion between continuity and differentiability is indeed confusion. While differentiability implies continuity, continuity guarantees nothing about differentiability.

Until well into the 1800s, there was a basic belief that all functions have derivatives, except possibly at a few isolated points such as one finds with the absolute value function, $|x|$, at $x = 0$. In 1806, Ampère tried to prove the general existence of derivatives. His proof is difficult to evaluate because it is not clear what implicit assumptions he was making about what constitutes a function. In 1839 with the publication of J. L. Raabe's calculus text, *Die Differential- und Integralrechnung*, the "theorem" that any continuous function is differentiable—with the possibility of at most finitely many exceptional points—started making its way into the standard textbooks.

Bolzano, Weierstrass, and Riemann knew this was wrong. By 1861 Riemann had introduced into his lectures the function

$$\sum_{n=1}^{\infty} \frac{\sin(n^2 x)}{n^2},$$

claiming that it is continuous at every x but not differentiable for infinitely many values of x. The convergence of this series is uniform (by the Weierstrass M-test with $M_n = 1/n^2$), and so it is continuous at every x. Nondifferentiability is harder to prove. It was not until 1916 that G. H. Hardy showed that in any finite interval, no matter how short, there will be infinitely many values of x for which the derivative does not exist. It was demonstrated in 1970 that there are also infinitely many values at which the derivative *does* exist. Riemann's example—while remarkable—does not go as far as nondifferentiability for all x.

The faith in the existence of derivatives is illustrated by the reaction to Hermann Hankel's 1870 paper "Untersuchungen über die unendlich oft oszillierenden und unstetigen Functionen" in which, among other things, he described a general method for creating continuous functions with infinitely many points of nondifferentiability. J. Hoüel applauded this result and expressed hope that it would change the current attitude in which "there is no mathematician today who would believe in the existence of continuous functions without derivatives."[1] Phillipe Gilbert pounced upon errors and omissions in Hankel's work and displayed them "so as to leave no doubt . . . about the inanity of the conclusions."[2]

But the tide had turned. Hankel responded with the observation that Riemann's example of an integrable function with infinitely many discontinuities implies that its integral,

$$F(x) = \int_0^x \left(\sum_{n=1}^{\infty} \frac{((nt))}{n^2} \right) dt,$$

is necessarily continuous at every x but cannot be differentiable at any of the infinitely many points where the integrand is not continuous. The real surprise came in 1872 when Karl Weierstrass showed the Berlin Academy the trigonometric series mentioned at the end of Chapter 1:

$$f(x) = \sum_{n=0}^{\infty} b^n \cos(a^n \pi x), \tag{6.80}$$

where a is an odd integer, b lies strictly between 0 and 1, and ab is strictly larger than $1 + 3\pi/2$. It is continuous at every value of x and differentiable at none. A flood of examples followed.

Proving Nondifferentiability

The continuity of Weierstrass's example is easy. We have uniform convergence from the M-test with $M_n = b^n$. To see how to prove that a function is not differentiable, we must first recall what it means to say that it *is* differentiable. If f is differentiable at x_0, then there is a number, denoted by $f'(x_0)$, for which

$$\left| \frac{f(x_1) - f(x_0)}{x_1 - x_0} - f'(x_0) \right| = |E(x_1, x_0)|$$

can be made as small as desired by taking x_1 sufficiently close to x_0. What is significant here is that we must be able to force E to be small not by how we choose x_1 but by how

---

[1] As quoted in Medvedev, *Scenes from the History of Real Functions*, p. 222.

[2] As quoted in Hawkins, *Lebesgue's Theory of Integration*, p. 45.

we bound it. There must be a response δ so that for all possible values of x_1 within δ of x_0, $E(x_1, x_0)$ is smaller than the allowed error.

To prove that f is not differentiable at x_0, we must show that no matter how we select a value for $f'(x_0)$ and how we select our response δ, there is at least one x_1 within δ of x_0 for which $E(x_1, x_0)$ is larger than the allowed error. One way of accomplishing this is to show that for any δ, there is always an x_1 within δ of x_0 for which

$$\left| \frac{f(x_1) - f(x_0)}{x_1 - x_0} \right|$$

is larger than any prespecified bound. If this ratio is unbounded inside every interval of the form $(x_0 - \delta, x_0 + \delta)$, then it cannot stay close to any single value of $f'(x_0)$.

We begin with a Fourier series of the form

$$f(x) = \sum_{n=0}^{\infty} b^n \cos(a^n \pi x),$$

where $0 < b < 1$ so that it converges uniformly and look for conditions on a and b that will imply that

$$\left| \frac{\sum_{n=0}^{\infty} b^n \cos(a^n \pi x_1) - \sum_{n=0}^{\infty} b^n \cos(a^n \pi x_0)}{x_1 - x_0} \right|$$

is unbounded as x_1 ranges over any interval of the form $(x_0 - \delta, x_0 + \delta)$. By Theorems 5.4 and 5.5, we can combine the summations and simplify this ratio to

$$\sum_{n=0}^{\infty} b^n \frac{\cos(a^n \pi x_1) - \cos(a^n \pi x_0)}{x_1 - x_0}.$$

We make two critical observations. First, given a, x_0, and a positive integer m, there will always be an integer N satisfying

$$1 \leq |N - a^m x_0| \leq 3/2. \tag{6.81}$$

Second, if we choose x_1 so that $a^m x_1 = N$, then $\cos(a^m \pi x_1)$ and $\cos(a^m \pi x_0)$ will have opposite signs. It follows that

$$\begin{aligned} \cos(a^m \pi x_1) - \cos(a^m \pi x_0) &= \cos(\pi N) - \cos[\pi N + \pi(a^m x_0 - N)] \\ &= (-1)^N \left\{ 1 + \cos[\pi(a^m x_0 - N - 1)] \right\}, \end{aligned} \tag{6.82}$$

where

$$1 + \cos[\pi(a^m x_0 - N - 1)] \geq 1. \tag{6.83}$$

If a is an odd integer and n is larger than m, then

$$\begin{aligned} \cos(a^n \pi x_1) - \cos(a^n \pi x_0) &= \cos(a^{n-m} N \pi) - \cos(a^n \pi x_0) \\ &= (-1)^N \left\{ 1 + \cos[a^{n-m} \pi(a^m x_0 - N - 1)] \right\}, \end{aligned} \tag{6.84}$$

and

$$1 + \cos[a^{n-m}\pi(a^m x_0 - N - 1)] \geq 0. \tag{6.85}$$

Equations (6.82–6.85) imply that all of the summands in

$$\sum_{n=0}^{\infty} b^n [\cos(a^n \pi x_1) - \cos(a^n \pi x_0)]$$

from the mth term on have the same sign and that the mth summand has absolute value greater than or equal to b^m:

$$\left| \sum_{n=m}^{\infty} b^n \frac{\cos(a^n \pi x_1) - \cos(a^n \pi x_0)}{x_1 - x_0} \right| \geq \frac{b^m}{|x_1 - x_0|}. \tag{6.86}$$

If we replace N by $a^m x_1$ in equation (6.81), we see that

$$|x_1 - x_0| \leq \frac{3}{2a^m}. \tag{6.87}$$

As long as we choose m large enough so that $3/2a^m < \delta$, there is such an x_1 inside our interval $(x_0 - \delta, x_0 + \delta)$. This upper bound on $|x_1 - x_0|$ combined with equation (6.86) tells us that

$$\left| \sum_{n=m}^{\infty} b^n \frac{\cos(a^n \pi x_1) - \cos(a^n \pi x_0)}{x_1 - x_0} \right| \geq \frac{2}{3}(ab)^m. \tag{6.88}$$

As long as ab is larger than 1, we can find an x_1 for which the tail of the series is as large as we wish.

We are not quite done. We must verify that the first m summands of our series do not cancel out the value of the tail. We need an upper bound on

$$\left| \sum_{n=0}^{m-1} b^n \frac{\cos(a^n \pi x_1) - \cos(a^n \pi x_0)}{x_1 - x_0} \right|.$$

The mean value theorem tells us that there is an x_2 between x_0 and x_1 for which

$$\frac{\cos(a^n \pi x_1) - \cos(a^n \pi x_0)}{x_1 - x_0} = -a^n \pi \sin(a^n \pi x_2). \tag{6.89}$$

Since the absolute value of the sine is bounded by 1, we see that

$$\left| \frac{\cos(a^n \pi x_1) - \cos(a^n \pi x_0)}{x_1 - x_0} \right| \leq a^n \pi, \tag{6.90}$$

and therefore

$$\left| \sum_{n=0}^{m-1} b^n \frac{\cos(a^n \pi x_1) - \cos(a^n \pi x_0)}{x_1 - x_0} \right| \leq \sum_{n=0}^{m-1} b^n \left| \frac{\cos(a^n \pi x_1) - \cos(a^n \pi x_0)}{x_1 - x_0} \right|$$

$$\leq \sum_{n=0}^{m-1} (ab)^n \pi \quad = \quad \pi \frac{(ab)^m - 1}{ab - 1}$$

$$< \pi \frac{(ab)^m}{ab - 1}. \tag{6.91}$$

If we choose ab so that

$$\frac{\pi}{ab - 1} < \frac{2}{3},$$

which is the same as saying that

$$ab > 1 + 3\pi/2, \tag{6.92}$$

then the absolute value of the sum of the first m terms will be a strictly smaller multiple of $(ab)^m$ than is the sum of the tail:

$$\left| \frac{1}{x_1 - x_0} \sum_{n=0}^{\infty} b^n [\cos(a^n \pi x_1) - \cos(a^n \pi x_0)] \right| > (ab)^m \left(\frac{2}{3} - \frac{\pi}{ab - 1} \right). \tag{6.93}$$

Since m is not bounded and ab is larger than 1, an x_1 exists for which this average rate of change is larger than any predetermined error.

Q.E.D.

As an example, let us take $b = 6/7$ and $a = 7$ so that $ab = 6 > 1 + 3\pi/2 \approx 5.7$. Given δ, we can choose any m for which $7^m > 3/2\delta$. We have demonstrated that there will be an x_1 within δ of x_0 for which

$$\left| \sum_{n=0}^{\infty} \left(\frac{6}{7} \right)^n \frac{\cos(7^n \pi x_1) - \cos(7^n \pi x_0)}{x_1 - x_0} \right| > 6^m \left(\frac{2}{3} - \frac{\pi}{5} \right) > 0.038 \times 6^m. \tag{6.94}$$

Even Weierstrass Could Be Wrong

Even after Weierstrass announced his example of a function that is continuous everywhere and differentiable nowhere, the question remained of whether a "nice" continuous function would have to be differentiable. The simplest additional condition would be to insist on monotonicity or piecewise monotonicity. Several people searched for a proof that a continuous monotonic function would have to be differentiable at all but finitely many points. Weierstrass responded with a continuous monotonic function that is not differentiable at any rational number. He then sought such a function that is not differentiable at any number. Though he never found one, he believed that they must exist.

Weierstrass was wrong. They do not exist. Henri Lebesgue would prove in 1903 that a continuous and monotonic function must be differentiable at *most* points. What he proved was that the set of points where the function is not differentiable must have measure zero, where *measure zero* is a technical restriction on the size of a set. It will be defined in the next chapter. The rational numbers have measure zero. Even if we include all *algebraic numbers*, numbers like $\sqrt{2}$ that are the roots of polynomials with rational coefficients, we still have a set of measure zero. It may seem like there are a lot of them, but in a sense that can be made precise, most real numbers are neither rational nor algebraic.

Exercises

The symbol (**M&M**) indicates that *Maple* and *Mathematica* codes for this problem are available in the **Web Resources** at **www.macalester.edu/aratra**.

6.4.1. Prove that if $f(a-0) \neq f(a) \neq f(a+0)$, then

$$F(x) = \int_0^x f(t)\,dt$$

cannot be differentiable at $x = a$.

6.4.2. (**M&M**) The exercises beginning here and continuing through exercise 6.4.10 develop and verify a standard example of an everywhere continuous, nowhere differentiable function. We begin with the function that assigns to the variable x the distance between x and the nearest integer. For example: $f(2.15) = 0.15$, $f(1.78) = 0.22$, $f(1/2) = 1/2$, $f(3) = 0$. Graph this function for $-2 \leq x \leq 2$. The function you get should look like the teeth of a saw.

6.4.3. (**M&M**) Graph the function

$$f_n(x) = \frac{1}{4^n} f(4^n x)$$

for $n = 2,\ 3,\ 4$ over the interval $-4^{1-n} \leq x \leq 4^{1-n}$.

6.4.4. (**M&M**) Define a new function F by

$$F(x) = \sum_{n=0}^{\infty} f_n(x). \tag{6.95}$$

This is the function that will be shown to be continuous but never differentiable. Let

$$S_N(x) = \sum_{n=0}^{N} f_n(x)$$

be the partial sums. Graph $S_2(x),\ S_3(x),\ S_4(x)$ for $-2 \leq x \leq 2$.

6.4.5. Prove that the series expansion for F given in equation (6.95) converges uniformly for all x. This implies that F must be a continuous function.

6.4.6. The first step in showing that f is never differentiable is to consider the real number line divided into intervals that are split at the half integers:

$$\begin{array}{ccccccccccccccc} & & & & & & & & & & & & & \\ \hline \end{array}$$

$$-1.5 \quad -1 \quad -0.5 \quad 0 \quad \;\; 0.5 \quad 1 \quad \;\; 1.5 \quad 2 \quad \;\; 2.5 \quad 3 \quad \;\; 3.5 \quad 4 \quad \;\; 4.5 \quad 5 \quad \;\; 5.5$$

Let m be any positive integer. If $4^m x$ and $4^m x + 1/4$ are in different intervals, then $4^m x$ and $4^m x - 1/4$ must be in the same interval. Define $\sigma_m = 1$ or -1 so that $4^m x$ and $4^m x + \sigma_m/4$ are in the same interval (if $4^m x$ is a half integer, then take $\sigma_m = 1$). Show that

$$f_m(x + \sigma_m/4^{m+1}) - f_m(x) = \pm \frac{\sigma_m}{4^{m+1}}. \tag{6.96}$$

6.4.7. The reason why equation (6.96) works is that x and $x + \sigma_m/4^{m+1}$ lie on the same edge of the same tooth in the graph of the function f_m. Prove that if $n \le m$, then x and $x + \sigma_m/4^{m+1}$ lie on the same edge of the same tooth in the graph of the function f_n, and then show how this implies that

$$f_n(x + \sigma_m/4^{m+1}) - f_n(x) = \pm \frac{\sigma_m}{4^{m+1}}, \quad n \le m. \tag{6.97}$$

6.4.8. Prove that if $n > m$, then $f_n(x + \sigma_m/4^{m+1}) = f_n(x)$, and so

$$f_n(x + \sigma_m/4^{m+1}) - f_n(x) = 0, \quad n > m. \tag{6.98}$$

6.4.9. Show that

$$\frac{F(x + \sigma_m/4^{m+1}) - F(x)}{\sigma_m/4^{m+1}} = \sum_{n=0}^{m} \pm 1 = \alpha_m, \tag{6.99}$$

where α_m is even if and only if m is odd.

6.4.10. If F is differentiable at x, then we can find a number—denoted by $F'(x)$—such that

$$\alpha_m = \frac{F(x + \sigma_m/4^{m+1}) - F(x)}{\sigma_m/4^{m+1}} = F'(x) - E(x + \sigma_m/4^{m+1}, x), \tag{6.100}$$

where $E(x + h, x)$ can be made arbitrarily small by taking h sufficiently small. Explain why no such $F'(x)$ can exist.

6.4.11. **M&M** Graph the partial sums

$$f_M(x) = \sum_{n=0}^{M} \left(\frac{6}{7}\right)^n \cos(7^n \pi x)$$

for x in $[-1/7^{M-1}, 1/7^{M-1}]$ with $M = 1, 2,$ and 3.

6.4.12. Let

$$f(x) = \sum_{n=0}^{\infty} \left(\frac{6}{7}\right)^n \cos(7^n \pi x).$$

What is $f(0)$? How many terms of this series would you have to take in order to be certain that you are within 0.01 of the correct value of $f(x)$? The fact that this series is uniformly convergent implies that there is an answer to this question that is independent of x.

6.4.13. Find a value of x_1 that lies within 0.001 of 0.5 and for which

$$\left| \sum_{n=0}^{\infty} \left(\frac{6}{7}\right)^n \frac{\cos(7^n \pi x_1) - \cos(7^n \pi \times 0.5)}{x_1 - 0.5} \right| > 1,000,000. \qquad (6.101)$$

7

Epilogue

For over a decade, Weierstrass was a voice crying in the wilderness, proclaiming the importance of uniform convergence and the need for ϵ–δ proofs. Few understood what he was saying. But he trained students who spread the seeds of his message. The publication of Riemann's treatise on trigonometric series became the catalyst for the acceptance of Weierstrass's position into the mathematical canon. By the 1870s, the mathematical world was abuzz with the questions that emerged from Riemann's work.

One of these questions was the uniqueness of the trigonometric representation: can two different trigonometric series be equal at every value of x? Fourier's proof of the uniqueness of his series rests on the interchange of integration and summation, of the ability to integrate an infinite sum by integrating each summand. Weierstrass, Riemann, and Dirichlet recognized the potential hazards of this approach. Weierstrass knew that it *was* legitimate when the series converged uniformly, but some of the most interesting Fourier series do not converge uniformly.

In 1870, Heinrich Heine introduced the notion of **uniform convergence in general**. A trigonometric series converges uniformly in general if there are a finite number of break points in the interval of the basic period so that on any closed interval that does not include a break point, the series converges uniformly. Fourier's cosine series for the function that is 1 on $(-1, 1)$ is not uniformly convergent, but it is uniformly convergent in general. Heine proved that if the series converges uniformly in general, then there is no other trigonometric series that represents that same function. The representation is unique. Mathematicians became aware of the subtleties of term-by-term integration and uniform convergence.

They also began to evince unease with Riemann's definition of the integral. In that same year, 1870, Hermann Hankel pointed out that if a function is integrable in Riemann's sense, then inside any open interval, no matter how small, there is at least one point where the function is continuous. In modern terminology, the set of points at which the function is continuous is **dense**. But there is no reason why a trigonometric series, even if it converges, need be continuous at any point. The search was on for a more general definition of integration.

The road to this redefinition of integration leads through set theory and point-set topology and involves such names as Cantor, Baire, and Borel. The new integral was announced in 1901 by Henri Lebesgue and became the subject of his doctoral thesis, published a year later. Whenever a function can be integrated in the Riemann sense, its integral exists and is the same whether one uses the Riemann or Lebesgue definition. The advantage of Lebesgue's definition is that it completely divorces integrability from continuity and so extends integrability to many more functions.

The Lebesgue Integral

The Lebesgue definition of the integral begins with the notion of the **measure** of a set. For an interval, whether open or closed, its measure is simply its length. The measure of a single point is 0. For other sets, the measure gets more interesting. It is not always defined, but when the measure of a set S exists, then it can be found by looking at all coverings of S. A **covering** of S is a countable collection of open intervals whose union contains every point in S. For each possible covering, we calculate the sum of the lengths of the intervals in the covering. When S is measurable, its measure $m(S)$ is the greatest lower bound (or *infimum*) of these calculations, taken over all possible coverings:

$$m(S) = \inf_{S \subseteq \cup I_n} \left(\sum m(I_n) \right).$$

To find the measure of the set of rational numbers between 0 and 1, we observe that we can order them in one-to-one correspondence with the integers:

$$\left(0, 1, \frac{1}{2}, \frac{1}{3}, \frac{2}{3}, \frac{1}{4}, \frac{3}{4}, \frac{1}{5}, \frac{2}{5}, \frac{3}{5}, \frac{4}{5}, \frac{1}{6}, \frac{5}{6}, \frac{1}{7}, \frac{2}{7}, \cdots \right).$$

Given any $\epsilon > 0$, we can create a covering out of an open interval of length $\epsilon/2$ containing 0, $\epsilon/4$ containing 1, $\epsilon/8$ containing 1/2, $\epsilon/16$ containing 1/3, and so on. The sum of these lengths is ϵ. The greatest lower bound of the sum of the lengths over all coverings of this set is 0.

The set of irrational numbers between 0 and 1 has measure 1. While it is possible to construct sets that are not measurable, they are very strange creatures indeed.

The **characteristic function** of a set S, χ_S, is defined to be 1 at any point in S and 0 at any point not in S:

$$\chi_S(x) = \begin{cases} 1, & x \in S, \\ 0, & x \notin S. \end{cases}$$

The Lebesgue integral starts with characteristic functions. The integral over the interval $[a, b]$ of χ_S is the measure of $S \cap [a, b]$:

$$\int_a^b \chi_S(x)\, dx = m\left(S \cap [a, b] \right).$$

If a function is a finite linear combination of characteristic functions, its integral is the appropriate linear combination of the integrals:

$$\int_a^b \left(\sum_{i=1}^n a_i \chi_{S_i}(x) \right) dx = \sum_{i=1}^n a_i \, m \left(S_i \cap [a, b] \right).$$

The hard part comes in showing that any reasonably nice function is the limit of linear combinations of characteristic functions and that the limit of the corresponding integrals is well defined.

Not everyone was happy with the direction analysis had taken in the last decades of the nineteenth century. In 1889, Henri Poincaré wrote

> So it is that we see the emergence of a multitude of bizarre functions that seem to do their best to resemble as little as possible those honest functions that serve a useful purpose. No longer continuous, or maybe continuous but not differentiable, etc. More than this, from a logical point of view, it is these strange functions that are the most common. The functions that one encounters without having searched for them and which follow simple laws now appear to be no more than a very special case. Only a small corner remains for them.

> In earlier times, when we invented a new function it was for the purpose of some practical goal. Today, we invent them expressly to show the flaws in our forefathers' reasoning, and we draw from them nothing more than that.

More succinctly, Hermite wrote to Stieltjes in 1893, "I turn away with fright and horror from this lamentable plague of functions that do not have derivatives."

Why?

Had analysis gone too far? Had it totally divorced itself from reality to wallow in a self-generated sea of miscreations and meaningless subtleties? Some good mathematicians may have been alarmed by the direction analysis was taking, but both of the quotes just given were taken out of context. Poincaré was not disparaging what analysis had become, but how it was taught. Hermite was not complaining of artificially created functions that lack derivatives, but of trigonometric series that he had encountered in his explorations of the Bernoulli polynomials. They were proving intractable precisely because they were not differentiable.

With characteristic foresight, Riemann put his finger on the importance of these studies:

> In fact, [the problem of the representability of a function by trigonometric series] was completely solved for all cases which present themselves in nature alone, because however great may be our ignorance about how the forces and states of matter vary in space and time in the infinitely small, we can certainly assume that the functions to which Dirichlet's research did not extend do not occur in nature.

> Nevertheless, those cases that were unresolved by Dirichlet seem worthy of attention for two reasons.

> The first is that, as Dirichlet himself remarked at the end of his paper, this subject stands in the closest relationship to the principles of the infinitesimal calculus

and can serve to bring these principles to greater clarity and certainty. In this connection its treatment has an immediate interest.

But the second reason is that the applicability of Fourier series is not restricted to physical researches; it is now also being applied successfully to one area of pure mathematics, number theory. And just those functions whose representability by a trigonometric series Dirichlet did not explore seem here to be important.

Riemann was intimately acquainted with the connection between infinite series and number theory. Legendre had conjectured in 1788 that if a and q are relatively prime integers, then there are infinitely many primes p for which $p - a$ is a multiple of q. As an example, there should be infinitely many primes p for which $p - 6$ is a multiple of 35. Dirichlet proved Legendre's conjecture by showing that if we sum the reciprocals of these primes,

$$\sum \frac{1}{p}, \quad p \text{ is prime and } q \text{ divides } p - a,$$

this series always diverges, irrespective of the choice of a and q (provided they have no common factor). Not only is there no limit to the number of primes of this form, they are in some sense quite common.

The methods that Dirichlet introduced to prove this result are extremely powerful and far-reaching. Riemann himself modified and extended them to suggest how it might be possible to prove that the number of primes less than or equal to x is asymptotically $x / \ln x$. The route he mapped out is tortuous, involving strange series and very subtle questions of convergence. It was not completely negotiated until the independent proofs of Jacques Hadamard and Charles de la Vallée Poussin in 1896. They needed all of the machinery of analysis that was available to them.

Analysis has continued to be a key component of modern number theory. It is more than a toy for the investigation of primes and Bernoulli numbers. It has emerged as an important tool for the study of a wide range of discrete systems with interesting structure. It sits at the heart of the modern methods used by Andrew Wiles and Richard Taylor to attack Fermat's last theorem. It plays a critical role in the theoretical constructs of modern physics. As Riemann foresaw, "just those functions whose representability by a trigonometric series Dirichlet did not explore seem here to be important."

Appendix A
Explorations of the Infinite

The four sections of Appendix A lead to a proof of Stirling's remarkable formula for the value of $n!$:

$$n! = n^n e^{-n} \sqrt{2\pi n} \; e^{E(n)}, \qquad\qquad (A.1)$$

where $\lim_{n\to\infty} E(n) = 0$, and this error term can be represented by the asymptotic series

$$E(n) \sim \frac{B_2}{1\cdot 2\cdot n} + \frac{B_4}{3\cdot 4\cdot n^3} + \frac{B_6}{5\cdot 6\cdot n^5} + \cdots, \qquad\qquad (A.2)$$

where B_1, B_2, B_3, ... are rational numbers known as the **Bernoulli numbers**. Note that we do not write equation (A.2) as an equality since this series does not converge to $E(n)$ (see section A.4).

In this first section, we follow John Wallis as he discovers an infinite product that is equal to π. While his formula is a terrible way to approximate π, it establishes the connection between $n!$ and π, explaining that curious appearance of π in equation (A.1). In section 2, we show how Jacob Bernoulli was led to discover his mysterious and pervasive sequence of numbers by his search for a simple formula for the sum of kth powers. We continue the applications of the Bernoulli numbers in section 3 where we follow Leonhard Euler's development of formulæ for the sums of reciprocals of even powers of the positive integers. It all comes together in section 4 when we shall prove this identity discovered by Abraham deMoivre and James Stirling.

A.1 Wallis on π

When Newton said, "If I have seen a little farther than others it is because I have stood on the shoulders of giants," one of those giants was John Wallis (1616–1703). Wallis taught at Cambridge before becoming Savilian Professor of Geometry at Oxford. His *Arithmetica Infinitorum*, published in 1655, derives the rule (found also by Fermat) for the integral of a fractional power of x:

$$\int_0^1 x^{m/n}\, dx = \frac{1}{1+m/n} = \frac{n}{m+n}. \tag{A.3}$$

We begin our development of Wallis's formula for π with the observation that π is the area of any circle with radius 1. If we locate the center of our circle at the origin, then the quarter circle in the first quadrant has area

$$\frac{\pi}{4} = \int_0^1 (1-x^2)^{1/2}\, dx. \tag{A.4}$$

This looks very much like the kind of integral Wallis had been studying. Any means of calculating this integral will yield a means of calculating π. Realizing that he could not attack it head on, Wallis looked for similar integrals that he could handle. His genius is revealed in his decision to look at

$$\int_0^1 (1-x^{1/p})^q\, dx.$$

When q is a small positive integer, we can expand the integrand:

$$\int_0^1 (1-x^{1/p})^0\, dx = \int_0^1 dx$$
$$= 1,$$

$$\int_0^1 (1-x^{1/p})^1\, dx = \int_0^1 (1-x^{1/p})\, dx$$
$$= 1 - \frac{p}{p+1}$$
$$= \frac{1}{p+1},$$

$$\int_0^1 (1-x^{1/p})^2\, dx = \int_0^1 (1 - 2x^{1/p} + x^{2/p})\, dx$$
$$= 1 - \frac{2p}{p+1} + \frac{p}{p+2}$$
$$= \frac{2}{(p+1)(p+2)},$$

$$\int_0^1 (1-x^{1/p})^3\, dx = \int_0^1 (1 - 3x^{1/p} + 3x^{2/p} - x^{3/p})\, dx$$
$$= 1 - \frac{3p}{p+1} + \frac{3p}{p+2} - \frac{p}{p+3}$$
$$= \frac{6}{(p+1)(p+2)(p+3)}.$$

A pattern is emerging, and it requires little insight to guess that

$$\int_0^1 (1 - x^{1/p})^4 \, dx = \frac{4!}{(p+1)(p+2)(p+3)(p+4)},$$

$$\int_0^1 (1 - x^{1/p})^5 \, dx = \frac{5!}{(p+1)(p+2)(p+3)(p+4)(p+5)},$$

$$\vdots$$

where $4! = 4 \cdot 3 \cdot 2 \cdot 1 = 24$, $5! = 5 \cdot 4 \cdot 3 \cdot 2 \cdot 1 = 120$, and so on.

These numbers should look familiar. When p and q are both integers, we get reciprocals of binomial coefficients,

$$\frac{q!}{(p+1)(p+2)\cdots(p+q)} = \frac{1}{\binom{p+q}{q}},$$

where

$$\binom{p+q}{q} = \binom{p+q}{p} = \frac{(p+q)!}{p! \, q!}.$$

This suggested to Wallis that he wanted to work with the reciprocals of his integrals:

$$f(p, q) = 1 \Big/ \int_0^1 (1 - x^{1/p})^q \, dx = \frac{(p+1)(p+2)\cdots(p+q)}{q!}. \tag{A.5}$$

We want to find the value of $f(1/2, 1/2) = 4/\pi$. Our first observation is that we can evaluate $f(p, q)$ for any p so long as q is a nonnegative integer. We use induction on q to prove equation (A.5). As shown above, when $q = 0$ we have $f(p, q) = 1/\int_0^1 dx = 1$. In exercise A.1.2, you are asked to prove that

$$\int_0^1 \left(1 - x^{1/p}\right)^q \, dx = \frac{q}{p+q} \int_0^1 \left(1 - x^{1/p}\right)^{q-1} \, dx. \tag{A.6}$$

With this recursion and the induction hypothesis that

$$f(p, q - 1) = \frac{(p+1)(p+2)\cdots(p+q-1)}{(q-1)!},$$

it follows that

$$f(p, q) = \left(\frac{p+q}{q}\right) \frac{(p+1)(p+2)\cdots(p+q-1)}{(q-1)!} = \frac{(p+1)(p+2)\cdots(p+q)}{q!}.$$

We also can use this recursion to find $f(1/2, 3/2)$ in terms of $f(1/2, 1/2)$:

$$f(1/2, 3/2) = \frac{1/2 + 3/2}{3/2} f(1/2, 1/2) = \frac{4}{3} f(1/2, 1/2).$$

We can now prove by induction (see exercise A.1.3) that

$$f\left(\frac{1}{2}, q - \frac{1}{2}\right) = \frac{(2q)(2q - 2)\cdots 4}{(2q - 1)(2q - 3)\cdots 3} \, f\left(\frac{1}{2}, \frac{1}{2}\right). \tag{A.7}$$

Table A.1. Values of $f(p,q)$ in terms of $\square = 4/\pi$.

$p\downarrow$ $q\rightarrow$	$-1/2$	0	1/2	1	3/2	2	5/2	3	7/2	4
$-1/2$	∞	1	$\frac{1}{2}\square$	$\frac{1}{2}$	$\frac{1}{3}\square$	$\frac{3}{8}$	$\frac{4}{15}\square$	$\frac{5}{16}$	$\frac{8}{35}\square$	$\frac{35}{128}$
0	1	1	1	1	1	1	1	1	1	1
1/2	$\frac{1}{2}\square$	1	$\square$	$\frac{3}{2}$	$\frac{4}{3}\square$	$\frac{15}{8}$	$\frac{8}{5}\square$	$\frac{35}{16}$	$\frac{64}{35}\square$	$\frac{315}{128}$
1	$\frac{1}{2}$	1	$\frac{3}{2}$	2	$\frac{5}{2}$	3	$\frac{7}{2}$	4	$\frac{9}{2}$	5
3/2	$\frac{1}{3}\square$	1	$\frac{4}{3}\square$	$\frac{5}{2}$	$\frac{8}{3}\square$	$\frac{35}{8}$	$\frac{64}{15}\square$	$\frac{105}{16}$	$\frac{128}{21}\square$	$\frac{1155}{128}$
2	$\frac{3}{8}$	1	$\frac{15}{8}$	3	$\frac{35}{8}$	6	$\frac{63}{8}$	10	$\frac{99}{8}$	15
5/2	$\frac{4}{15}\square$	1	$\frac{8}{5}\square$	$\frac{7}{2}$	$\frac{64}{15}\square$	$\frac{63}{8}$	$\frac{128}{15}\square$	$\frac{231}{16}$	$\frac{512}{35}\square$	$\frac{3003}{128}$
3	$\frac{5}{16}$	1	$\frac{35}{16}$	4	$\frac{105}{16}$	10	$\frac{231}{16}$	20	$\frac{429}{16}$	35
7/2	$\frac{8}{35}\square$	1	$\frac{64}{35}\square$	$\frac{9}{2}$	$\frac{128}{21}\square$	$\frac{99}{8}$	$\frac{512}{35}\square$	$\frac{429}{16}$	$\frac{1024}{35}\square$	$\frac{6435}{128}$
4	$\frac{35}{128}$	1	$\frac{315}{128}$	5	$\frac{1155}{128}$	15	$\frac{3003}{128}$	35	$\frac{6435}{128}$	70

What if p is a nonnegative integer or half-integer? The binomial coefficient is symmetric in p and q,

$$\binom{p+q}{q} = \binom{p+q}{p} = \frac{(p+q)!}{p!\,q!}.$$

It is only a little tricky to verify that for *any* values of p and q, we also have that $f(p,q) = f(q,p)$ (see exercise A.1.4). This can be used to prove that when p and q are positive integers (see exercise A.1.5),

$$f\left(p-\frac{1}{2}, q-\frac{1}{2}\right) = \frac{2\cdot 4\cdot 6\cdots(2p+2q-2)}{3\cdot 5\cdots(2p-1)\cdot 3\cdot 5\cdots(2q-1)}\, f\left(\frac{1}{2},\frac{1}{2}\right). \quad (A.8)$$

Wallis could now construct a table of values for $f(p,q)$, allowing $\square$ to stand for $f(1/2, 1/2) = 4/\pi$ (see Table A.1.).

We see that any row in which p is a positive integer is increasing from left to right, and it is reasonable to expect that the row $p = 1/2$ is also increasing from left to right (see exericse A.1.6 for the proof). Recalling that $\square = 4/\pi$, this implies a string of inequalities:

$$1 < \frac{4}{\pi} < \frac{3}{2} < \frac{16}{3\pi} < \frac{15}{8} < \frac{32}{5\pi} < \frac{35}{16} < \frac{256}{35\pi} < \frac{315}{128}.$$

These, in turn, yield inequalities for $\pi/2$:

$$\frac{4}{3} < \frac{\pi}{2} < 2,$$

$$\frac{64}{45} < \frac{\pi}{2} < \frac{16}{9},$$

$$\frac{256}{175} < \frac{\pi}{2} < \frac{128}{75},$$

$$\frac{16384}{11025} < \frac{\pi}{2} < \frac{2048}{1225}.$$

It is easier to see what is happening with these inequalities if we look at our string of inequalities in terms of the ratios that led us to find them in the first place:

$$1 < \frac{4}{\pi} < \frac{3}{2} < \frac{4}{3}\cdot\frac{4}{\pi} < \frac{3}{2}\cdot\frac{5}{4} < \frac{4}{3}\cdot\frac{6}{5}\cdot\frac{4}{\pi} < \frac{3}{2}\cdot\frac{5}{4}\cdot\frac{7}{6} < \cdots.$$

In general, we have that

$$\frac{3\cdot5\cdot7\cdots(2n-1)}{2\cdot4\cdot6\cdots(2n-2)} < \frac{4\cdot6\cdot8\cdots(2n)}{3\cdot5\cdot7\cdots(2n-1)}\frac{4}{\pi} < \frac{3\cdot5\cdot7\cdots(2n+1)}{2\cdot4\cdot6\cdots(2n)}. \qquad \text{(A.9)}$$

This yields a general inequality for $\pi/2$ that we can make as precise as we want by taking n sufficiently large:

$$\frac{2^2\cdot4^2\cdot6^2\cdots(2n)^2}{1\cdot3^2\cdot5^2\cdots(2n-1)^2\cdot(2n+1)} < \frac{\pi}{2} < \frac{2^2\cdot4^2\cdot6^2\cdots(2n-2)^2\cdot(2n)}{1\cdot3^2\cdot5^2\cdots(2n-1)^2}. \qquad \text{(A.10)}$$

As n gets larger, these bounds on $\pi/2$ approach each other. Their ratio is $2n/(2n+1)$ which approaches 1. Wallis therefore concluded that

$$\frac{\pi}{2} = \frac{2}{1}\cdot\frac{2}{3}\cdot\frac{4}{3}\cdot\frac{4}{5}\cdot\frac{6}{5}\cdot\frac{6}{7}\cdots. \qquad \text{(A.11)}$$

Note that this product alternately grows and shrinks as we take more terms.

Exercises

The symbol (M&M) indicates that *Maple* and *Mathematica* codes for this problem are available in the **Web Resources** at **www.macalester.edu/aratra**.

A.1.1.   (M&M) Consider Wallis's infinite product for $\pi/2$ given in equation (A.11).

a. Show that the product of the kth *pair* of fractions is $4k^2/(4k^2-1)$ and therefore

$$\pi = 2\prod_{k=1}^{\infty}\frac{4k^2}{4k^2-1}.$$

b. How many terms of this product are needed in order to approximate π to 3-digit accuracy?

c. We can improve our accuracy if we average the upper and lower bounds. Prove that this average is

$$\frac{2^2 \cdot 4^2 \cdots (2n) \cdot (2n + 1/2)}{1 \cdot 3^2 \cdots (2n - 1)^2 \cdot (2n + 1)}.$$

How large a value of n do we need in order to approximate π to 3-digit accuracy?

A.1.2. Prove equation (A.6).

A.1.3. Use equation (A.6) to prove equation (A.7) by induction on q.

A.1.4. Prove that

$$\int_0^1 \left(1 - x^{1/p}\right)^q \, dx = \int_0^1 \left(1 - x^{1/q}\right)^p \, dx, \qquad (A.12)$$

and therefore $f(p, q) = f(q, p)$.

A.1.5. Using the results from exercises A.1.3 and A.1.4, prove equation (A.8) when p and q are positive integers.

A.1.6. Prove that if p is positive and $q_1 > q_2$, then

$$\left(1 - x^{1/p}\right)^{q_1} < \left(1 - x^{1/p}\right)^{q_2},$$

for all x between 0 and 1, and therefore $f(p, q_1) > f(p, q_2)$. Prove that if p is negative and $q_1 > q_2$, then $f(p, q_1) < f(p, q_2)$.

A.1.7. **(M&M)** We have seen that as long as p or q is an integer, we have that $f(p, q) = \binom{p+q}{q}$. This suggests a way of defining binomial coefficients when neither p nor q are integers. We would expect the value of $\binom{1}{1/2}$ to be $f(1/2, 1/2) = 4/\pi$. Using your favorite computer algebra system, see what value it assigns to $\binom{1}{1/2}$. How should we define $\binom{a}{b}$ for arbitrary real numbers a and b?

A.1.8. When p and q are integers, we have the relationship of Pascal's triangle,

$$f(p, q) = f(p, q - 1) + f(p - 1, q). \qquad (A.13)$$

Does this continue to hold true when p and q are not both integers? Either give an example where it does not work or prove that it is always true.

A.1.9. Show that if we use the row $p = -1/2$ to approximate $\pi/2$:

$$1 > \frac{4}{2\pi} > \frac{1}{2} > \frac{4}{3\pi} > \frac{3}{8} > \frac{16}{15\pi} > \frac{5}{16} > \frac{32}{35\pi} > \frac{35}{128} > \cdots,$$

then we get the same bounds for $\pi/2$.

A.1.10. What bounds do we get for $\pi/2$ if we use the row $p = 3/2$?

A.1.11. What bounds do we get for $\pi/2$ if we use the diagonal:

$$1 < \frac{4}{\pi} < 2 < \frac{32}{3\pi} < 6 < \frac{512}{15\pi} < 20 < \frac{4094}{35\pi} < 70 < \cdots?$$

A.1.12. As far as you can, extend the table of values for $f(p, q)$ into negative values of p and q.

A.1.13. (M&M) Use the method of this section to find an infinite product that approaches the value of

$$f(2/3, 1/3) = 1 \Big/ \int_0^1 \left(1 - x^{3/2}\right)^{1/3} dx \ .$$

Compare this to the value of $\binom{1}{1/3}$ given by your favorite computer algebra system.

A.2 Bernoulli's Numbers

Johann and Jacob Bernoulli were Swiss mathematicians from Basel, two brothers who played a critical role in the early development of calculus. The elder, Jacob Bernoulli, died in 1705. Eight years later, his final masterpiece was published, *Ars Conjectandi*. It laid the foundations for the study of probability and included an elegant solution to an old problem: to find formulas for the sums of powers of consecutive integers. He bragged that with it he had calculated the sum of the tenth powers of the integers up to one thousand,

$$1^{10} + 2^{10} + 3^{10} + \cdots + 1000^{10},$$

in "half a quarter of an hour."

The formula for the sum of the first $k - 1$ integers is

$$1 + 2 + 3 + 4 + \cdots + (k - 1) = \frac{k^2}{2} - \frac{k}{2}. \tag{A.14}$$

The proof is simple:

$$
\begin{array}{cccccc}
 1 & + & 2 & + \cdots + (k - 2) & + (k - 1) & \\
 + (k - 1) & + (k - 2) & + \cdots + & 2 & + \quad 1 & \\
 \hline
 = \quad k & + \quad k & + \cdots + & k & + \quad k & = (k - 1)k.
\end{array}
$$

No one knows who first discovered this formula. Its origins are lost in the mists of time. Even the formula for the sum of squares is ancient,

$$1^2 + 2^2 + 3^2 + 4^2 + \cdots + (k - 1)^2 = \frac{k^3}{3} - \frac{k^2}{2} + \frac{k}{6}. \tag{A.15}$$

It was known to and used by Archimedes, but was probably even older. It took quite a bit longer to find the formula for the sum of cubes. The earliest reference that we have to this formula connects it to the work of Aryabhata of Patna (India) around the year 500 A.D. The formula for sums of fourth powers was discovered by ibn Al-Haytham of Baghdad about 1000 A.D. In the early 14th century, Levi ben Gerson of France found a general formula for arbitary kth powers, though the central idea which draws on patterns in the binomial coefficients can be found in other contemporary and earlier sources: *Al-Bahir fi'l Hisab* (*Shining Treatise on Calculation*) written by al-Samaw'al in 1144 in what is now Iraq, *Siyuan Yujian* (*Jade Mirror of the Four Unknowns*) written by Zhu Shijie in 1303 in China, and *Ganita Kaumudi* written by Narayana Pandita in 1356 in India.

> **Web Resource:** To see a derivation and proof of this historic formula for the sum of kth powers based on properties of binomial coefficients, go to **Binomial coefficients and sums of kth powers**.

Jacob Bernoulli may have been aware of the binomial coefficient formula, but that did not stop him from finding his own. He had a brilliant insight. The new integral calculus gave efficient tools for calculating limits of summations that today we call Riemann sums. Perhaps it could be turned to the task of finding formulas for other types of sums.

The Bernoulli Polynomials

Jacob Bernoulli looked for polynomials, $B_1(x)$, $B_2(x)$, $\ldots$, for which

$$1 + 2 + \cdots + (k-1) = \int_0^k B_1(x)\, dx,$$

$$1^2 + 2^2 + \cdots + (k-1)^2 = \int_0^k B_2(x)\, dx,$$

$$1^3 + 2^3 + \cdots + (k-1)^3 = \int_0^k B_3(x)\, dx,$$

$$\vdots$$

$$1^n + 2^n + \cdots + (k-1)^n = \int_0^k B_n(x)\, dx.$$

Such a polynomial must satisfy the equation

$$\int_k^{k+1} B_n(x)\, dx = k^n. \tag{A.16}$$

In fact, for each positive integer n, there is a unique monic[1] polynomial of degree n that satisfies this equation *for all values of k*, not only when k is a positive integer. It is easiest to see why this is so by means of an example. We shall set

$$B_3(x) = x^3 + a_2 x^2 + a_1 x + a_0$$

and show that there exist unique values for a_2, a_1, and a_0.

---

[1] The coefficient of x^n is 1.

Substituting this polynomial for $B_3(x)$ in equation (A.16), we see that

$$
\begin{aligned}
k^3 &= \int_k^{k+1} \left(x^3 + a_2 x^2 + a_1 x + a_0 \right) dx \\
&= \frac{1}{4}(k+1)^4 + \frac{a_2}{3}(k+1)^3 + \frac{a_1}{2}(k+1)^2 + a_0(k+1) - \frac{1}{4}k^4 - \frac{a_2}{3}k^3 - \frac{a_1}{2}k^2 - a_0 k \\
&= k^3 + \left(\frac{3}{2} + a_2 \right) k^2 + (1 + a_2 + a_1)k + \left(\frac{1}{4} + \frac{a_2}{3} + \frac{a_1}{2} + a_0 \right).
\end{aligned}
\tag{A.17}
$$

The coefficients of the different powers of k must be the same on both sides:

$$
0 = \frac{3}{2} + a_2,
\tag{A.18}
$$

$$
0 = 1 + a_2 + a_1,
\tag{A.19}
$$

$$
0 = \frac{1}{4} + \frac{a_2}{3} + \frac{a_1}{2} + a_0.
\tag{A.20}
$$

These three equations have a unique solution: $a_2 = -3/2$, $a_1 = 1/2$, $a_0 = 0$, and so

$$
B_3(x) = x^3 - \frac{3x^2}{2} + \frac{x}{2}.
\tag{A.21}
$$

Integrating this polynomial from 0 to k, we obtain the formula for the sum of cubes:

$$
\begin{aligned}
1^3 + 2^3 + \cdots + (k-1)^3 &= \int_0^1 \left(x^3 - \frac{3x^2}{2} + \frac{x}{2} \right) dx \\
&= \frac{k^4}{4} - \frac{k^3}{2} + \frac{k^2}{4}.
\end{aligned}
$$

The first two Bernoulli polynomials are

$$
B_1(x) = x - \frac{1}{2},
\tag{A.22}
$$

$$
B_2(x) = x^2 - x + \frac{1}{6}.
\tag{A.23}
$$

We now make an observation that will enable us to construct $B_{n+1}(x)$ from $B_n(x)$. If we differentiate both sides of equation (A.16) with respect to k, we get:

$$
B_n(k+1) - B_n(k) = nk^{n-1}.
\tag{A.24}
$$

This implies that

$$
\begin{aligned}
n \left[0^{n-1} + 1^{n-1} + 2^{n-1} + \cdots + (k-1)^{n-1} \right] \\
= [B_n(1) - B_n(0)] + [B_n(2) - B_n(1)] + [B_n(3) - B_n(2)] \\
+ \cdots + [B_n(k) - B_n(k-1)] \\
= B_n(k) - B_n(0),
\end{aligned}
\tag{A.25}
$$

and therefore

$$\frac{B_n(k) - B_n(0)}{n} = 1^{n-1} + 2^{n-1} + \cdots + (k-1)^{n-1} = \int_0^k B_{n-1}(x)\, dx. \quad (A.26)$$

Our recursive formula is

$$B_n(x) = n \int_0^x B_{n-1}(t)\, dt \quad + \quad B_n(0). \qquad (A.27)$$

Given that $B_4(0) = -1/30$, we can find $B_4(x)$ by integrating $B_3(x)$, multiplying by 4, and then adding $-1/30$:

$$B_4(x) = 4\left(\frac{x^4}{4} - \frac{x^3}{2} + \frac{x^2}{4}\right) - \frac{1}{30} = x^4 - 2x^3 + x^2 - \frac{1}{30}.$$

If we know the constant term in each polynomial: $B_1(0) = -1/2$, $B_2(0) = 1/6$, $B_3(0) = 0, \ldots$, then we can successively construct as many Bernoulli polynomials as we wish. These constants are called the **Bernoulli numbers**:

$$B_1 = \frac{-1}{2}, \quad B_2 = \frac{1}{6}, \quad B_3 = 0, \quad B_4 = \frac{-1}{30}, \ldots$$

A Formula for $B_n(x)$

We can do even better. Recalling that $B_1(x) = x + B_1$ and repeatedly using equation (A.27), we see that

$$B_2(x) = 2 \int_0^x (t + B_1)\, dt + B_2$$
$$= x^2 + 2B_1 x + B_2,$$
$$B_3(x) = 3 \int_0^x (t^2 + 2B_1 t + B_2)\, dt + B_3$$
$$= x^3 + 3B_1 x^2 + 3B_2 x + B_3,$$
$$B_4(x) = 4 \int_0^x (t^3 + 3B_1 t^2 + 3B_2 t + B_3)\, dt + B_4$$
$$= x^4 + 4B_1 x^3 + 6B_2 x^2 + 4B_3 x + B_4,$$
$$\vdots$$

A pattern is developing. Our coefficients are precisely the coefficients of the binomial expansion. Pascal's triangle has struck again. Once we see it, it is easy to verify by induction (see exercise A.2.3) that

$$B_n(x) = x^n + nB_1 x^{n-1} + \frac{n(n-1)}{2!} B_2 x^{n-2} + \frac{n(n-1)(n-2)}{3!} B_3 x^{n-3}$$
$$+ \cdots + nB_{n-1} x + B_n. \qquad (A.28)$$

The only problem left is to find an efficient means of determining the Bernoulli numbers.

A Recursive Formula for B_n

We turn to equation (A.24), set $k = 0$, and assume that n is larger than one:

$$B_n(1) - B_n(0) = n \cdot 0^{n-1} = 0. \tag{A.29}$$

We use equation (A.28) to evaluate $B_n(1)$:

$$0 = B_n(1) - B_n$$

$$= \left[1 + nB_1 + \frac{n(n-1)}{2!} B_2 + \cdots + \frac{n(n-1)}{2!} B_{n-2} + nB_{n-1} + B_n \right] - B_n, \tag{A.30}$$

$$B_{n-1} = \frac{-1}{n} \left(1 + nB_1 + \frac{n(n-1)}{2!} B_2 + \cdots + \frac{n(n-1)}{2!} B_{n-2} \right). \tag{A.31}$$

It follows that

$$B_5 = -\frac{1}{6} \left(1 + 6 \cdot \frac{-1}{2} + 15 \cdot \frac{1}{6} + 20 \cdot 0 + 15 \cdot \frac{-1}{30} \right)$$

$$= 0,$$

$$B_6 = -\frac{1}{7} \left(1 + 7 \cdot \frac{-1}{2} + 21 \cdot \frac{1}{6} + 35 \cdot 0 + 35 \cdot \frac{-1}{30} + 21 \cdot 0 \right)$$

$$= \frac{1}{42}.$$

Continuing, we obtain

$$B_7 = 0, \quad B_8 = \frac{-1}{30}, \quad B_9 = 0,$$

$$B_{10} = \frac{5}{66}, \quad B_{11} = 0, \quad B_{12} = \frac{-691}{2730}.$$

Bernoulli's Calculation

Equipped with equation (A.26) and the knowledge of $B_1, B_2, \ldots, B_{10}$, we can find the formula for the sum of the first $k - 1$ tenth powers:

$$\sum_{i=1}^{k-1} i^{10} = \frac{1}{11} [B_{11}(k) - B_{11}]$$

$$= \frac{1}{11} (k^{11} + 11B_1 k^{10} + 55B_2 k^9 + 165B_3 k^8 + 330B_4 k^7$$

$$+ 462B_5 k^6 + 462B_6 k^5 + 330B_7 k^4 + 165B_8 k^3$$

$$+ 55B_9 k^2 + 11B_{10} k)$$

$$= \frac{1}{11} k^{11} - \frac{1}{2} k^{10} + \frac{5}{6} k^9 - k^7 + k^5 - \frac{1}{2} k^3 + \frac{5}{66} k. \tag{A.32}$$

Since it is much easier to take powers of $1000 = 10^3$ than of 1001, let us add $1000^{10} = 10^{30}$ to the sum of the tenth powers of the integers up to 999:

$$1^{10} + 2^{10} + 3^{10} + \cdots + 1000^{10}$$
$$= 10^{30} + \frac{1}{11} 10^{33} - \frac{1}{2} 10^{30} + \frac{5}{6} 10^{27} - 10^{21} + 10^{15} - \frac{1}{2} 10^9 + \frac{5}{66} 10^3.$$

This is a simple problem in arithmetic:

```
        1 00000 00000 00000 00000 00000 00000 .00
      + 90 90909 09090 90909 09090 90909 09090 .90
      -    50000 00000 00000 00000 00000 00000 .00
      +          83 33333 33333 33333 33333 33333 .33
      -             10 00000 00000 00000 00000 .00
      +              1 00000 00000 00000 .00
      -                     5000 00000 .00
      +                           75 .75
        ─────────────────────────────────────────
        91 40992 42414 24243 42424 19242 42500
```

Seven and a half minutes is plenty of time.

Fermat's Last Theorem

The Bernoulli numbers will make appearances in each of the next two sections. Once they were discovered, mathematicians kept finding them again, and again, and again. One of the more surprising places that they turn up is in connection with Fermat's last theorem.

After studying Pythagorean triples, triples of positive integers satisfying

$$x^2 + y^2 = z^2,$$

Pierre de Fermat pondered the question of whether such triples could exist when the exponent was larger than 2. He came to the conclusion that no such triples exist, but never gave a proof. It should be noted that if there is no solution to

$$x^n + y^n = z^n,$$

then there can be no solution to

$$x^{mn} + y^{mn} = z^{mn},$$

because if $x = a$, $y = b$, $z = c$ were a solution to the second equation, then $x = a^m$, $y = b^m$, $z = c^m$ would be a solution to the first. If we want to prove Fermat's statement, then it is enough to prove that there are no solutions when $n = 4$ and no solutions when n is an odd prime.

The case $n = 4$ can be handled by methods described by Fermat. Euler essentially proved the case $n = 3$ in 1753. His proof was flawed, but his approach was correct. Fermat's "theorem" for $n = 5$ came in pieces. Sophie Germain (1776–1831), one of the first women to publish mathematics, showed that if a solution exists, then either x, y, or z *must* be divisible by 5. Gustav Lejeune Dirichlet made his mark on the mathematical scene when, in 1825 at the age of 20, he proved that the variable divisible by 5 cannot be even. In the same year, Adrien Marie Legendre, then in his 70's, picked up Dirichlet's analysis and carried it forward to show the general impossibility of the case $n = 5$. Gabriel Lamé settled $n = 7$ in 1839.

In 1847, Ernst Kummer proved that there is no solution in positive integers to

$$x^p + y^p = z^p$$

whenever p is a **regular prime**. The original definition of a regular prime is well outside the domain of this book, but Kummer found a simple and equivalent definition:

> An odd prime p is **regular** if and only if it does not divide the numerator of any of the Bernoulli numbers: $B_2, B_4, B_6, \ldots, B_{p-3}$.

The prime 11 is regular. Up to $p - 3 = 8$, the numerators are all 1. The prime 13 is regular. So is 17. And 19. And 23. Unfortunately, not all primes are regular. The prime 37 is not, nor is 59 or 67. Methods using Bernoulli numbers have succeeded in proving Fermat's last theorem for all primes below 4,000,000. The proof by Andrew Wiles and Richard Taylor uses a very different approach.

Exercises

The symbol (**M&M**) indicates that *Maple* and *Mathematica* codes for this problem are available in the **Web Resources** at **www.macalester.edu/aratra**.

A.2.1. (**M&M**) Find the polynomials $B_5(x)$, $B_6(x)$, $B_7(x)$, and $B_8(x)$.

A.2.2. (**M&M**) Graph the eight polynomials $B_1(x)$ through $B_8(x)$. Describe the symmetries that you see. Prove your guess about the *symmetry* of $B_n(x)$ for arbitrary n.

A.2.3. Prove equation (A.28) by induction on n.

A.2.4. Prove that

$$B_n(1 - x) = (-1)^n B_n(x) \tag{A.33}$$

provided $n \geq 1$.

A.2.5. Prove that

$$B_{2n+1} = 0 \tag{A.34}$$

provided that $n \geq 1$.

A.2.6. Show how to use Bernoulli polynomials and hand calculations to find the sum
$$1^8 + 2^8 + 3^8 + \cdots + 1000^8.$$

A.2.7. Show how to use Bernoulli polynomials and hand calculations to find the sum
$$1^{10} + 2^{10} + 3^{10} + \cdots + 1{,}000{,}000^{10}.$$

A.2.8. (M&M) Explore the factorizations of the numerators and of the denominators of the Bernoulli numbers. What conjectures can you make?

A.2.9. (M&M) Find all primes less than 100 that are not regular. Show that 691 is not a regular prime.

A.3 Sums of Negative Powers

Jacob Bernoulli and his brother Johann were also interested in the problem of summing negative powers of the integers. The first such case is the harmonic series,

$$1 + \frac{1}{2} + \frac{1}{3} + \cdots + \frac{1}{k},$$

which by then was well understood. The next case involves the sums of the reciprocals of the squares:

$$1 + \frac{1}{2^2} + \frac{1}{3^2} + \frac{1}{4^2} + \cdots .$$

We observe that this seems to approach a finite limit. The sum up to $1/100^2$ is 1.63498. Up to $1/1000^2$ it is 1.64393. In fact, the Bernoullis knew that it must converge because

$$\frac{1}{n^2} < \frac{1}{n(n-1)} = \frac{1}{n-1} - \frac{1}{n},$$

and so

$$\sum_{n=1}^{N} \frac{1}{n^2} < 1 + \sum_{n=2}^{N} \left(\frac{1}{n-1} - \frac{1}{n} \right)$$

$$= 1 + \left(1 - \frac{1}{2} \right) + \left(\frac{1}{2} - \frac{1}{3} \right)$$

$$+ \cdots + \left(\frac{1}{N-2} - \frac{1}{N-1} \right) + \left(\frac{1}{N-1} - \frac{1}{N} \right)$$

$$= 2 - \frac{1}{N}.$$

The sum of the reciprocals of the squares must converge and it must converge to something less than 2. What is the actual value of its limit?

It was around 1734 that Euler discovered that the value of this infinite sum is, in fact, $\pi^2/6$. His proof stretched even the credulity of his contemporaries, but it is worth giving to show the spirit of mathematical discovery. The fact that $\pi^2/6 = 1.64493\ldots$ is very close to the expected value was convincing evidence that it must be correct.

Consider the power series expansion of $\sin(x)/x$:

$$\frac{\sin x}{x} = 1 - \frac{x^2}{3!} + \frac{x^4}{5!} + \cdots . \tag{A.35}$$

We know that this function has roots at $\pm\pi$, $\pm 2\pi$, $\pm 3\pi$, $\ldots$, and so we should be able to factor it:

$$\frac{\sin x}{x} = \left(1 - \frac{x}{\pi}\right)\left(1 + \frac{x}{\pi}\right)\left(1 - \frac{x}{2\pi}\right)\left(1 + \frac{x}{2\pi}\right)\cdots$$

$$= \left(1 - \frac{x^2}{\pi^2}\right)\left(1 - \frac{x^2}{2^2\pi^2}\right)\left(1 - \frac{x^2}{3^2\pi^2}\right)\cdots . \tag{A.36}$$

We compare the coefficient of x^2 in equations (A.35) and (A.36) and see that

$$\frac{-1}{6} = -\left(\frac{1}{\pi^2} + \frac{1}{2^2\pi^2} + \frac{1}{3^2\pi^2} + \cdots\right),$$

or equivalently,

$$\frac{\pi^2}{6} = 1 + \frac{1}{2^2} + \frac{1}{3^2} + \frac{1}{4^2} + \cdots . \tag{A.37}$$

Comparing the coefficients of x^4 and doing a little bit of work, we can also find the formula for the sum of the reciprocals of the fourth powers:

$$\frac{1}{120} = \frac{1}{1^2 \cdot 2^2\pi^4} + \frac{1}{1^2 \cdot 3^2\pi^4} + \frac{1}{1^2 \cdot 4^2\pi^4} + \cdots + \frac{1}{2^2 \cdot 3^2\pi^4} + \frac{1}{2^2 \cdot 4^2\pi^4} + \cdots$$

$$= \sum_{1 \le j < k < \infty} \frac{1}{j^2 k^2 \pi^4},$$

$$\frac{\pi^4}{120} = \sum_{1 \le j < k < \infty} \frac{1}{j^2 k^2}. \tag{A.38}$$

If we square both sides of equation (A.37) and separate the pieces of the resulting product, we see that

$$\frac{\pi^4}{36} = \left(1 + \frac{1}{2^2} + \frac{1}{3^2} + \cdots\right)\left(1 + \frac{1}{2^2} + \frac{1}{3^2} + \cdots\right)$$

$$= \sum_{1 \le j < k < \infty} \frac{1}{j^2 k^2} + \sum_{1 \le j = k < \infty} \frac{1}{j^2 k^2} + \sum_{\infty > j > k \ge 1} \frac{1}{j^2 k^2}$$

$$= 2 \sum_{1 \le j < k < \infty} \frac{1}{j^2 k^2} + \sum_{k=1}^{\infty} \frac{1}{k^4}.$$

Using the result from equation (A.38), we obtain our formula:

$$\frac{\pi^4}{36} = 2\frac{\pi^4}{120} + \sum_{k=1}^{\infty} \frac{1}{k^4},$$

$$\frac{\pi^4}{90} = 1 + \frac{1}{2^4} + \frac{1}{3^4} + \frac{1}{4^4} + \cdots . \tag{A.39}$$

One can—and Euler did—continue this to find formulas for the sums of the reciprocals of the other even powers,

$$\sum_{k=1}^{\infty} \frac{1}{k^6}, \quad \sum_{k=1}^{\infty} \frac{1}{k^8}, \quad \sum_{k=1}^{\infty} \frac{1}{k^{10}}, \quad \cdots$$

In 1740, Euler discovered a formula that covered *all* of these cases.

A Generating Function

A problem that is not unreasonable at first glance is that of finding a power series expansion for $1/(1 - e^x)$. It looks as if it should be quite straightforward. Expand as a geometric series, use the power series for the exponential function, then rearrange (note that we need to define $0! = 1$):

$$\frac{1}{1 - e^x} = 1 + e^x + e^{2x} + e^{3x} + \cdots$$

$$= 1 + \sum_{k=1}^{\infty} e^{kx}$$

$$= 1 + \sum_{k=1}^{\infty} \left(1 + kx + \frac{(kx)^2}{2!} + \frac{(kx)^3}{3!} + \cdots \right)$$

$$= 1 + \sum_{k=1}^{\infty} \sum_{n=0}^{\infty} \frac{(kx)^n}{n!}$$

$$= 1 + \sum_{n=0}^{\infty} \frac{x^n}{n!} \sum_{k=1}^{\infty} k^n \quad \cdots \quad \textbf{whoa!}$$

Something is wrong. We are getting infinite coefficients. We need to back up.

The constant term in the power series expansion should be the value of the function at $x = 0$. There is our problem. If we set $x = 0$ in our original function, we get a zero in the denominator. We can get rid of the zero in the denominator if we multiply the numerator by x. The function we should try to expand is

$$\frac{x}{1 - e^x}.$$

We check what happens as x approaches 0 and see that we get -1. So far so good. It would be nice if the constant were $+1$ instead, so we change the sign of the denominator. We are looking for the coefficients in the power series expansion:

$$\frac{x}{e^x - 1} = 1 + a_1 x + a_2 x^2 + a_3 x^3 + \cdots. \tag{A.40}$$

The fact that we have multiplied by $-x$ is not going to make our original argument work, but this power series should exist. There is little choice but to compute the coefficients, the a_n, by brute force. We could do it by using Taylor's formula, but those derivatives quickly become very messy. Instead, we shall multiply both sides of equation (A.40) by $e^x - 1$, expanded as a power series, and then equate the coefficients of comparable powers of x to solve for the a_n:

$$x = (e^x - 1)(1 + a_1 x + a_2 x^2 + a_3 x^3 + \cdots)$$

$$= \left(x + \frac{x^2}{2!} + \frac{x^3}{3!} + \frac{x^4}{4!} + \cdots\right)(1 + a_1 x + a_2 x^2 + a_3 x^3 + \cdots)$$

$$= x + \left(\frac{1}{2!} + a_1\right) x^2 + \left(\frac{1}{3!} + \frac{a_1}{2!} + a_2\right) x^3 + \left(\frac{1}{4!} + \frac{a_1}{3!} + \frac{a_2}{2!} + a_3\right) x^4 + \cdots.$$

$$(A.41)$$

We obtain an infinite sequence of equations which we can solve for a_n:

$$0 = \frac{1}{2} + a_1 \qquad \Longrightarrow a_1 = \frac{-1}{2},$$

$$0 = \frac{1}{6} + \frac{a_1}{2} + a_2 \qquad \Longrightarrow a_2 = \frac{1}{12},$$

$$0 = \frac{1}{24} + \frac{a_1}{6} + \frac{a_2}{2} + a_3 \Longrightarrow a_3 = 0.$$

We continue in this manner:

$$a_4 = \frac{-1}{720}, \quad a_5 = 0, \quad a_6 = \frac{1}{30240}, \quad a_7 = 0, \quad a_8 = \frac{-1}{1209600}, \quad \cdots.$$

If you do not yet see what is happening, try multiplying each a_n by $n!$:

$$1 \cdot a_1 = \frac{-1}{2}, \quad 2! \cdot a_2 = \frac{1}{6}, \quad 3! \cdot a_3 = 0, \quad 4! \cdot a_4 = \frac{-1}{30},$$

$$5! \cdot a_5 = 0, \quad 6! \cdot a_6 = \frac{1}{42}, \quad 7! \cdot a_7 = 0, \quad 8! \cdot a_8 = \frac{-1}{30}.$$

The Bernoulli numbers!

Once we see them, it is not hard to prove that they are really there. Equation (A.41) implies that

$$0 = \frac{1}{n!} + \frac{a_1}{(n-1)!} + \frac{a_2}{(n-2)!} + \frac{a_3}{(n-3)!} + \cdots + \frac{a_{n-2}}{2!} + \frac{a_{n-1}}{1!}. \qquad (A.42)$$

We multiply both sides by $n!$:

$$0 = 1 + n a_1 + \frac{n(n-1)}{2!} (2! \cdot a_2) + \frac{n(n-1)(n-2)}{3!} (3! \cdot a_3)$$

$$+ \cdots + \frac{n(n-1)}{2!} [(n-2)! \cdot a_{n-2}] + n [(n-1)! \cdot a_{n-1}]. \qquad (A.43)$$

This is *precisely* the recursion that we saw in equation (A.30) for the Bernoulli numbers. We have proven that

$$\frac{x}{e^x - 1} = 1 + \sum_{n=1}^{\infty} B_n \frac{x^n}{n!}. \qquad (A.44)$$

Euler's Analysis

Once he had realized equation (A.44), Euler was off and running. One of the things that it shows is that $x/(e^x - 1)$ is *almost* an even function: B_n is zero whenever n is odd and larger than 1. If we add $x/2$ to both sides, we knock out the single odd power of x and obtain an even function:

$$1 + \sum_{m=1}^{\infty} B_{2m} \frac{x^{2m}}{(2m)!} = \frac{x}{e^x - 1} + \frac{x}{2} = \frac{2x + xe^x - x}{2(e^x - 1)} = \frac{x(e^x + 1)}{2(e^x - 1)}. \tag{A.45}$$

We replace x by $2t$:

$$1 + \sum_{m=1}^{\infty} B_{2m} \frac{(2t)^{2m}}{(2m)!} = \frac{t(e^{2t} + 1)}{e^{2t} - 1}$$

$$= t \frac{e^t + e^{-t}}{e^t - e^{-t}}$$

$$= t \coth t, \tag{A.46}$$

where $\coth t$ is the hyperbolic cotangent of t. Euler knew that

$$\sin z = \frac{e^{iz} - e^{-iz}}{2i} = -i \sinh iz, \tag{A.47}$$

$$\cos z = \frac{e^{iz} + e^{-iz}}{2} = \cosh iz, \tag{A.48}$$

and so he saw that

$$z \cot z = z \frac{\cosh iz}{-i \sinh iz}$$

$$= iz \coth iz$$

$$= 1 + \sum_{m=1}^{\infty} B_{2m} \frac{(2iz)^{2m}}{(2m)!}$$

$$= 1 + \sum_{m=1}^{\infty} (-1)^m B_{2m} \frac{(2z)^{2m}}{(2m)!}. \tag{A.49}$$

Euler knew of another expansion for $z \cot z$. Recognizing that the denominator of $\cot z = \cos z / \sin z$ is zero whenever $z = k\pi$, k any integer, he had found an infinite partial fraction decomposition of the cotangent (see exercise A.3.16):

$$\cot z = \cdots + \frac{1}{z + 2\pi} + \frac{1}{z + \pi} + \frac{1}{z} + \frac{1}{z - \pi} + \frac{1}{z - 2\pi} + \frac{1}{z - 3\pi} + \cdots$$

$$= \frac{1}{z} + \sum_{k=1}^{\infty} \left(\frac{1}{z + k\pi} + \frac{1}{z - k\pi} \right)$$

$$= \frac{1}{z} + 2 \sum_{k=1}^{\infty} \frac{z}{z^2 - k^2\pi^2}$$

$$= \frac{1}{z} - 2 \sum_{k=1}^{\infty} \frac{z}{k^2\pi^2 - z^2}. \tag{A.50}$$

If we multiply by z, we get an alternate expression for $z \cot z$:

$$z \cot z = 1 - 2 \sum_{k=1}^{\infty} \frac{z^2}{k^2 \pi^2 - z^2}$$

$$= 1 - 2 \sum_{k=1}^{\infty} \frac{z^2/k^2\pi^2}{1 - z^2/k^2\pi^2}$$

$$= 1 - 2 \sum_{k=1}^{\infty} \left(\frac{z^2}{k^2\pi^2} + \frac{z^4}{k^4\pi^4} + \frac{z^6}{k^6\pi^6} + \cdots \right)$$

$$= 1 - 2 \sum_{k=1}^{\infty} \sum_{m=1}^{\infty} \frac{z^{2m}}{k^{2m}\pi^{2m}}$$

$$= 1 - 2 \sum_{m=1}^{\infty} \frac{z^{2m}}{\pi^{2m}} \left(1 + \frac{1}{2^{2m}} + \frac{1}{3^{2m}} + \frac{1}{4^{2m}} + \cdots \right). \tag{A.51}$$

Comparing the coefficients of z^{2m} in equations (A.49) and (A.51), we see that

$$(-1)^m \frac{B_{2m}2^{2m}}{(2m)!} = \frac{-2}{\pi^{2m}} \left(1 + \frac{1}{2^{2m}} + \frac{1}{3^{2m}} + \frac{1}{4^{2m}} + \cdots \right), \tag{A.52}$$

or equivalently

$$1 + \frac{1}{2^{2m}} + \frac{1}{3^{2m}} + \frac{1}{4^{2m}} + \cdots = (-1)^{m+1} \frac{(2\pi)^{2m}}{2 \cdot (2m)!} B_{2m}. \tag{A.53}$$

Euler had them all, provided the exponent was even:

$$\sum_{n=1}^{\infty} \frac{1}{n^2} = \frac{(2\pi)^2}{4} \cdot \frac{1}{6} = \frac{\pi^2}{6}, \tag{A.54}$$

$$\sum_{n=1}^{\infty} \frac{1}{n^4} = \frac{(2\pi)^4}{2 \cdot 24} \cdot \frac{1}{30} = \frac{\pi^4}{90}, \tag{A.55}$$

$$\sum_{n=1}^{\infty} \frac{1}{n^6} = \frac{(2\pi)^6}{2 \cdot 720} \cdot \frac{1}{42} = \frac{\pi^6}{945}, \tag{A.56}$$

$$\vdots$$

The function $\sum n^{-s}$ which Euler had shown how to evaluate when s is a positive even integer would come to play a very important role in number theory. Today it is called the **zeta function**:

$$\zeta(s) = \sum_{n=1}^{\infty} \frac{1}{n^s}, \quad s > 1.$$

It can be defined for all complex values of s except $s = 1$. When Riemann laid out his prescription for a proof that the number of primes less than or equal to x is asymptotically $x / \ln x$, he conjectured that all of the nonreal roots of $\zeta(s)$ lie on the line of s's with real part 1/2. This is known as the Riemann hypothesis. It says a great deal about the error that is introduced when $x/\ln x$ is used to approximate the prime counting function. It is still unproven.

If the Exponent is Odd?

If the exponent is odd, it appears that there is no simple formula. The most that can be said, and this was only proved in 1978 by Roger Apéry, is that

$$\sum_{n=1}^{\infty} \frac{1}{n^3}$$

is definitely *not* a rational number.

Exercises

The symbol (**M&M**) indicates that *Maple* and *Mathematica* codes for this problem are available in the **Web Resources** at **www.macalester.edu/aratra**.

A.3.1. (**M&M**) Calculate

$$\sum_{n=1}^{100} \frac{1}{n^2} \quad \text{and} \quad \sum_{n=1}^{1000} \frac{1}{n^2}.$$

The first differs from $\pi^2/6$ by about 1/100, the second from $\pi^2/6$ by about 1/1000. This suggests that

$$\frac{1}{N} + \sum_{n=1}^{N} \frac{1}{n^2}$$

should be a pretty good approximation to $\pi^2/6$. Test this hypothesis for different values of N, including at least $N = 5000$ and $N = 10000$.

A.3.2. (**M&M**) Calculate

$$\sum_{n=1}^{N} \frac{1}{n^4} \quad \text{and} \quad \sum_{n=1}^{N} \frac{1}{n^6}$$

for $N = 100, 500$, and 1000. Compare your results with the predicted values of $\pi^4/90$ and $\pi^6/945$, respectively. Are you willing to believe Euler's result? In each case, what is the approximate size of the error?

A.3.3. Prove that if k is larger than 1, then

$$\int_{N+1}^{\infty} \frac{dx}{x^k} < \sum_{n=N+1}^{\infty} \frac{1}{n^k} < \int_{N}^{\infty} \frac{dx}{x^k}. \tag{A.57}$$

Use these bounds to prove that $\sum_{n=1}^{N} 1/n^k$ differs from $\sum_{n=1}^{\infty} 1/n^k$ by an amount that lies between

$$\frac{1}{(k-1)(N+1)^{k-1}} \quad \text{and} \quad \frac{1}{(k-1)N^{k-1}}.$$

A.3.4. Set $x = \pi/2$ in equation (A.36) and see what identity you get. Does it look familiar? It should.

A.3.5. Set $x = \pi/3$ in equation (A.36) and see what identity you get. What happens if you set $x = \pi/4$?

A.3.6. Comparing the coefficients of x^6 in equations (A.35) and (A.36) tells us that

$$\frac{\pi^6}{7!} = \sum_{1 \le j < k < l < \infty} \frac{1}{j^2 k^2 l^2}. \tag{A.58}$$

Use this fact together with equations (A.37) and (A.38) to prove that

$$\sum_{k=1}^{\infty} \frac{1}{k^6} = \frac{\pi^6}{945}.$$

A.3.7. Consider the aborted derivation on page 286. Remember that any equality involving infinite series must, in general, carry a restriction on those x's for which it is valid. What are the restrictions that need to go with each equality? Where precisely does the argument go wrong?

A.3.8. (M&M) Graph the polynomials

$$y = 1 + \sum_{n=1}^{N} B_n \frac{x^n}{n!}$$

for $N = 4$, 6, 8, 10, and 12. Compare these to the graph of $x/(e^x - 1)$. Describe what you see. Where does it appear that this series converges?

A.3.9. We observe that $\sum_{n=1}^{\infty} n^{-2m} = 1 + 2^{-2m} + \cdots$ is always larger than 1. Use this fact and equation (A.53) to prove that

$$|B_{2m}| > \frac{2 \cdot (2m)!}{(2\pi)^{2m}}. \tag{A.59}$$

Evaluate this lower bound for B_{20}, B_{40}, and B_{100}. Do these numbers stay small or do they get large? Express the lower bound in scientific notation with six digits of accuracy.

A.3.10. Show that $\lim_{n \to \infty} \zeta(n) = 1$. Use this fact and the formula for B_n implied by equation (A.53) to find the interval of convergence of the series $1 + \sum_{n=1}^{\infty} B_n x^n/n!$. Explain your analysis of convergence at the endpoints.

A.3.11. (**M&M**) Taylor's theorem tells us that B_n must be the nth derivative of $x/(e^x - 1)$ evaluated at $x = 0$. Verify that this is correct when $n = 1, 2, 3$, and 4 by finding the derivatives.

A.3.12. Use the power series expansions of e^x, $\cos x$, and $\sin x$ to prove that

$$e^{ix} = \cos x + i \sin x. \tag{A.60}$$

A.3.13. Use equation (A.60) to prove equations (A.47) and (A.48).

A.3.14. (**M&M**) Graph the polynomials

$$y = 1 + \sum_{m=1}^{N} (-1)^m B_{2m} \frac{(2z)^{2m}}{(2m)!}$$

for $N - 2, 4$, and 6. Compare these to the graph of $z \cot z$. Describe what you see. Estimate the radius of convergence for this series.

A.3.15. Determine the interval of convergence for the series in exercise A.3.14. Show the work that supports your answer.

A.3.16. We assume that $\cot z$ has a partial fraction decomposition. This means that there are constants, a_k, such that

$$\cot z = \sum_{-\infty < k < \infty} \frac{a_k}{z - k\pi}.$$

To find the values of the a_k, we multiply both sides by $\sin z$,

$$\cos z = \sum_{-\infty < k < \infty} a_k \frac{\sin z}{z - k\pi},$$

and then take the limit as z approaches $m\pi$. Show that

$$\frac{\sin z}{z - k\pi}$$

approaches 0 if $m \neq k$ and that it approaches $\cos m\pi = (-1)^m$ if $m = k$. Finish the proof that $a_m = 1$.

A.3.17. (**M&M**) Graph the functions

$$R_N(z) = \frac{1}{z} + 2 \sum_{k=1}^{N} \frac{z}{z^2 - k^2\pi^2}$$

for $N = 3, 6, 9$, and 12. Compare these to the graph of $\cot z$. Describe what you see. Where does it appear that this series converges? Plot the differences $\cot(z) - R_N(z)$ for various values of N and find a reasonable approximation, in terms of N and z, to this error function. Test the validity of your approximation for $N = 1000$.

A.3.18. What are the exact values of

$$\sum_{n=1}^{\infty} \frac{1}{n^8} \quad \text{and} \quad \sum_{n=1}^{\infty} \frac{1}{n^{10}},$$

expressed as a power of π times a rational number?

A.3.19. Prove that B_{2m} and B_{2m+2} always have opposite sign.

A.3.20. (**M&M**) Apéry proved that $\sum n^{-3}$ is not a rational number. We still do not know if it can be written as π^3 times a rational number. Calculate

$$\sum_{n=1}^{N} \frac{1}{n^3}$$

for large values of N (at least 1000) and estimate the size of your error (see exercise A.3.3).

A.4 The Size of *n*!

An accurate approximation to $n!$ was discovered in 1730 in a collaboration between Abraham de Moivre (1667–1754) and James Stirling (1692–1770). de Moivre was a French Protestant. He and his parents had fled to London after the revocation of the Edict of Nantes in 1685. Despite his brilliance, he was always a foreigner and never obtained an academic appointment. He struggled throughout his life to support himself on the meager income earned as a tutor. Stirling was a Jacobite and in 1716, a year after the Jacobite rebellion, was expelled from Oxford for refusing to swear an oath of allegiance to the king. Because of his politics, he too was denied an academic position.

Even though it was a joint effort, the formula that we will find is called Stirling's formula. This is primarily de Moivre's own fault. When he published his result he gave Stirling credit for finding the constant, but his language was sufficiently imprecise that the attribution of the constant to Stirling could easily be misread as crediting him with the entire identity. In any event, Stirling's name does deserve to be attached to this identity because it was the fruit of both their efforts.

Our first task is to turn $n!$ into a summation so we can use an integral approximation. This is easily accomplished by taking the natural logarithm:

$$\ln(n!) = \sum_{k=1}^{n} \ln k.$$

We can bound this above and below by integrals:

$$\int_1^n \ln x \, dx < \sum_{k=1}^{n} \ln k < \int_0^{n-1} \ln(x+1)\, dx + \ln n,$$

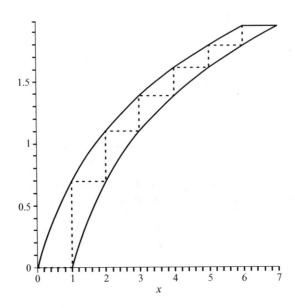

FIGURE A.1. Graphs of $\ln(x+1)$ and $\ln x$ bounding the step function $\ln\lfloor x\rfloor$.

where the value of $\sum_{k=1}^{n} \ln k$ is represented by the area under the staircase in Figure A.1. Evaluating these integrals, we see that

$$n \ln n - n + 1 < \ln(n!) < n \ln n - n + 1 + \ln n,$$

$$e\left(\frac{n}{e}\right)^{n} < \quad n! \quad < ne\left(\frac{n}{e}\right)^{n}.$$

This gives us a pretty good idea of how fast $n!$ grows, but because the summands are increasing, our upper and lower bounds get further apart as n increases, unlike the situation when we estimated the rate of the growth of the harmonic series.

A Trick for Approximating Summations

To get a better approximation, we use a trick that is part of the repertoire of number theory where there are many summations that need to be approximated. We rewrite $\ln k$ as the integral of $1/x$ from $x = 1$ to $x = k$ and then interchange the integral and the summation:

$$\sum_{k=1}^{n} \ln k = \sum_{k=1}^{n} \int_{1}^{k} \frac{dx}{x}$$

$$= \int_{1}^{n} \frac{\sum_{k=\lfloor x\rfloor+1}^{n} 1}{x} \, dx$$

$$= \int_{1}^{n} \frac{n - \lfloor x\rfloor}{x} \, dx. \qquad (A.61)$$

We now split this integral into two pieces:

$$\int_{1}^{n} \frac{n - \lfloor x\rfloor}{x} \, dx = \int_{1}^{n} \frac{n - x + 1/2}{x} \, dx + \int_{1}^{n} \frac{x - \lfloor x\rfloor - 1/2}{x} \, dx$$

$$= n \ln n - n + 1 + \frac{1}{2} \ln n + \int_{1}^{n} \frac{x - \lfloor x\rfloor - 1/2}{x} \, dx. \qquad (A.62)$$

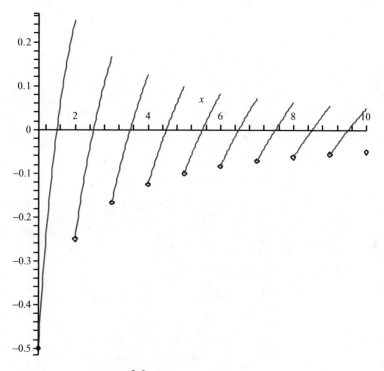

FIGURE A.2. Graph of $(x - \lfloor x \rfloor - 1/2)/x$.

The integrand in the last line has a graph that oscillates about and approaches the x-axis (see Figure A.2). The limit of this integral as n approaches infinity exists because

$$\left| \int_n^\infty \frac{x - \lfloor x \rfloor - 1/2}{x} \, dx \right| < \left| \int_n^{n+1/2} \frac{x - \lfloor x \rfloor - 1/2}{x} \, dx \right|$$

$$= \left| \int_0^{1/2} \frac{x - 1/2}{n + x} \, dx \right|$$

$$= \frac{-1}{2} + \left(n + \frac{1}{2} \right) \left(\ln \left(n + \frac{1}{2} \right) - \ln n \right),$$

(A.63)

which approaches 0 as n goes to infinity (see exercise A.4.1).

We have proven that

$$\ln(n!) = n \ln n - n + \frac{1}{2} \ln n + 1 + \int_1^\infty \frac{x - \lfloor x \rfloor - 1/2}{x} \, dx + E(n),$$

(A.64)

where

$$E(n) = - \int_n^\infty \frac{x - \lfloor x \rfloor - 1/2}{x} \, dx$$

approaches 0 as n approaches infinity. Equivalently,

$$n! = C \left(\frac{n}{e} \right)^n \sqrt{n} \, e^{E(n)}$$

(A.65)

where

$$\ln C = 1 + \int_1^\infty \frac{x - \lfloor x \rfloor - 1/2}{x}\, dx.$$

What is the value of C?

Evaluating C

Wallis's formula comes to our aid in a very slick evaluation. Stirling's formula implies that

$$\frac{(2n)!}{n! \cdot n!} = \frac{C(2n/e)^{2n}\sqrt{2n}\; e^{E(2n)}}{C^2(n/e)^{2n}n\; e^{2E(n)}}$$

$$= \frac{2^{2n}}{C}\sqrt{\frac{2}{n}}\; e^{E(2n)-2E(n)}. \tag{A.66}$$

We solve for C and then do a little rearranging:

$$C = \frac{2^{2n}(n!)^2}{(2n)!} \frac{\sqrt{2}}{\sqrt{n}} e^{E(2n)-2E(n)}$$

$$= \frac{(2 \cdot 4 \cdot 6 \cdots 2n)^2}{1 \cdot 2 \cdot 3 \cdots 2n} \frac{2}{\sqrt{2n}} e^{E(2n)-2E(n)}$$

$$= \frac{2 \cdot 4 \cdot 6 \cdots 2n}{1 \cdot 3 \cdot 5 \cdots (2n-1)} \frac{2}{\sqrt{2n}} e^{E(2n)-2E(n)}$$

$$= \frac{2 \cdot 4 \cdot 6 \cdots (2n-2) \cdot \sqrt{2n}}{1 \cdot 3 \cdot 5 \cdots (2n-1)} 2\, e^{E(2n)-2E(n)}. \tag{A.67}$$

Looking back at Wallis's work, we see from equation (A.10) that

$$\frac{2 \cdot 4 \cdot 6 \cdots (2n-2) \cdot \sqrt{2n}}{1 \cdot 3 \cdot 5 \cdots (2n-1)}$$

approaches $\sqrt{\pi/2}$ as n gets large. That means that the right side of equation (A.67) approaches $2\sqrt{\pi/2} = \sqrt{2\pi}$ as n approaches infinity. Since the left side is independent of n, the constant C must actually equal $\sqrt{2\pi}$. We have proven **Stirling's formula**:

$$n! = n^n e^{-n} \sqrt{2\pi n}\; e^{E(n)}, \tag{A.68}$$

where $E(n)$ is an error that approaches 0 as n gets large.

The Asymptotic Series for $E(n)$

Not long after deMoivre and Stirling published their formula for $n!$ in 1730, both Leonard Euler and Colin Maclaurin realized that something far more general was going on, a formula for approximating arbitrary series that today is called the Euler–Maclaurin formula. Euler wrote to Stirling in 1736 describing this general formula. Stirling wrote back in 1738 saying that Colin Maclaurin had also discovered this result. Euler's proof was published in 1738, Maclaurin's in 1742. Because it takes very little extra work, we shall develop the asymptotic series for $E(n)$ in the more general context of the Euler-Maclaurin formula.

We want to find a formula for $\sum_{k=1}^{n} f(k)$ where f is an analytic function for $x > 0$. We set

$$S(n) = \sum_{k=1}^{n} f(k)$$

and assume that S also can be defined for all $x > 0$ so that it is analytic. By Taylor's formula, we have that

$$S(n + x) = S(n) + S'(n)x + \frac{S''(n)}{2!} x^2 + \frac{S'''(n)}{3!} x^3 + \cdots.$$

We set $x = -1$ and observe that

$$f(n) = S(n) - S(n - 1) = S'(n) - \frac{S''(n)}{2!} + \frac{S'''(n)}{3!} - \frac{S^{(4)}(n)}{4!} + \cdots.$$

We want to invert this and write $S'(n)$ in terms of f and its derivatives at n. In principle, this is doable because

$$f'(n) = S''(n) - \frac{S'''(n)}{2!} + \frac{S^{(4)}(n)}{3!} - \frac{S^{(5)}(n)}{4!} + \cdots$$

$$f''(n) = S'''(n) - \frac{S^{(4)}(n)}{2!} + \frac{S^{(5)}(n)}{3!} - \frac{S^{(6)}(n)}{4!} + \cdots$$

$$f'''(n) = S^{(4)}(n) - \frac{S^{(5)}(n)}{2!} + \frac{S^{(6)}(n)}{3!} - \frac{S^{(7)}(n)}{4!} + \cdots$$

$$\vdots$$

In other words, we want to find the constants $a_1, a_2, a_3, \ldots$ such that

$$S'(n) = f(n) + a_1 f'(n) + a_2 f''(n) + a_3 f'''(n) + \cdots. \tag{A.69}$$

We substitute the expansions of the derivatives of f in terms of the derivatives of S into equation (A.69). This tells us that

$$S'(n) = \sum_{j=1}^{\infty} (-1)^{j-1} \frac{S^{(j)}(n)}{j!} + \sum_{k=1}^{\infty} a_k \sum_{j=k+1}^{\infty} (-1)^{j-k-1} \frac{S^{(j)}(n)}{(j-k)!}$$

$$= S'(n) + \sum_{j=2}^{\infty} (-1)^{j-1} S^{(j)}(n) \left(1 + \sum_{k=1}^{j-1} (-1)^k \frac{a_k}{(j-k)!} \right). \tag{A.70}$$

This will be true if and only if

$$1 - \frac{a_1}{(j-1)!} + \frac{a_2}{(j-2)!} - \frac{a_3}{(j-3)!} + \cdots + (-1)^{j-1} \frac{a_{j-1}}{1!} = 0, \quad j \geq 2.$$

This equation should look familiar. Except for the sign changes, it is exactly the equality that we saw in equation (A.42) on page 287, an equality uniquely satisfied by the Bernoulli numbers divided by the factorials. In our case,

$$a_k = (-1)^k \frac{B_k}{k!}.$$

Since we know that $B_{2m+1} = 0$ for $m \geq 1$, we have shown that

$$S'(x) = f(x) + \sum_{k=1}^{\infty} (-1)^k \frac{B_k}{k!} f^{(k)}(x)$$

$$= f(x) + \frac{1}{2} f'(x) + \sum_{m=1}^{\infty} \frac{B_{2m}}{(2m)!} f^{(2m)}(x). \tag{A.71}$$

We do need to keep in mind that these have all been formal manipulations, what Cauchy referred to as "explanations drawn from algebraic technique." This derivation should be viewed as suggestive. In no sense is it a proof. In particular, there is no guarantee that this series converges.

Nevertheless, even when the series does not converge, it does provide useful approximations. If we now integrate each side of equation (A.71) from $x = 1$ to $x = n$ and then add $S(1) = f(1)$, we get the **Euler–Maclaurin formula**.

Theorem A.1 (Euler–Maclaurin Formula). *Let f be an analytic function for $x > 0$, then, provided the series converge, we have that*

$$\sum_{k=1}^{n} f(k) = \int_{1}^{n} f(x)\,dx + \frac{1}{2} f(n) + \sum_{m=1}^{\infty} \frac{B_{2m}}{(2m)!} f^{(2m-1)}(n)$$

$$+ \frac{1}{2} f(1) - \sum_{m=1}^{\infty} \frac{B_{2m}}{(2m)!} f^{(2m-1)}(1). \tag{A.72}$$

When we set $f(x) = \ln x$ and use the fact that we know that the constant term is $\ln(2\pi)/2$, this becomes **Stirling's Formula**:

$$\ln(n!) = \sum_{k=1}^{n} \ln k = n \ln n - n + \frac{1}{2} \ln n + \frac{1}{2} \ln(2\pi) + E(n), \tag{A.73}$$

where $E(n)$ can be approximated by the asymptotic series,

$$E(n) \sim \sum_{m=1}^{\infty} \frac{B_{2m}}{(2m)(2m-1)n^{2m-1}}. \tag{A.74}$$

Difficulties

Does the fact that the constant term is $\ln(\sqrt{2\pi})$ mean that

$$1 - \left(\frac{B_2}{1 \cdot 2} + \frac{B_4}{3 \cdot 4} + \frac{B_6}{5 \cdot 6} + \cdots \right) = \ln\left(\sqrt{2\pi}\right) = .9189385\ldots ?$$

Hardly. If we try summing this series, we find that it does not approach anything. The first few Bernoulli numbers are small, but as we saw in the last section, they start to grow. They

Table A.2. Partial sums of de Moivre's series.

N	$1 - \sum_{m=1}^{N} B_{2m}/2m(2m-1)$
1	0 .9166667
2	0 .9194444
3	0 .9186508
4	0 .9192460
5	0 .9184043
6	0 .9203218
7	0 .9139116
8	0 .9434622
9	0 .763818
10	2 .15625
11	−11 .2466
12	145 .602
13	−2047 .5
14	34061 .3
15	−657411 .0

grow faster than $2(2m)!/(2\pi)^{2m}$. Table A.2. lists the partial sums of

$$1 - \sum_{m=1}^{N} \frac{B_{2m}}{2m(2m-1)}.$$

The first few values look good—up to $N = 4$ they seem to be approaching $\ln \sqrt{2\pi}$—but then they begin to move away and very quickly the series is lurching out of control.

What about the error function:

$$E(n) \sim \frac{B_2}{1 \cdot 2n} + \frac{B_4}{3 \cdot 4n^3} + \frac{B_6}{5 \cdot 6n^5} + \cdots;$$

does it converge? In exercises A.4.2 and A.4.3, the reader is urged to experiment with this series. What you should see is that no matter how large n is, eventually this series will start to oscillate with increasing swings. *But that does not mean that it is useless.* If you take the first few terms, say the first two, then

$$n^n e^{-n} \sqrt{2\pi n}\, e^{1/(12n)-1/(360n^3)}$$

is a better approximation to n! than just

$$n^n e^{-n} \sqrt{2\pi n}.$$

Something very curious is happening. As we take more terms, the approximation keeps getting better *up to some point*, and then it starts to get worse as the series moves into its uncontrolled swings. This is what we mean by an **asymptotic series**. Even though it does not converge, it does give an approximation to the quantity in question. How many terms of the asymptotic series should you take? That depends on n. As n gets larger, you can go farther. Infinite series do strange things.

Exercises

The symbol $\boxed{\textbf{M\&M}}$ indicates that *Maple* and *Mathematica* codes for this problem are available in the **Web Resources** at **www.macalester.edu/aratra**.

A.4.1. Show that

$$\frac{-1}{2} + \left(n + \frac{1}{2}\right)\left(\ln\left(n + \frac{1}{2}\right) - \ln n\right)$$

$$= \frac{-1}{2} + \frac{1}{2}\ln\left(1 + \frac{1}{2n}\right) + \frac{1}{2}\ln\left(\left[1 + \frac{1}{2n}\right]^{2n}\right).$$

Use this identity to prove (see equation (A.63)) that

$$\lim_{n\to\infty} \int_n^\infty \frac{x - \lfloor x\rfloor - 1/2}{x}\, dx = 0.$$

A.4.2. $\boxed{\textbf{M\&M}}$ Evaluate

$$n^n e^{-n} \sqrt{2\pi n}\, e^{1/(12n) - 1/(360n^3)}$$

for $n = 5, 10, 20, 50,$ and 100 and compare it to $n!$.

A.4.3. $\boxed{\textbf{M\&M}}$ To see how many terms of the asymptotic series we should take, find the summand in the asymptotic series that is closest to zero and stop at that term. For each of the values $n = 5, 10, 20, 50,$ and 100, find which summand is the smallest in absolute value. Estimate the function of n that describes how many terms of the asymptotic series should be taken for any given n. How accurately does this approximate n! when the number of terms is chosen optimally?

A.4.4. Use the approximation

$$|B_{2m}| \approx \frac{2(2m)!}{(2\pi)^{2m}}$$

to check your estimate from exercise A.4.3.

A.4.5. $\boxed{\textbf{M\&M}}$ Using the Euler–Maclaurin formula with $f(x) = 1/x$ gives us an approximation for the harmonic series. Show that the constant term of the Euler–Maclaurin formula is

$$\frac{1}{2} + \sum_{m=1}^{\infty} \frac{B_{2m}}{2m}.$$

Determine how useful this is in approximating the value of Euler's γ.

A.4.6.   (**M&M**) Use the Euler–Maclaurin formula to show that

$$\sum_{k=1}^{n} \frac{1}{k} = \ln n + \frac{1}{2n} + \gamma - H(n)$$

where $H(n)$ can be approximated by the asymptotic series

$$H(n) \sim \sum_{m=1}^{\infty} \frac{B_{2m}}{2m \, n^{-2m}}.$$

For each of the values $n = 5, 10, 20, 50$, and 100, find which summand is the smallest in absolute value. Estimate the function of n that describes how many terms of the asymptotic series should be taken for any given n. How accurately does this approximate the harmonic series when the number of terms is chosen optimally?

Appendix B
Bibliography

Birkhoff, Garrett, *A Source Book in Classical Analysis*, Harvard University Press, Cambridge, MA, 1973.

Bonnet, Ossian, "Remarques sur quelques intégrales définies," *Journal de Mathématiques Pures et Appliquées*, vol. **14**, August 1849, pages 249–256.

Borwein, J. M., P. B. Borwein, and D. H. Bailey, "Ramanujan, Modular Equations, and Approximations to Pi or How to Compute One Billion Digits of Pi," *The American Mathematical Monthly*, vol. **96**, no. 3, March 1989, pages 201–219.

Cauchy, Augustin-Louis, *Cours d'Analyse de l'École Royale Polytechnique*, series 2, vol. **3** in *Œuvres complètes d'Augustin Cauchy*, Gauthier-Villars, Paris, 1897.

Cauchy, Augustin-Louis, *Leçons sur le calcul différentiel*, series 2, vol. **4** in *Œuvres complètes d'Augustin Cauchy*, Gauthier-Villars, Paris, 1899.

Cauchy, Augustin-Louis, *Résumé des Leçons données a l'École Royale Polytechnique sur le calcul infinitésimal*, series 2, vol. **4** in *Œuvres complètes d'Augustin Cauchy*, Gauthier-Villars, Paris, 1899.

Dijksterhuis, E. J., *Archimedes*, translated by C. Dikshoorn, Princeton University Press, Princeton, 1987.

Dirichlet, G. Lejeune, *Werke*, reprinted by Chelsea, New York, 1969.

Dunham, William, *Journey through Genius: the great theorems of mathematics*, John Wiley & Sons, New York, 1990.

Edwards, C. H., Jr., *The Historical Development of the Calculus*, Springer–Verlag, New York, 1979.

Euler, Leonhard, *Introduction to Analysis of the Infinite*, books I & II, translated by John D. Blanton, Springer–Verlag, New York, 1988.

Gauss, Carl Friedrich, *Werke*, vol. **3**, Königlichen Gesellschaft der Wissenschaften, 1876.

Grabiner, Judith V., *The Origins of Cauchy's Rigorous Calculus*, MIT Press, Cambridge, MA, 1981.

Grattan-Guinness, Ivor, *Convolutions in French Mathematics, 1800–1840*, vols. I, II, III, Birkhäuser Verlag, Basel, 1990.

Grattan-Guinness, Ivor, *The Development of the Foundations of Mathematical Analysis from Euler to Riemann*, MIT Press, Cambridge, MA, 1970.

Grattan-Guinness, Ivor, *Joseph Fourier, 1768–1830*, MIT Press, Cambridge, MA, 1972.

Hawkins, Thomas, *Lebesgue's theory of integration: its origins and development*, 2nd edition, Chelsea, New York, 1975.

Hermite, Charles and Thomas Jan Stieltjes, *Correspondance d'Hermite et de Stieltjes*, B. Baillaud and H. Bourget, eds., Gauthier-Villars, Paris, 1903–1905.

Kaczor, W. J., and M. T. Nowak, *Problems in Mathematical Analysis*, vols. I, II, III, Student Mathematical Library vols. 4, 12, 21, American Mathematical Society, Providence, RI, 2000–2003.

Kline, Morris, *Mathematical Thought from Ancient to Modern Times*, Oxford, 1972.

Lacroix, S. F., *An Elementary Treatise on the Differential and Integral Calculus*, translated by Babbage, Peacock, and Herschel with appendix and notes, J. Deighton and Sons, Cambridge, 1816.

Lacroix, S. F., *Traité Élémentaire de Calcul Différentiel et de Calcul Intégral*, 4th edition, Bachelier, Paris, 1828.

Medvedev, Fyodor A., *Scenes from the History of Real Functions*, translated by Roger Cooke, Birkhäuser Verlag, Basel, 1991.

Olsen, L., A new proof of Darboux's theorem, *American Mathematical Monthly*, vol. 111 (2004), pp. 713–715.

Poincaré, Henri, "La Logique et l'Intuition dans la Science Mathématique et dans l'Enseignement," *L'Ensiegnement mathématique*, vol. **1** (1889), pages 157–162.

Preston, Richard, "The Mountains of Pi," *The New Yorker*, March 2, 1992, pages 36–67.

Riemann, Bernhard, *Gesammelte Mathematische Werke*, reprinted with comments by Raghavan Narasimhan, Springer–Verlag, New York, 1990.

Rudin, Walter, *Principles of Mathematical Analysis*, 3rd edition, McGraw-Hill, New York, 1976.

Serret, J.-A., *Calcul Différentiel et Intégral*, 4th edition, Gauthier-Villars, Paris, 1894.

Struik, D. J., *A Source Book in Mathematics 1200–1800*, Princeton University Press, Princeton, 1986.

Truesdell, C., "The Rational Mechanics of flexible or elastic bodies 1638–1788," *Leonardi Euleri Opera Omnia*, series 2, volume **11**, section 2, Orell Füssli Turici, Switzerland, 1960.

Van Vleck, Edward B., "The influence of Fourier's series upon the development of mathematics, *Science*, N.S. vol. **39**, 1914, pages 113–124.

Weierstrass, Karl Theodor Wilhelm, *Mathematische werke von Karl Weierstrass*, 7 volumes, Mayer & Muller, Berlin, 1894–1927.

Whittaker, E. T., and G. N. Watson, *A Course of Modern Analysis*, 4th ed., Cambridge University Press, Cambridge, 1978.

Appendix C
Hints to Selected Exercises

Exercises which can also be found in Kaczor and Nowak are listed at the start of each section following the symbol $\boxed{\textbf{KN}}$. The significance of $3.1.2 = \text{II}:2.1.1$ is that exercise 3.1.2 in this book can be found in Kaczor and Nowak, volume II, problem 2.1.1.

2.1.6 Use the fact that $1 + x + x^2 + \cdots + x^{k-1} = (1 - x^k)/(1 - x)$.

2.1.8 If you stop at the kth term, how far away are the partial sums that have more terms?

2.2.1 Use the fact that $1 + x + x^2 + \cdots + x^{k-1} = (1 - x^k)/(1 - x)$.

2.2.4 Take the first $3k + 3$ terms and rewrite this finite summation as $(1 + 2^{-3} + 2^{-6} + \cdots + 2^{-3k}) + (2^{-1} + 2^{-4} + 2^{-7} + \cdots + 2^{-(3k+1)}) - (2^{-2} + 2^{-5} + 2^{-8} + \cdots + 2^{-(3k+2)})$.

2.2.6 Use the work from exercise 2.2.5.

2.2.8 Find an expression in terms of r and s for a partial sum of a rearranged series that uses the first r positive summands and the first s negative summands. Show that you can get as close as desired to the target value provided only that r and s are sufficiently close, regardless of their respective sizes.

2.3.4 Take pairs of terms and assume that regrouping of the summands is allowed.

2.3.5 Take the tangent of each side and use the formula

$$\tan(x + y) = \frac{\tan x + \tan y}{1 - \tan x \, \tan y}.$$

2.3.8 Explain what happens when you take $a = -1$ in equation (2.20).

2.4.10 Begin by separating the summands according to the total number of digits in the denominator:

$$\left(\frac{1}{1} + \frac{1}{2} + \cdots + \frac{1}{8}\right)$$

$$+ \left(\frac{1}{10} + \frac{1}{11} + \cdots + \frac{1}{18} + \frac{1}{20} + \cdots + \frac{1}{88}\right)$$

$$+ \left(\frac{1}{100} + \frac{1}{101} + \cdots + \frac{1}{888}\right)$$

$$+ \cdots + \left(\frac{1}{10^k} + \frac{1}{10^k + 1} + \cdots + \frac{1}{8(10^{k+1} - 1)/9}\right)$$

$$+ \cdots .$$

a. There are 8 summands in the first pair of parentheses. Show that there are $72 = 8 \cdot 9$ in the second, $648 = 8 \cdot 9^2$ in the third, $5832 = 8 \cdot 9^3$ in the fourth, and that in general there are $8 \cdot 9^k$ in the $k + 1$st. Hint: what digits are you allowed to place in the first position? in the second? in the third?

b. Each summand in a given pair of parentheses is less than or equal to the first term. Show that the sum of the terms in the $k + 1$st parentheses is strictly less than $8 \cdot 9^k / 10^k$, and thus our series is bounded by

$$\frac{8}{1} + \frac{8 \cdot 9}{10} + \frac{8 \cdot 9^2}{10^2} + \frac{8 \cdot 9^3}{10^3} + \cdots .$$

c. Evaluate the geometric series given above.

2.4.14 Show that

$$1 + \frac{1}{3} + \cdots + \frac{1}{2n - 1} = \left(1 + \frac{1}{2} + \frac{1}{3} + \cdots + \frac{1}{2n}\right) - \frac{1}{2}\left(1 + \frac{1}{2} + \frac{1}{3} + \cdots + \frac{1}{n}\right).$$

2.4.16 Show that $\displaystyle\sum_{m=n}^{\infty} \frac{1}{m^2} < \frac{1}{n^2} + \int_n^{\infty} \frac{dx}{x^2}$.

2.4.18 Work with the fraction of the road that you have covered. The first step takes you 1/2000th of the way, the next step 1/4000th, the third 1/6000th.

2.5.3 Integration by parts.

2.5.15 Is c the same for all values of n?

2.6.5 How can you use the fact that e^{-1/x^2} has all of its derivatives equal to 0 at $x = 0$?

(KN) 3.1.2 = II:2.1.1, 3.1.3 = II:2.1.2, 3.1.4 = II:2.1.3, 3.1.5 = II:2.1.4, 3.1.6 = II:2.1.5, 3.1.15 = II:2.1.8, 3.1.16 = II:2.1.10b, 3.1.17 = II:2.1.9b, 3.1.18 = II:2.1.12, 3.1.19 = II:2.1.13, 3.1.20 = II:2.1.13.

3.1.2 For those functions with $|x|$, consider $x > 0$ and $x < 0$ separately. Use the definition of the derivative at $x = 0$. For functions with $\lfloor x \rfloor$, consider $x \notin \mathbb{Z}$ separately. Use the definition of the derivative at $x \in \mathbb{Z}$.

3.1.3 $\log_x a = (\ln a)/(\ln x)$.

3.1.4 Consider the transition points: Is the function continuous there? If it is, rely on the definition of the derivative.

3.1.15 $xf(a) - af(x) = (x - a)f(a) - a(f(x) - f(a))$. The same trick will work in part (b).

3.1.16 Rewrite the fraction as $\dfrac{f(x)e^x - f(0)e^0}{x - 0} \div \dfrac{f(x)\cos x - f(0)\cos 0}{x - 0}$.

3.1.19 (b) Consider $f(x) = x^2 \sin(1/x)$, $x \neq 0$.

3.1.20 Rewrite

$$\frac{f(x_n) - f(z_n)}{x_n - z_n} = \frac{f(x_n) - f(a)}{x_n - a} \cdot \frac{x_n - a}{x_n - z_n} + \frac{f(z_n) - f(a)}{z_n - a} \cdot \frac{a - z_n}{x_n - z_n}.$$

Show that this must lie between $\dfrac{f(x_n) - f(a)}{x_n - a}$ and $\dfrac{f(z_n) - f(a)}{z_n - a}$. Why doesn't this approach work when x_n and z_n lie on the same side of a?

3.2.9 Show that there is a k between 0 and Δx for which

$$\frac{f(x_0 + 2\Delta x) - 2f(x_0 + \Delta x) + f(x_0)}{\Delta x^2} = \frac{f'(x_0 + 2k) - f'(x_0 + k)}{k}.$$

Define $g(h) = f'(x_0 + k + h)$, so that

$$\frac{f(x_0 + 2\Delta x) - 2f(x_0 + \Delta x) + f(x_0)}{\Delta x^2} = \frac{g(k) - g(0)}{k}.$$

Use the generalized mean value theorem a second time.

(KN) 3.3.4 = II:1.2.1, 3.3.5 = II:1.2.2, 3.3.6 = II:1.2.3, 3.3.7 = II:1.2.4, 3.3.8 = II:1.3.3, 3.3.9 = II:1.3.4, 3.3.10 = II:1.3.7, 3.3.11 = II:1.3.10, 3.3.12 = II:1.3.11, 3.3.13 = II:1.3.12, 3.3.14 = II:1.2.6, 3.3.15 = II:1.2.7, 3.3.34 = II:2.1.23.

3.3.3 What fractions in $(\sqrt{2} - 1, \sqrt{2} + 1)$ have denominators ≤ 5?

3.3.4 Where is $\sin x = 0$?

3.3.6 For rational numbers, $f(p/q) = p/(q + 1)$. What is the difference between p/q and $f(p/q)$?

3.3.8 Apply the intermediate value theorem to the function g defined by $g(x) = f(x) - x$.

3.3.11 Consider $g(x) = f(x + 1) - f(x)$, $0 \leq x \leq 1$.

3.3.12 $f(2) - f(0) = (f(2) - f(1)) + (f(1) - f(0))$.

3.3.13 Start by explaining why $f(i + 1) - f(i)$ cannot be strictly positive for all integer values of $i \in [0, n - 1]$.

3.3.14 Consider separately the cases $x^2 \in \mathbb{N}$, $x^2 \notin \mathbb{N}$.

3.3.15 Consider separately the cases $x \in \mathbb{N}$, $x \notin \mathbb{N}$.

3.3.17

$$\begin{aligned}
|\sin(x + h) - \sin x| &= |(\sin x)(\cos h - 1) + (\cos x)(\sin h)| \\
&\leq |\sin x| \cdot |\cos h - 1| + |\cos x| \cdot |\sin h| \\
&\leq |\cos h - 1| + |\sin h|.
\end{aligned}$$

Graph $|\cos h - 1| + |\sin h|$ and find an interval containing $h = 0$ where this function is less than 0.1.

3.3.18

$$|(x + h)^2 - x^2| = |2xh + h^2|$$
$$= |h| \cdot |2x + h|$$
$$\leq |h| \cdot |2 + h|.$$

3.3.22 Use the power series for $\ln(1 + x)$ to show that $\ln(1 + x) < x$ for $x > 0$, and therefore if $a > b > 0$ then

$$\ln a - \ln b = \ln\left(1 + \frac{a - b}{b}\right) < \frac{a - b}{b}.$$

3.3.27 Since f is continuous on any interval that does not contain 0, you only need to prove that if $c_1 \leq 0 \leq c_2$ and if A is between $f(c_1)$ and $f(c_2)$, then there is some c, $c_1 < c < c_2$, for which $f(c) = A$.

3.3.28 When does a small change in x result in a change in $f(x)$ that cannot be made arbitrarily small?

3.3.33 $\dfrac{f(x)}{g(x)} - \dfrac{f(c)}{g(c)} = \dfrac{f(x)}{g(x)} \cdot \dfrac{g(c) - g(x)}{g(c)} + \dfrac{f(x) - f(c)}{g(c)}.$

(KN) $3.4.6 = \text{I:}1.1.7\text{--}12$, $3.4.23 = \text{II:}2.1.24$, $3.4.24 = \text{II:}2.1.25$, $3.4.25 = \text{II:}2.1.26$, $3.4.26 = \text{II:}2.1.27$, $3.4.27 = \text{II:}2.1.28$, $3.4.28 = \text{II:}2.1.29$, $3.4.29 = \text{II:}1.2.17$, $3.4.30 = \text{II:}2.2.1$.

3.4.1 The function cannot be continuous.

3.4.2 The domain cannot be a closed, bounded interval.

3.4.4 Prove the contrapositive. Explain why if A and B have opposite signs, then $|A - B| \geq |A|$.

3.4.5 What exactly is the technical statement that corresponds to this condition? For *every* pair (ϵ, δ), what must exist? What happens for very large values of ϵ? Does this technical statement of existence make sense as the definition of a vertical asymptote?

3.4.6 (h) If you hold n constant, what value of m, $1 \leq m \leq 2n - 1$ maximizes this expression? (j) How close can this expression get to 1? (n) Find the minimum value of $x/y + 4y/x$ in the first quadrant. (o) Set $m = kn$ and find the values of k that maximize, minimize the resulting expression. (r) Find the maximum value of $xy/(1 + x + y)$ in the first quadrant.

3.4.11 Given the sequences $x_1 \leq x_2 \leq \cdots \leq x_k \leq \cdots < \cdots \leq y_k \leq \cdots \leq y_2 \leq y_1$, let c be the least upper bound of $\{x_1, x_2, x_3, \ldots\}$. Prove that $c \in [x_k, y_k]$ for every k.

3.4.12 Let S be the set of all x for which $a \leq x < x_2$ and $g(x) \geq g(x_2)$. If S is not empty, then it is bounded and so has a least upper bound, call it $B \leq x_2$. Note that B may or may not be in S.

   a. Use the continuity of g to prove that $g(B) \geq g(x_2)$.
   b. Use the fact that we can make $|g'(x_2) - (g(x_2) - g(x))/(x_2 - x)|$ as small as we wish by taking x sufficiently close to x_2 to prove that $B < x_2$.

c. Use the fact that $B < x_2$, $g(B) \geq g(x_2)$, and $g'(x) \geq 0$ to prove that there are elements of S that are strictly larger than B. This implies that B is not an upper bound and so S must be the empty set.

3.4.13 Assume that we can find a pair (x_1, x_2), $a \leq x_1 < x_2 \leq b$, for which $f(x_1) > f(x_2)$. It follows that there is a positive number α such that

$$\frac{f(x_2) - f(x_1)}{x_2 - x_1} < -\alpha < 0.$$

3.4.16 If $|\sin(1/c) - c^{-1} \cos(1/c)| > 1$, then c cannot be in the range of g.

3.4.19 Let $c = e^{(-8n+1)\pi/4}$, $n \in \mathbb{N}$, and try to find an x for which $\sin(\ln c) + \cos(\ln c) = \sin(\ln x)$. What are other values of $c \in (0, 1)$ that do not correspond to any value of x?

3.4.20 Recall Theorem 3.4.

3.4.22 Start by proving that between any two real roots of P there must be at least one real root of P'. If a polynomial P has a root of order $n > 1$ at $x = a$, then $P(x) = (x - a)^n Q(x)$ where Q is a polynomial, $Q(a) \neq 0$. The derivative $P'(x) = n(x - a)^{n-1} Q(x) + (x - a)^n Q'(x)$ has a root of order $n - 1$ at $x = a$.

3.4.24 If f has a local maximum at $x = c$, then $f'_-(c) \geq 0$ (why?). Let $d = \sup\{x \mid f(x) > f(c)/2\}$. Show that $f'_-(d) \leq 0$. Complete the proof.

3.4.25 Use the idea that helped us prove the mean value theorem.

3.4.26 Use the result of exercise 3.4.25.

3.4.27 Use the result of exercise 3.4.26.

3.4.28 Let $c = \inf\{x \in (a, b) \mid f(x) = 0\}$. Why is this set non-empty?

3.4.30 Consider $f(x)e^{\alpha x}$.

(KN) $3.5.2 = \text{II}2.3.6$, $3.5.3 = \text{II}:2.3.7$, $3.5.4 = \text{II}:2.2.11$, $3.5.17 = \text{II}:2.3.8$, $3.5.18 = \text{II}:2.3.34$.

3.5.1 Prove the contrapositive.

3.5.2 Consider negative as well as positive values of x.

3.5.4 Consider derivatives.

3.5.11 Rewrite the limit as

$$\lim_{x \to 0} \frac{e^{-1/x^2}}{x} = \lim_{x \to 0} \frac{x^{-1}}{e^{1/x^2}}.$$

3.5.17 Differentiate each side of

$$[f(x) - f(0)] g'(\theta(x)) = [g(x) - g(0)] f'(\theta(x))$$

with respect to x, collect the terms that involve $\theta'(x)$ on one side, divide both sides by x, and then take the limit of each side as $x \to 0^+$.

3.5.18 Rewrite $f(x)^{g(x)}$ as $e^{g(x) \ln(f(x))}$.

(KN) $4.1.16 = \text{I}:3.4.10a$.

4.1.1 In this case we know that the partial sum to n terms differs from the value of the series by exactly $(1/2)^n/(1 - 1/2) = 1/2^{n-1}$.

4.1.3 For an alternating series with summands whose absolute values are decreasing toward zero, the partial sum approximation differs from the target value by at most the absolute value of the next term.

4.1.7 A function f is even if and only if $f(-x) = f(x)$.

4.1.14 Are the hypotheses of the Alternating Series Test satisfied?

4.1.15 Combine consecutive summands with the same sign.

4.1.16 Write out enough terms that you get a feel for ϵ_n. Combine consecutive summands with the same sign.

(KN) $4.2.4 = \text{I:2.2.50}, 4.2.5 = \text{I:3.2.1}, 4.2.6 = \text{I:3.4.1}, 4.2.7 = \text{I:3.4.13}, 4.2.29 = \text{I:3.2.17}.$

4.2.1 How do you know that for all n sufficiently large, $|a_n| < 1$?

4.2.2 Use the definition of convergence. To what value does this series converge? Show that given any $\epsilon > 0$, there is some N so that all of the partial sums past the Nth differ from this value by less than ϵ.

4.2.4 (a) The arctangent function is bounded. (d) To test for absolute convergence, combine consecutive pairs of terms. (f) Show that $n/(n+1)^2 > 1/(n+3)$.

4.2.5 (a) $\sqrt{n^2 + 1} - \sqrt[3]{n^3 + 1} = n(1 + 1/n^2)^{1/2} - n(1 + 1/n^3)^{1/3}$. Use a Taylor polynomial approximation. (b) Show that $\lim_{n \to \infty}(n/(n+1))^{n+1} = \lim_{n \to \infty}(1 + 1/n)^{-n-1} = 1/e$. (c) Use a Taylor polynomial approximation. (f) Use the root test.

4.2.6 (b) When does the rational function of a have absolute value less than 1? (c) Use the root test.

4.2.8 Show that if $n \geq N$, then $|a_n| \leq |a_N| \alpha^{n-N}$ and so

$$\sqrt[n]{|a_n|} \leq \alpha \sqrt[n]{|a_N|/\alpha^N}.$$

4.2.22 Prove and then use the fact that for $k \geq 2$:

$$\frac{1}{k \ln(k \ln 2)} > \frac{1}{2k \ln k}.$$

4.2.24 Show that $n^{1+(\ln \ln n + \ln \ln \ln n)/\ln n} = n(\ln n)(\ln \ln n)$.

4.2.28 Use Stirling's formula in place of the factorials.

4.2.29 (a) $2^{n/2} > 2n$ for $n > 8$.

(KN) $4.3.1 = \text{I:3.3.2}, 4.3.2 = \text{I:3.3.3}, 4.3.3 = \text{I:3.3.6}, 4.3.4 = \text{I:3.3.7}, 4.3.18 = \text{I:2.4.11},$
$4.3.20 = \text{I:2.4.15}, 4.3.21 = \text{I:2.4.19}, 4.3.22 = \text{I:2.4.20}, 4.3.23 = \text{I:2.4.26}, 4.3.24 = \text{II:1.2.18}, 4.3.25 = \text{II:1.2.19}, 4.3.26 = \text{II:1.2.20}.$

4.3.1 (a) Using the limit ratio test, we have absolute convergence if

$$1 > \lim_{n \to \infty} \frac{(n+1)^3 |x|^{n+1}}{n^3 |x|^n} = |x|.$$

Check for convergence at $x = +1$ and at $x = -1$. (d) Using the lim sup root test, we have absolute convergence when

$$1 > \varlimsup_{n \to \infty} \left| (2 + (-1)^n)x \right| = 3 |x|.$$

Check for convergence at $x = 1/3$ and at $x = -1/3$. (f) Rewrite this summation so that the power of x is the index of summation: $\sum_{n=1}^{\infty} 2^n x^{n^2} = \sum_{m=1}^{\infty} a_m x^m$ where $a_m = 2^{\sqrt{m}}$ if m is a perfect square, $a_m = 0$ if m is not a perfect square. Now use the lim sup root test. (h) Use the lim sup root test and remember that $\lim_{n \to \infty}(1 + 1/n)^n = e$.

4.3.2 (b) Use the lim sup root test. This converges absolutely if

$$1 > \varlimsup_{n \to \infty} \left(\frac{n}{n+1}\right)^{1/n} \left|\frac{2x+1}{x}\right| = \left|\frac{2x+1}{x}\right|.$$

Check what happens when $|(2x + 1)/x| = 1$, i.e. when $2x + 1 = x$ and when $2x + 1 = -x$.

4.3.3 (a) This implies that $\lim_{n \to \infty} |a_n x^n|^{1/n} = |x| \lim_{n \to \infty} L^{1/n} n^{-\alpha/n}$.

4.3.4 (a) Since the radius of convergence is R, we know that $\varlimsup_{n \to \infty} \sqrt[n]{|a_n x^n|} = 1/R$. It follows that $\varlimsup_{n \to \infty} \sqrt[n]{|2^n a_n x^n|} = 2/R$. (c) Use Stirling's formula.

4.3.6 Do the summands approach 0 when $|x|$ equals the radius of convergence?

4.3.7 Use Stirling's formula.

4.3.8 Use Stirling's formula.

4.3.11 Use the ratio test.

4.3.12 This is a hypergeometric series.

4.3.13 Show that

$$1 \cdot 3 \cdot 5 \cdots (2n - 1) = \frac{(2n)!}{2 \cdot 4 \cdot 6 \cdots 2n} = \frac{(2n)!}{2^n \cdot n!}.$$

If we ignore $F(n)$ in equation (4.15), how close is this approximation when $n = 10? = 20? = 100?$

4.3.14 Either use the result of exercise 4.3.13 together with Stirling's formula, or use the fact that $\lim_{k \to \infty}(1 + 1/k)^k = e$.

4.3.18 (a) Let $\alpha = p/q$ where $\gcd(p, q) = 1$. The answer is in terms of q.

4.3.20 First show that it is enough to prove the last two inequalities. Use the equivalent definition on the lim sup found in exercise 4.3.19.

4.3.21 Use the result from exercise 4.3.20.

4.3.23 Show that it is enough to prove the last inequality. Let

$$A = \varlimsup_{n \to \infty} \frac{a_{n+1}}{a_n}.$$

choose an $\epsilon > 0$ and a response N such that for all $n \geq N$, $a_{n+1}/a_n < A + \epsilon$. Show that for $n \geq N$, $a_n < A_N (A + \epsilon)^{n-N}$. Take the limit as n approaches infinity of the nth root of this upper bound.

4.4.2 Use equation (4.27).

4.4.4 At $x = 1/2$,

$$\left|\sum_{k=1}^{n}(-1)^{k-1} \cos[(2k - 1)\pi/4]\right| = \left|\frac{1 - (-1)^n \cos(\pi n/2)}{2 \cos(\pi/4)}\right| \leq \sqrt{2}.$$

By equation (4.23), $|T_n - T_m| \leq 2\sqrt{2}/(2m + 1)$.

4.4.12 (c) Use Dirichlet's test with $b_k = c_k R^k$ and $a_k = (x/R)^k = e^{ik\theta}$.

5.1.1 The regrouped series has initial term 2/3 and ratio 1/9.

5.1.10 Choose *any* ten terms from the original series to be the first ten terms of the rearranged series. Can the remaining terms be arranged so that the resulting series converges to the target value? Does it matter in what order we put the first ten terms?

5.2.1 Show that the function represented by this series is not continuous.

5.2.4 For all $x \in [-\pi, \pi]$, this is an alternating series and therefore the sum of the first N terms differs from the value of the series by an amount whose absolute value is less than $|x|^{2N+1}/(2N+1)!$.

5.2.7

$$-1 + \frac{1}{4} - \frac{1}{9} + \frac{1}{16} - \frac{1}{25} + \frac{1}{36} - \cdots$$

$$= -\left(1 + \frac{1}{4} + \frac{1}{9} + \frac{1}{16} + \cdots\right) + 2\left(\frac{1}{4} + \frac{1}{16} + \frac{1}{36} + \cdots\right).$$

5.2.8 What is the power series expansion of $\ln(1-x)$?

5.2.9 Use the partial sums

$$S_n(x) = \sum_{k=1}^{n} \frac{x^k}{k^2}$$

and the fact that

$$|\text{Li}_2(1) - \text{Li}_2(x)| \leq |\text{Li}_2(1) - S_n(1)| + |S_n(1) - S_n(x)| + |S_n(x) - \text{Li}_2(x)|.$$

KN 5.3.2 = II:3.2.29.

5.3.1 Consider functions for which $f_k'(x) = 0$ for all k and all x.

5.3.2 Show that it converges at $x = 0$. Explain why it is enough to show that for any N and any x,

$$\sum_{n=N+1}^{\infty} \frac{1}{n^2 + x^2} \leq \sum_{n=N+1}^{\infty} \frac{1}{n^2},$$

and then explain why this is true.

5.3.4 Use equation (4.28) from page 166.

5.3.6 Show that

$$\frac{x^2}{(1 + kx^2)(1 + (k-1)x^2)} = \frac{1}{1 + (k-1)x^2} - \frac{1}{1 + kx^2}.$$

5.3.7 Show that $|G(x) - G_n(x)| = |\sin x|/(1 + nx^2)$. Given $\epsilon > 0$, find $x_0 > 0$ so that $|x| \leq x_0$ implies that $|\sin x|/(1 + nx^2) \leq |\sin x| \leq |\sin x_0| < \epsilon$. For this value of x_0, find an N so that $n \geq N$ and $|x| \geq x_0$ implies that $|\sin x|/(1 + nx^2) \leq 1/(1 + nx^2) \leq 1/(nx_0^2) < \epsilon$. Explain why this proves that the convergence is uniform.

(KN) $5.4.1 = \text{II}:3.2.2$, $5.4.12 = \text{II}:3.2.14$.

5.4.1 (a) For each n, find the supremum of $\{n^2 x^2 e^{-n^2 |x|} \mid x \in \mathbb{R}\}$. (c) Show that for any n, there is an $x > 0$ for which the nth summand is equal to 2^n and all of the summands beyond the nth are ≥ 0. Explain why you cannot have uniform convergence if this is true. (f) Show that $\arctan x + \arctan x^{-1} = \pi/2$, and therefore the nth summand is equal to $\arctan(1/(n^2(1+x^2)))$. Explain why this is less than or equal to $1/(n^2(1+x^2))$.

5.4.2 Use the fact that $a_1 + 2a_2 x + 3a_3 x^2 + \cdots$ converges uniformly and absolutely on $(0, R)$.

5.4.5 Find the values of N that are responses to ϵ at $x = a$, over the open interval (a, b), and at $x = b$.

5.4.6 Consider summands that are not continuous at $x = a$ or $x = b$.

5.4.10 Show that for any $n \geq 2$:

$$\sum_{k=2}^{\infty} \frac{\sin k\pi/n}{\ln k} \geq \sum_{k=2}^{2n} \frac{\sin k\pi/n}{\ln k}$$

$$= \frac{-\sin \pi/n}{\ln(n+1)} + \sum_{k=2}^{n} \sin(k\pi/n) \left(\frac{1}{\ln k} - \frac{1}{\ln(k+n)} \right)$$

$$\geq \frac{-\sin \pi/n}{\ln(n+1)} + \sum_{k=2}^{n} \sin(k\pi/n) \frac{\ln 2}{(\ln n)(\ln 2n)}$$

$$= \frac{f(n) \ln 2}{(\ln n)(\ln 2n)} - \sin(\pi/n) \left(\frac{1}{\ln(n+1)} + \frac{\ln 2}{(\ln n)(\ln 2n)} \right),$$

where

$$f(n) = \sin(\pi/n) + \sin(2\pi/n) + \sin(3\pi/n) + \cdots + \sin(n\pi/n).$$

Use equation (5.64) to show that

$$f(n) = \frac{\sin(\pi/n)}{1 - \cos(\pi/n)}.$$

5.4.12 (c) Show that $2 \sin(n^2 x) \sin(nx) = \cos(n(n-1)x) - \cos(n(n+1)x)$.
(d) Rewrite the summation as

$$\sum_{n=1}^{\infty} \frac{\sin(nx)}{n} \left(\arctan(nx) - \frac{\pi}{2} \right) + \frac{\pi}{2} \sum_{n=1}^{\infty} \frac{\sin(nx)}{n}.$$

In the first sum, let $b_n(x) = \pi/2 - \arctan(nx)$. (e) Rewrite the summation as

$$\sum_{n=1}^{\infty} \frac{(-1)^{n+1}}{n^{x-a/2}} n^{-a/2}.$$

5.4.13 Consider $\sum_{k=1}^{\infty} x^k - x^{k-1}$ on $[0, 1]$.

6.1.4 Show that $f(x) = f(-x)$, $g(x) = -g(-x)$.

6.1.5 If f is even and g is odd, then $F(-x) = f(x) - g(x)$.

6.1.6 If the Fourier series converges at $x = 0$, then $\sum_{k=1}^{\infty} a_k$ converges, and therefore the partial sums of $\sum_{k=1}^{\infty} a_k$ are bounded.

6.1.9 Find an algebraic expression for this function on $(-1, 1)$.

6.1.10 Uniform convergence means that you are allowed to interchange integration and infinite summation.

6.1.18 Change variables using $t = \alpha + \beta - u$ and let $h(u) = g(\alpha + \beta - u)$. Show that h is nonnegative and increasing on $[\alpha, \beta]$.

6.2.2 Where is the graph of $y = x^3 - 2x^2 + x$ increasing? Where is it decreasing? Where is the slope steepest?

6.2.4 Fix $\epsilon > 0$. Put a bound on the error contributed by using an approximating sum over the interval $[0, \epsilon]$. Use the fact that $\sin(1/x)$ is continuous on the interval $[\epsilon, 1]$.

6.2.6 Use the mean value theorem.

6.2.7 We need a bounded, differentiable function whose derivative is not bounded.

6.2.11 Use the definition of differentiability. You must show that

$$\lim_{x \to x_0} \left| \frac{\int_{x_0}^{x} f(t)\, dt}{x - x_0} - f(x_0) \right| = 0.$$

6.2.12 Consider Theorem 3.14.

(KN) $6.3.8 = \text{III}:1.1.7, 6.3.9 = \text{III}:1.1.6, 6.3.10 = \text{III}:1.1.14.$

6.3.2 Show that given any $\sigma > 0$, there is a response δ so that for any partition with subintervals of length $< \delta$, the variation is less than σ.

6.3.7 Fix a variation σ. Can we limit the sum of the lengths of the intervals on which the variation exceeds σ?

6.3.8 Where is this function discontinuous? How large is the variation at the points of discontinuity?

6.3.10 (d) $1/(n + k) = (1/2n)(2/(1 + k/n))$. (f) First show that the function of n is equal to $e^{\sum_{k=1}^{n}(1/n)\ln(1+k/n)}$.

6.3.11 The summation is an approximation using a partition with infinitely many intervals of the form $[q^{n+1}, q^n]$. Show that for any $\epsilon > 0$, we can find a Riemann sum with intervals of length less than $1 - q$ that differs from our infinite summation by less than ϵ.

6.3.13 $\displaystyle\sum_{n=1}^{\infty} \frac{1}{(2n-1)^2} = \sum_{n=1}^{\infty} \frac{1}{n^2} - \sum_{n=1}^{\infty} \frac{1}{(2n)^2}.$

6.3.20 Note that at points of discontinuity, the function decreases. Otherwise, it is an increasing function. Show that if we approximate $f(x)$ with $\sum_{n=1}^{100} (\!(nx)\!)/n^2$, then we are within $1/200$ of the correct value. Now explain why it follows that if $0 \le x < y \le 1$, then

$$f(y) - f(x) < \sum_{n=1}^{100} \frac{ny - nx}{n^2} + \frac{1}{100} < (y - x)(\ln(101) + \gamma) + \frac{1}{100}.$$

6.3.16 Write $f(x) = f_N(x) + R_N(x)$ where

$$R_N(x) = \sum_{n=N+1}^{\infty} \frac{((nx))}{n^2}.$$

Given an $\epsilon > 0$, the task is to show how to find a response δ such that for $0 < v < \delta$,

$$|f_N(x + v) + R_N(x + v) - f_N(x + 0) - R_N(x + 0)| < \epsilon.$$

6.3.19 Let $\sigma(k)$ be the sum of the divisors of k, and set $k = 2^a k_1$ where k_1 is odd. Show that $\psi(k) = (2^{a+1} - 3)\sigma(k_1)$. It is known that

$$\overline{\lim} \frac{\sigma(k)}{k \ln \ln k} = e^\gamma.$$

This is Gronwall's Theorem, published in 1913.

6.3.20 Show that

$$g(1/5) = \frac{1}{5} \sum_{n=0}^{\infty} \frac{1}{5n + 1} + \frac{2}{5} \sum_{n=0}^{\infty} \frac{1}{5n + 2} - \frac{2}{5} \sum_{n=0}^{\infty} \frac{1}{5n + 3} - \frac{1}{5} \sum_{n=0}^{\infty} \frac{1}{5n + 4}$$

$$= \sum_{n=0}^{\infty} \frac{125n^2 + 125n + 26}{5(5n + 1)(5n + 2)(5n + 3)(5n + 4)}.$$

6.3.21 Let $x = p/q$, $\gcd(p, q) = 1$. Let $m = \lfloor (q - 1)/2 \rfloor$. Show that

$$g(x) = \sum_{k=1}^{m} \frac{((kx))}{k} + \sum_{k=-m}^{m} ((kx)) \sum_{n=1}^{\infty} \frac{1}{qn + k}$$

$$= \sum_{k=1}^{m} ((kx)) \left(\frac{1}{k} - 2k \sum_{n=1}^{\infty} \frac{1}{q^2 n^2 - k^2} \right).$$

6.4.1 Consider Theorem 3.14.

6.4.10 Use the fact that α_m is an integer that is odd when m is even and even when m is odd.

A.1.2 Start with

$$\int_0^1 (1 - x^{1/p})^q = \int_0^1 (1 - x^{1/p})^{q-1} \, dx - \int_0^1 (1 - x^{1/p})^{q-1} x^{1/p} \, dx$$

$$= \int_0^1 (1 - x^{1/p})^{q-1} \, dx$$

$$+ p \int_0^1 (1 - x^{1/p})^{q-1} \left(\frac{-x^{(1-p)/p}}{p} \right) x \, dx,$$

and then use integration by parts on the second integral.

A.1.4 Use the substitution $u = (1 - x^{1/p})^q$.

A.1.8 Using equations (A.6) and (A.12), we see that

$$f(p, q) + \frac{p + q}{q} f(p, q - 1) = \frac{p + q}{p} f(p - 1, q).$$

A.1.12 The values are undefined when p or q is a negative integer, but it is defined for other negative values of p and q.

A.1.13 Show that $f(2/3, k) = (5 \cdot 8 \cdot 11 \cdots (3k + 2))/(3 \cdot 6 \cdot 9 \cdots (3k))$. Show that $f(2/3, 1/3 + k) = f(2/3, 1/3) (6 \cdot 9 \cdot 12 \cdots (3k + 3))/(4 \cdot 7 \cdot 10 \cdots (3k + 1))$.

A.2.4 Use the fact that equation (A.16) defines $B_n(x)$. Show that
$\int_k^{k+1} B_n(1 - x) \, dx = (-1)^n \int_k^{k+1} B_n(x) \, dx$.

A.2.5 Use equation (A.33) from the previous exercise and equation (A.29).

A.3.6 Use the fact that
$$\sum_{j=1}^{\infty} \frac{1}{j^2} \sum_{k=1}^{\infty} \frac{1}{k^4} = \sum_{j \neq k} \frac{1}{j^2 \, k^4} + \sum_{k=1}^{\infty} \frac{1}{k^6}.$$

Find the coefficients of the summmations on the right side:
$$\left(\sum_{j=1}^{\infty} \frac{1}{j^2} \right)^3 = ? \times \sum_{1 \leq j < k < l < \infty} \frac{1}{j^2 k^2 l^2} + ? \times \sum_{j \neq k} \frac{1}{j^2 \, k^4} + ? \times \sum_{k=1}^{\infty} \frac{1}{k^6}.$$

A.3.9 Rather than trying to evaluate 100!, find
$$A = \ln \left(\frac{2 \cdot (2m)!}{(2\pi)^{2m}} \right) = \left(\sum_{n=1}^{2m} \ln n \right) - (2m - 1) \ln 2 - (2m) \ln \pi.$$

Use the observation that
$$\frac{2 \cdot (2m)!}{(2\pi)^{2m}} = e^A = 10^{A/\ln 10}.$$

A.3.10 Show that $\zeta(n) < 1 + \int_1^{\infty} x^{-n} dx = 1 + 1/(n - 1)$.

A.3.19 Use equation (A.53).

Index